Probability and Mathematical Statistics (Continued)

WILLIAMS • Diffusions, Markov Processes, and Martingales, Volume I: Foundations
ZACKS • Theory of Statistical Inference

Applied Probability and Statistics

ANDERSON, AUQUIER, HAUCK, OAKES, VANDAELE, and WEISBERG • Statistical Methods for Comparative Studies
ARTHANARI and DODGE • Mathematical Programming in Statistics
BAILEY • The Elements of Stochastic Processes with Applications to the Natural Sciences
BAILEY • Mathematics, Statistics and Systems for Health
BARNETT and LEWIS • Outliers in Statistical Data
BARTHOLOMEW • Stochastic Models for Social Processes, *Second Edition*
BARTHOLOMEW and FORBES • Statistical Techniques for Manpower Planning
BECK and ARNOLD • Parameter Estimation in Engineering and Science
BELSLEY, KUH, and WELSCH • Regression Diagnostics: Identifying Influential Data and Sources of Collinearity
BENNETT and FRANKLIN • Statistical Analysis in Chemistry and the Chemical Industry
BHAT • Elements of Applied Stochastic Processes
BLOOMFIELD • Fourier Analysis of Time Series: An Introduction
BOX • R. A. Fisher, The Life of a Scientist
BOX and DRAPER • Evolutionary Operation: A Statistical Method for Process Improvement
BOX, HUNTER, and HUNTER • Statistics for Experimenters: An Introduction to Design, Data Analysis, and Model Building
BROWN and HOLLANDER • Statistics: A Biomedical Introduction
BROWNLEE • Statistical Theory and Methodology in Science and Engineering, *Second Edition*
BURY • Statistical Models in Applied Science
CHAMBERS • Computational Methods for Data Analysis
CHATTERJEE and PRICE • Regression Analysis by Example
CHERNOFF and MOSES • Elementary Decision Theory
CHOW • Analysis and Control of Dynamic Economic Systems
CLELLAND, BROWN, and deCANI • Basic Statistics with Business Applications, *Second Edition*
COCHRAN • Sampling Techniques, *Third Edition*
COCHRAN and COX • Experimental Designs, *Second Edition*
CONOVER • Practical Nonparametric Statistics, *Second Edition*
CORNELL • Experiments with Mixtures: Designs, Models and The Analysis of Mixture Data
COX • Planning of Experiments
DANIEL • Biostatistics: A Foundation for Analysis in the Health Sciences, *Second Edition*
DANIEL • Applications of Statistics to Industrial Experimentation
DANIEL and WOOD • Fitting Equations to Data: Computer Analysis of Multifactor Data, *Second Edition*
DAVID • Order Statistics, *Second Edition*
DEMING • Sample Design in Business Research
DODGE and ROMIG • Sampling Inspection Tables, *Second Edition*
DRAPER and SMITH • Applied Regression Analysis, *Second Edition*
DUNN • Basic Statistics: A Primer for the Biomedical Sciences, *Second Edition*
DUNN and CLARK • Applied Statistics: Analysis of Variance and Regression
ELANDT-JOHNSON • Probability Models and Statistical Methods in Genetics
ELANDT-JOHNSON and JOHNSON • Survival Models and Data Analysis

continued on back

Statistical Methods for Rates and Proportions

Statistical Methods for Rates and Proportions

Second Edition

JOSEPH L. FLEISS

Division of Biostatistics, School of Public Health, Columbia University

JOHN WILEY & SONS New York • Chichester • Brisbane • Toronto • Singapore

Library of Congress Cataloging in Publication Data:

Fleiss, Joseph L.
Statistical methods for rates and proportions.

(Wiley series in probability and mathematical statistics)
Includes bibliographies and indexes.
1. Analysis of variance. 2. Sampling
(Statistics) 3. Biometry. I. Title.

QA279.F58 1981 519.5′352 80-26382
ISBN 0-471-06428-9

Printed in the United States of America

10 9 8

TO ISABEL

Preface

The need for a revised edition became apparent a few years after the publication of the first edition. Reviewers, researchers, teachers, and students cited some important topics that were absent, treated too briefly, or not presented in their most up-to-date form. In the meantime, the field of applied statistics continued to develop, and new results were obtained that deserved citation and illustration.

Of the several topics I had omitted from the first edition, the most important was the construction of confidence intervals. In the revision, interval estimation is treated almost as extensively as hypothesis testing. In fact, the close connection between the two is pointed out in the new Section 1.4. The reader will find there, in the new Section 5.6, and elsewhere realizations of the warning I gave in the Preface to the first edition that a properly constructed confidence interval is frequently more complicated than simply the point estimate plus or minus a multiple of its standard error.

Another important topic missing from the first edition was the planning of comparative studies with unequal sample sizes. This is treated in the new Section 3.4.

Several other topics not covered in the first edition are covered here. The Fisher-Irwin "exact" test for a fourfold table is described in the new Section 2.2. Attributable risk, an important indicator of the effect of exposure to a risk factor, is discussed in the new Sections 5.7 and 6.4. The Cornfield method for making inferences about the odds ratio is presented in the new Sections 5.5 and 5.6.

A number of topics touched on superficially or not dealt with adequately in the first edition have, I hope, now been covered properly. The analysis of data from a two-period crossover study is described in an expansion of Section 7.2. A more appropriate method for analyzing data from a study of matched pairs when the response variable is qualitatively ordered is presented in Section 8.2. The comparison of proportions from several matched samples in Section 8.4 has been expanded to include the case of quantitatively ordered samples. A method for comparing data from several fourfold tables that has been found capable of yielding erroneous results has been relegated to the section (now Section 10.7) on methods to be avoided.

Developments in statistics since the appearance of the first edition are reflected in most sections and every chapter of the revision. The determination of sample sizes is brought up to date in Section 3.2; the corresponding table in the Appendix (Table A.3) has been revised accordingly. Some recently proposed alternatives to simple randomization in clinical studies are discussed in two new sections, 4.3 and 7.3. The presentation of ridit analysis in Section 9.4 has been revised in the light of recent research. The effects and control of misclassification in both variables in a fourfold table are considered in Sections 11.3 and 12.2. The new Chapter 13, which is an expansion and updating of the old Section 12.2, presents recent results on the measurement of interrater agreement for categorical data. Some recent insights into indirect standardization are cited in Sections 14.3 and 14.5.

The emphasis continues to be on, and the examples continue to be from, the health sciences. The selection of illustrative material was determined by the field I know best, not by the field I necessarily consider the most important.

The revision is again aimed at research workers and students who have had at least a year's course in applied statistics, including chi square and correlation. Many of the problems that conclude the chapters have been revised. Several new problems have been added.

Several of my colleagues and a few reviewers urged me to include the solutions to at least some of the numerical problems. I have decided to provide the solutions to all of them. Teachers who wish to assign these problems for homework or on examinations may do so simply by changing some of the numerical values.

The mathematical prerequisites continue to be a knowledge of high school algebra and an ability to take logarithms and extract square roots. All methods presented can be applied using only a desktop or pocket calculator. As a consequence, the book does not present the powerful but mathematically complicated methods of log linear or logistic regression analysis for high order cross-classification tables. The texts by D. R. Cox (*The analysis of binary data*, Methuen, London, 1970) and by Y. M. M. Bishop, S. E. Fienberg, and P. W. Holland (*Discrete multivariate analysis: Theory and practice*, M.I.T. Press, Cambridge, Mass., 1975) are excellent references at a somewhat advanced mathematical level. Two more recent short monographs (B. S. Everitt, *The analysis of contingency tables*, Halsted

Press, New York, 1977 and S. E. Fienberg, *The analysis of cross-classified categorical data*, M.I.T. Press, Cambridge, Mass., 1977) provide less mathematically advanced reviews of these topics.

Professors Agnes Berger, John Fertig, Bruce Levin, and Patrick Shrout of Columbia University and Professor Gary Simon of the State University of New York at Stony Brook reviewed draft copies of the revision and made many helpful suggestions. Professors Berger, Fertig, and Simon were especially critical, and offered advice that I took seriously but did not always follow.

Most helpful of all were the students who took my course on the analysis of categorical data the last couple of years at the Columbia University School of Public Health, and the students who took my course on advanced statistical methods in epidemiology in the 1978 Graduate Summer Session in Epidemiology at the University of Minnesota School of Public Health. They served as experimental subjects without informed consent as I tried out various approaches to the presentation of the new (and old) material. Students who took my course in the 1980 Graduate Summer Session in Epidemiology saw draft copies of the revision and pointed out several typographical errors I had made. I thank them all.

Ms. Blanche Agdern patiently and carefully typed the several drafts of the revision. Ms. Beatrice Shube, my editor at Wiley, was always supportive and a ready source of advice and encouragement. My wife Isabel was a constant source of inspiration and reinforcement when the going got tough.

The new table of sample sizes was generated by a program run at the computer center of the New York State Psychiatric Institute. The publishers of the *American Journal of Epidemiology*, *Biometrics*, and the *Journal of Chronic Diseases* kindly gave me permission to use published data.

JOSEPH L. FLEISS

New York, New York
December 1980

Preface to the First Edition

This book is concerned solely with comparisons of qualitative or categorical data. The case of quantitative data is treated in the many books devoted to the analysis of variance. Other books have restricted attention to categorical data (such as A. E. Maxwell, *Analysing qualitative data*, Methuen, London, 1961, and R. G. Francis, *The rhetoric of science: A methodological discussion of the two-by-two table*, University of Minnesota Press, Minneapolis, 1961), but an updated monograph seemed overdue. A recent text (D. R. Cox, *The analysis of binary data*, Methuen, London, 1970) is at once more general than the present book in that it treats categorical data arising from more complicated study designs and more restricted in that it does not treat such topics as errors of misclassification and standardization of rates.

Although the ideas and methods presented here should be useful to anyone concerned with the analysis of categorical data, the emphasis and examples are from the disciplines of clinical medicine, epidemiology, psychiatry and psychopathology, and public health. The book is aimed at research workers and students who have had at least a year's course in applied statistics, including a thorough grounding in chi square and correlation. Most chapters conclude with one or more problems. Some call for the proof of an algebraic identity. Others are numerical, designed either to have the reader apply what he has learned or to present ideas mentioned only in passing in the text.

No more complicated mathematical techniques than the taking of logarithms and the extraction of square roots are required to apply the methods described. This means that anyone with only high school algebra, and with only a desktop calculator, can apply the methods presented. It also means, however, that analyses requiring matrix inversion or other complicated mathematical techniques (e.g., the analysis of multiple contingency tables) are not described. Instead, the reader is referred to appropriate sources.

The estimation of the degree of association or difference assumes equal importance with the assessment of statistical significance. Except where the formulas are excessively complicated, I present the standard error of almost every measure of association or difference given in the text. The standard errors are used to test hypotheses about the corresponding parameters, to compare the precision of different methods of estimation, and to obtain a weighted average of a number of independent estimates of the same parameter.

I have tried to be careful in giving both sides of various arguments that are still unresolved about the proper design of studies and analysis of data. Examples are the use of matched samples and the measurement of association. Inevitably, my own biases have probably affected how I present the opposing arguments.

In two instances, however, my bias is so strong that I do not even permit the other side to be heard. I do not find confidence intervals to be useful, and therefore do not discuss interval estimation at all. The reader who finds a need for confidence intervals will have to refer to some of the cited references for details. He will find, by the way, that the proper interval is almost always more complicated than simply the point estimate plus or minus a multiple of its standard error.

The second instance is my bias against the Bayesian approach to statistical inference. See W. Edwards, H. Lindman, and L. J. Savage, Bayesian statistical inference for psychological research, *Psychol. Rev.*, **70**, 193–242, 1963, for a description of the Bayesian approach to data in psychology; and J. Cornfield, A Bayesian test of some classical hypotheses —with applications to sequential clinical trials, *J. Am. Stat. Assoc.*, **61**, 577–594, 1966, for a description of that approach to data in medicine. I believe that the kind of thinking described in Chapter 3, especially in Section 3.1, provides an adequate alternative to the Bayesian approach.

It is with gratitude that I acknowledge the advice, criticism, and encouragement of Professors John Fertig, Mervyn Susser, and Andre Varma of Columbia University and of Dr. Joseph Zubin of the Biometrics Research unit of the New York State Department of Mental Hygiene. Dr. Gary Simon of Princeton University and Professor W. Edwards Deming of New York University reviewed the manuscript and pointed out a number of errors I had made in an earlier draft. Needless to say, I take full responsibility for any and all errors that remain.

My wife Isabel was a constant source of inspiration as well as an invaluable editorial assistant.

The major portion of the typing was admirably performed by Vilma Rivieccio. Additional typing, collating, and keypunching were ably carried out by Blanche Agdern, Rosalind Fruchtman, Cheryl Keller, Sarah Lichtenstaedter, and Edith Pons.

My work was supported in part by grant DR 00793 from the National Institute of Dental Research (John W. Fertig, Ph.D., Principal Investigator) and in part by grant MH 08534 from the National Institute of Mental Health (Robert L. Spitzer, M.D., Principal Investigator). Except when noted otherwise, the tables in the Appendix were generated by programs run on the computers of the New York State Psychiatric Institute and of Rockland State Hospital.

I thank Professor E. S. Pearson and the Biometrika Trustees for permission to quote from Tables 1, 4, and 8 of *Biometrika tables for statisticians, Vol. I*, edited by E. S. Pearson and H. O. Hartley; John Wiley & Sons for permission to use Tables A.1 to A.3 of *Statistical inference under order restrictions* by R. E. Barlow, D. J. Bartholomew, J. M. Bremner, and H. D. Brunk; Van Nostrand Reinhold Co. for permission to quote data from *Smoking and Health*; the Institute of Psychiatry of the University of London for permission to quote data from *Psychiatric diagnosis in New York and London* by J. E. Cooper et al.; and Sir Austin Bradford Hill and Oxford University Press for permission to quote from *Statistical methods in clinical and preventive medicine*.

I also thank the editors of the following journals for permission to use published data: the *American Journal of Public Health*, the *American Statistician*, *Biometrics*, the *Journal of Laboratory and Clinical Medicine*, the *Journal of the National Cancer Institute*, the *Journal of Psychiatric Research*, and *Psychometrika*.

JOSEPH L. FLEISS

New York, New York
June 1972

Contents

CHAPTER

1. An Introduction to Applied Probability 1

1.1. Notation and Definitions 1
1.2. The Evaluation of a Screening Test 4
1.3. Biases Resulting from the Study of Selected Samples 8
1.4. Inferences About a Single Proportion 13
Problems 15
References 17

2. Assessing Significance in a Fourfold Table 19

2.1. Methods for Generating a Fourfold Table 20
2.2. "Exact" Analysis of a Fourfold Table 24
2.3. Yates' Correction for Continuity 26
2.4. One-Tailed Versus Two-Tailed Tests 27
2.5. A Simple Confidence Interval for the Difference Between Two Independent Proportions 29
2.6. An Alternative Critical Ratio Test 30
Problems 31
References 32

3. Determining Sample Sizes Needed to Detect a Difference Between Two Proportions 33

3.1. Specifying a Difference Worth Detecting 34
3.2. The Mathematics of Sample Size Determination 38
3.3. Using the Sample Size Tables 42

3.4. Unequal Sample Sizes 44
3.5. Some Additional Comments 46
Problems 46
References 48

4. How to Randomize 50

4.1. Selecting a Simple Random Sample 51
4.2. Randomization in a Clinical Trial 52
4.3. Variations on Simple Randomization 53
References 55

5. Sampling Method I: Naturalistic or Cross-Sectional Studies 56

5.1. Some Hypothetical Data 57
5.2. Measures of Association Derived from χ^2 58
5.3. Other Measures of Association: The Odds Ratio 61
5.4. Some Properties of the Odds Ratio and its Logarithm 64
5.5. Testing Hypotheses About the Odds Ratio 67
5.6. Confidence Intervals for the Odds Ratio 71
5.7. Attributable Risk 75
Problems 78
References 80

6. Sampling Method II: Prospective and Retrospective Studies 83

6.1. Prospective Studies 83
6.2. Retrospective Studies 87
6.3. Criticisms of the Odds Ratio 90
6.4. Estimating Attributable Risk from Retrospective Studies 93
6.5. The Retrospective Approach Versus the Prospective Approach 95
Problems 97
References 98

7. Sampling Method III: Controlled Comparative Trials 100

7.1. The Simple Comparative Trial 101
7.2. The Two-Period Crossover Design 104

7.3. Alternatives to Simple Randomization 105
Problems 108
References 109

8. The Analysis of Data from Matched Samples 112

8.1. Matched Pairs: Dichotomous Outcome 113
8.2. Matched Pairs: More than Dichotomous Outcome 119
8.3. The Case of Multiple Matched Controls 123
8.4. The Comparison of m Matched Samples 126
8.5. Advantages and Disadvantages of Matching 133
Problems 134
References 135

9. The Comparison of Proportions from Several Independent Samples 138

9.1. The Comparison of m Proportions 138
9.2. Gradient in Proportions: Samples Quantitatively Ordered 143
9.3. Gradient in Proportions: Samples Qualitatively Ordered 147
9.4. Ridit Analysis 150
Problems 156
References 158

10. Combining Evidence from Fourfold Tables 160

10.1. The Construction and Interpretation of Some Chi Square Tests 161
10.2. Combining the Logarithms of Odds Ratios 165
10.3. Method Due to Cornfield and Gart 168
10.4. The Mantel-Haenszel Method 173
10.5. A Comparison of the Three Procedures 175
10.6. Alternatives to Matching 176
10.7. Methods to be Avoided 178
Problems 185
References 186

11. The Effects of Misclassification Errors 188

11.1. An Example of the Effects of Misclassification 188
11.2. The Algebra of Misclassification 193

11.3. The Algebra of Misclassification: Both Variables in Error 196
Problems 198
References 199

12. The Control of Misclassification Error 201

12.1. Statistical Control for Error 201
12.2. Probabilistic Control for Error 204
12.3. The Experimental Control of Error 205
Problems 209
References 210

13. The Measurement of Interrater Agreement 211

13.1. The Case of Two Raters 212
13.2. Multiple Ratings per Subject 225
13.3. Further Applications 232
Problems 234
References 234

14. The Standardization of Rates 237

14.1. Reasons for and Warnings Against Standardization 239
14.2. Indirect Standardization 240
14.3. A Feature of Indirect Standardization 243
14.4. Direct Standardization 244
14.5. Some Other Summary Indices 247
14.6. Adjustment for Two Factors 249
Problems 253
References 254

Appendix Tables 257

Answers to Numerical Problems 295

Author Index 305

Subject Index 311

Statistical Methods for Rates and Proportions

CHAPTER 1

An Introduction to Applied Probability

Some elements of applied probability theory are needed to appreciate fully and to manipulate the different kinds of rates that arise in research. Thus the clearest and most suggestive interpretation of a rate is as a probability —as a measure of the likelihood that a specified event occurs to, or that a specified characteristic is possessed by, a typical member of a population. An important use of probabilities is in estimating the number of individuals, out of a sample of size n, who have the characteristic under consideration. If P is the probability that an individual possesses the characteristic, the *expected number* having the characteristic is simply nP.

Section 1.1 presents notation and some important definitions. The theory of Section 1.1 is applied in Section 1.2 to the evaluation of a screening test, and in Section 1.3 to the study of the bias possible in making inferences from selected samples. Section 1.4 is devoted to methods for testing hypotheses about and constructing confidence intervals for single probabilities or proportions.

1.1. NOTATION AND DEFINITIONS

In this book, the terms probability, relative frequency, proportion, and rate are used synonymously. If A denotes the event that a randomly selected individual from a population has a defined characteristic (e.g., has arteriosclerotic heart disease), then $P(A)$ denotes the proportion of all people who have the characteristic. For the given example $P(A)$ is the probability that a randomly selected individual has arteriosclerotic heart disease, or, in the terminology of vital statistics, the case rate for arteriosclerotic heart disease.

One can go only so far with overall rates, however. Of greater usefulness usually are so-called *specific rates:* the rate of the defined characteristic specific for age, race, sex, occupation, and so on. What is known in epidemiology and vital statistics as a specific rate is known in probability theory as a *conditional probability*. The notation is

$P(A|B)$ = probability that a randomly selected individual has characteristic A, given that he has characteristic B, or *conditional* on his having characteristic B.

If, in our example, we denote by B the characteristic of being aged 65–74, then $P(A|B)$ is the conditional probability that a person has arteriosclerotic heart disease, given that he is aged 65–74. In the terminology of vital statistics, $P(A|B)$ is the rate of arteriosclerotic heart disease specific to people aged 65–74.

Let $P(B)$ represent the proportion of all people who possess characteristic B, and let $P(A \text{ and } B)$ represent the proportion of all people who possess both characteristic A and characteristic B. Then, by definition, provided $P(B) \neq 0$,

$$P(A|B) = \frac{P(A \text{ and } B)}{P(B)}. \tag{1.1}$$

Similarly, provided $P(A) \neq 0$,

$$P(B|A) = \frac{P(A \text{ and } B)}{P(A)}. \tag{1.2}$$

By the *association* of two characteristics we mean that when a person has one of the characteristics, say B, his chances of having the other are affected. By the *independence* or lack of association of two characteristics we mean that the fact that a person has one of the characteristics does not affect his chances of having the other. Thus, if A and B are independent, then the rate at which A is present specific to people who possess B, $P(A|B)$, is equal to the overall rate at which A is present, $P(A)$. By (1.1), this implies that

$$\frac{P(A \text{ and } B)}{P(B)} = P(A),$$

or

$$P(A \text{ and } B) = P(A)P(B). \tag{1.3}$$

Equation 1.3 is often taken as the definition of independence, instead of the equivalent statement

$$P(A|B) = P(A).$$

A heuristic justification of (1.1) is the following. Let N denote the total number of people in the population; N_A the number of people who have characteristic A; N_B the number of people who have characteristic B; and N_{AB} the number of people who have both characteristics. It is then clear that

$$P(A) = \frac{N_A}{N},$$

$$P(B) = \frac{N_B}{N},$$

and

$$P(A \text{ and } B) = \frac{N_{AB}}{N}.$$

By $P(A|B)$ we mean the proportion out of all people who have characteristic B who also have characteristic A, so that both the numerator and the denominator of $P(A|B)$ must be specific to B. Thus

$$P(A|B) = \frac{N_{AB}}{N_B}. \tag{1.4}$$

If we now divide the numerator and denominator of (1.4) by N, we find that

$$P(A|B) = \frac{N_{AB}/N}{N_B/N} = \frac{P(A \text{ and } B)}{P(B)}.$$

Equation 1.2 may be derived similarly:

$$P(B|A) = \frac{N_{AB}}{N_A} = \frac{N_{AB}/N}{N_A/N} = \frac{P(A \text{ and } B)}{P(A)}.$$

Equations 1.1 and 1.2 are connected by means of *Bayes' theorem:*

$$P(B|A) = \frac{P(A|B)P(B)}{P(A)}. \tag{1.5}$$

Equation 1.5 follows from the definition (1.2) of $P(B|A)$ and from the fact, seen by multiplying both sides of (1.1) by $P(B)$, that $P(A \text{ and } B) = P(A|B)P(B)$.

1.2. THE EVALUATION OF A SCREENING TEST

A frequent application of Bayes' theorem is in evaluating the performance of a diagnostic test intended for use in a screening program. Let B denote the event that a person has the disease in question; $\bar{B}$ the event that he does not have the disease; A the event that he gives a positive response to the test; and $\bar{A}$ the event that he gives a negative response. Suppose that the test has been applied to a sample of B's, that is, to a sample of people with the disease, and to a sample of $\bar{B}$'s, that is, to a sample of people without the disease.

The results of this trial of the screening test may by represented by the two conditional probabilities $P(A|B)$ and $P(A|\bar{B})$. $P(A|B)$ is the conditional probability of a positive response given that the person has the disease; the larger $P(A|B)$ is, the more *sensitive* the test is. $P(A|\bar{B})$ is the conditional probability of a positive response given that the person is free of the disease; the smaller $P(A|\bar{B})$ is [equivalently, the larger $P(\bar{A}|\bar{B})$ is], the more *specific* the test is. These definitions of a test's sensitivity and specificity are due to Yerushalmy (1947).

Of greater concern than the test's sensitivity and specificity, however, are the error rates to be expected if the test is actually used in a screening program. If a positive result is taken to indicate the presence of the disease, then the false positive rate, say P_{F+}, is the proportion of people, among those responding positive, who are actually free of the disease, or $P(\bar{B}|A)$. By Bayes' theorem,

$$P_{F+} = P(\bar{B}|A) = \frac{P(A|\bar{B})P(\bar{B})}{P(A)} = \frac{P(A|\bar{B})[1 - P(B)]}{P(A)}, \tag{1.6}$$

since $P(\bar{B}) = 1 - P(B)$.

The false negative rate, say P_{F-}, is the proportion of people, among those responding negative on the test, who nevertheless have the disease, or $P(B|\bar{A})$. Again by Bayes' theorem,

$$P_{F-} = P(B|\bar{A}) = \frac{P(\bar{A}|B)P(B)}{P(\bar{A})} = \frac{[1 - P(A|B)]P(B)}{1 - P(A)}, \tag{1.7}$$

since $P(\bar{A}|B) = 1 - P(A|B)$ and $P(\bar{A}) = 1 - P(A)$.

We still need the overall rates $P(A)$ and $P(B)$ in order to evaluate these two error rates. Actually, we only need $P(B)$, for the following reason. Note that

$$P(A) = \frac{N_A}{N} = \frac{N_{AB} + N_{A\bar{B}}}{N} = \frac{N_{AB}}{N} + \frac{N_{A\bar{B}}}{N}. \tag{1.8}$$

In (1.8), N_{AB} denotes the number of people who have the disease and respond positive, and $N_{A\bar{B}}$ denotes the number of people who are free of the disease and respond positive. Multiplying and dividing the first of the two terms on the right-hand side of (1.8) by N_B, the number of people with the disease, we find that

$$\frac{N_{AB}}{N} = \frac{N_{AB}}{N_B}\frac{N_B}{N} = P(A|B)P(B). \tag{1.9}$$

Similarly, by multiplying and dividing the second term by $N_{\bar{B}}$, the number of people without the disease, we find that

$$\frac{N_{A\bar{B}}}{N} = \frac{N_{A\bar{B}}}{N_{\bar{B}}}\frac{N_{\bar{B}}}{N} = P(A|\bar{B})P(\bar{B}). \tag{1.10}$$

Substituting the expressions from (1.9) and (1.10) in (1.8), we find that

$$P(A) = P(A|B)P(B) + P(A|\bar{B})P(\bar{B}). \tag{1.11}$$

This equation is a special case of the familiar result that an overall rate—$P(A)$—is a weighted average of specific rates—$P(A|B)$ and $P(A|\bar{B})$—with the weights being the proportions of people in the specific categories—$P(B)$ and $P(\bar{B})$.

Since $P(\bar{B}) = 1 - P(B)$, (1.11) becomes

$$\begin{aligned} P(A) &= P(A|B)P(B) + P(A|\bar{B})(1 - P(B)) \\ &= P(A|\bar{B}) + P(B)\left[P(A|B) - P(A|\bar{B})\right]. \end{aligned} \tag{1.12}$$

Substitution of (1.12) in (1.6) yields, as the expression for the false positive rate,

$$P_{F+} = \frac{P(A|\bar{B})(1 - P(B))}{P(A|\bar{B}) + P(B)\left[P(A|B) - P(A|\bar{B})\right]}. \tag{1.13}$$

Substitution of (1.12) in (1.7) yields, as the expression for the false negative rate,

$$P_{F-} = \frac{(1 - P(A|B))P(B)}{1 - P(A|\bar{B}) - P(B)\left[P(A|B) - P(A|\bar{B})\right]}. \tag{1.14}$$

Analyses of the false positive and false negative rates associated with screening and diagnostic tests have been performed by Cochrane and

Holland (1971) and by Galen and Gambino (1975) for a variety of medical disorders. We see from (1.13) and (1.14) that, in general, the two error rates are functions of the proportions $P(A|B)$ and $P(A|\bar{B})$, which may be estimated from the results of a trial of the screening test, and of the overall case rate $P(B)$, for which an accurate estimate is rarely available. Nevertheless, a range of likely values for the error rates may be determined as in the following example.

Suppose that the test is applied to a sample of 1000 people known to have the disease and to a sample of 1000 people known not to have the disease. Suppose that this trial resulted in the frequencies shown in Table 1.1.

Table 1.1. Results of a trial of a screening test

	Test Result		
Disease Status	$+(A)$	$-(\bar{A})$	Total
Present (B)	950	50	1000
Absent ($\bar{B}$)	10	990	1000

We would then have the estimate of sensitivity,

$$P(A|B) = 950/1000 = .95,$$

and the estimate of specificity,

$$P(\bar{A}|\bar{B}) = 990/1000 = .99,$$

a pair of probabilities indicating a test that is both sensitive and specific to the disease being studied.

Substitution of these two probabilities in (1.13) gives, as the value for the false positive rate,

$$\begin{aligned} P_{F+} &= \frac{.01(1-P(B))}{.01+P(B)(.95-.01)} = \frac{.01(1-P(B))}{.01+.94P(B)} \\ &= \frac{1-P(B)}{1+94P(B)}, \end{aligned} \tag{1.15}$$

the final expression resulting by multiplying both the numerator and the denominator of the preceding expression by 100. Substitution in (1.14) gives, as the value for the false negative rate,

$$P_{F-} = \frac{(1-.95)P(B)}{1-.01-P(B)(.95-.01)} = \frac{.05P(B)}{.99-.94P(B)}$$
$$= \frac{5P(B)}{99-94P(B)}. \tag{1.16}$$

Table 1.2 gives the error rates associated with various values of $P(B)$, the overall case rate. Rarely will the case rate exceed 1% of the population.

Table 1.2. Error rates associated with screening test

$P(B)$	False Positive (P_{F+})	False Negative (P_{F-})
1/million	.9999	0
1/100,000	.9991	0
1/10,000	.9906	.00001
1/1000	.913	.00005
1/500	.840	.00010
1/200	.677	.00025
1/100	.510	.00051

If the disease is not too prevalent—if it affects, say, less than 1% of the population—the false negative rate will be quite small, but the false positive rate will be rather large. From one point of view, the test is a successful one: since P_{F-} is less than 5/10,000, therefore, of every 10,000 people who respond negative and are thus presumably given a clean bill of health, no more than five should actually have been informed they were ill. From another point of view, the test is a failure: since P_{F+} is greater than 50/100, therefore, of every 100 people who respond positive and thus presumably are told they have the disease or are at least advised to undergo further tests, more than 50 will actually be free of the disease.

The final decision whether or not to use the test will depend on the seriousness of the disease and on the cost of further tests or treatment. Because the false positive rate is so great, however, it would be hard to justify using this screening test for any but the most serious diseases. Since the proportions $P(A|B)$ and $P(\bar{A}|\bar{B})$ considered in this example are better than those associated with most existing screening tests, it is disquieting to realize that their false positive rates are probably greatly in excess of the 50 out of 100 that we found as a lower limit here.

One method for reducing the false positive or false negative rate associated with a diagnostic screening procedure (but thereby increasing its cost) is to repeat the test a number of times and to declare the final result positive if the subject responds positive to each administration of the test or if he responds positive to a majority of the administrations. Parts *b* and *c* of Problem 1.2 illustrate the improved performance of a screening procedure when a test is administered twice. Sandifer, Fleiss, and Green (1968) have shown that, for some disorders, a better rule is to administer the test three times and to declare the final result positive if the subject responds positive to at least two of the three administrations. Only those subjects who respond positive to one of the first two administrations and negative to the other will have to be tested a third time. Those who respond positive to both of the first two administrations would be declared positive, and those who respond negative to both would be declared negative.

A more accurate but more complex assessment of the performance of a screening procedure than the above is possible when disease severity is assumed to vary and not, as here, merely to be present or absent. The appropriate analysis was originally proposed by Neyman (1947) and later extended by Greenhouse and Mantel (1950) and by Nissen-Meyer (1964). The reader is referred to these papers and to a review article by McNeil, Keeler, and Adelstein (1975) for details.

It must be noted that universally accepted definitions do not exist for the false positive and false negative rate. Galen and Gambino (1975) and Rogan and Gladen (1978), for example, use the definitions adopted here. Others use these terms to refer simply to the complements of the test's specificity and sensitivity, a convention that is wasteful of precious terminology. When discussing what is here termed the false positive rate, they refer to "the complement of the predictive value of a positive test result." Similarly, our false negative rate is their "complement of the predictive value of a negative test result." These locutions are, to say the least, awkward.

1.3. BIASES RESULTING FROM THE STUDY OF SELECTED SAMPLES

The first clues to the association between diseases and antecedent conditions frequently come from the study of such selected samples as hospitalized patients and autopsy cases. Because not all subjects are equally likely to end up in the study samples, bias may result when the associations found in the selected samples are presumed to apply in the population at large.

A classic example of this kind of bias occurs in a study by Pearl (1929). A large number of autopsy cases were cross-classified by the presence or absence of cancer and by the presence or absence of tuberculosis. A negative association between these two diseases was found, that is, tuberculosis was less frequent in autopsy cases with cancer than in cases without cancer. Pearl inferred that the same negative association should apply to live patients, and in fact acted on the basis of this inference by conceiving a study to treat terminal cancer patients with tuberculin (the protein of the tubercle bacillus) in the anticipation that the cancer would be arrested. What Pearl ignored is that, unless all deaths are equally likely to be autopsied, it is improper to extrapolate to live patients an association found for autopsied cases. It is even possible that there would be no association for live patients but, due to the differential selection of patients for autopsy, a strong association for autopsied cases.

The same kind of bias, known as Berkson's fallacy after the person who first studied it in detail (Berkson, 1946, 1955), is possible whenever the chances of a subject's being included in the study sample vary. Berkson's fallacy has also been studied by Mainland (1953; 1963, pp. 117–124), White (1953), Mantel and Haenszel (1959), and Brown (1976). Curiously, the actual existence of the fallacy was not demonstrated empirically until a report by Roberts, Spitzer, Delmore, and Sackett (1978). It is illustrated below in data from Sackett (1979).

Of a random sample of 2784 individuals interviewed in the community, 257 had been hospitalized during the prior six months. Data on the presence or absence of respiratory disease and on the presence or absence of disease of bones and organs of movement (abbreviated locomotor disease) for the subsample that had been hospitalized are presented in Table 1.3.

Table 1.3. Locomotor disease by respiratory disease in hospitalized subsample

Respiratory Disease	Locomotor Disease			
	Present	Absent	Total	Proportion with Locomotor Disease
Present	5	15	20	$.25 = p_1$
Absent	18	219	237	$.08 = p_2$
Total	23	234	257	$.09 = p$

There is clearly an association between whether a hospitalized patient has or does not have respiratory disease and whether he has or does not have locomotor disease: the proportion of patients with locomotor disease among those who have respiratory disease, $p_1 = .25$, is over three times the proportion of patients with locomotor disease among those who do not have respiratory disease, $p_2 = .08$. Would it, however, be correct to conclude that the two kinds of disease were associated in the community?

Not necessarily. In fact, these two characteristics—respiratory disease and locomotor disease—are essentially independent in the community. As shown in Table 1.4, the rates of locomotor disease are virtually the same in people with and without respiratory disease.

Table 1.4. Locomotor disease by respiratory disease in general population

Respiratory Disease	Locomotor Disease			
	Present	Absent	Total	Proportion with Locomotor Disease
Present	17	207	224	.08
Absent	184	2376	2560	.07
Total	201	2583	2784	.07

The source of the apparent paradox lies in variations across the four possible kinds of people (both diseases present, only respiratory disease present, only locomotor disease present, neither disease present) in the likelihood of being hospitalized. It may be checked by comparing corresponding cell frequencies in Tables 1.3 and 1.4 that the hospitalization rate for people with both diseases, 29%, was about three times each of the rates for the other kinds of people, which ranged from 7 to 10%.

Berkson's fallacy is always possible when admission rates for people with different combinations of factors vary and can reasonably be ruled out only when the disease or diseases in question almost always require care (e.g., leukemia and some other cancers). The algebra underlying the phenomenon is as follows. Let B denote the event that a person has one of the two characteristics under study (in this case, respiratory disease), and $\bar{B}$ the event that he does not. Let $P(B)$ denote the proportion of all people in the community who have the characteristic and let $P(\bar{B}) = 1 - P(B)$ denote the proportion of all people who do not.

Let A denote the event that a person has the other characteristic under study (in this case, locomotor disease) and $\bar{A}$ the event that he does not, and let $P(A)$ and $P(\bar{A}) = 1 - P(A)$ denote the corresponding proportions. Let $P(A \text{ and } B)$ denote the proportion of all people in the community who have both characteristics, and assume that the two characteristics are independent in the community. Thus by (1.3), $P(A \text{ and } B) = P(A)P(B)$.

Let H denote the event that a person from the community is hospitalized for some reason or other. Define $P(H \mid B \text{ and } A)$ as the proportion, out of all people who have both characteristics, who are hospitalized; $P(H \mid B \text{ and } \bar{A})$ as the proportion, out of all people who have one characteristic but not the other, who are hospitalized; and define $P(H|\bar{B}$ and $A)$ and $P(H|\bar{B}$ and $\bar{A})$ similarly. Our problem is to evaluate, in terms of these probabilities,

$$p_1 = P(A|B \text{ and } H),$$

that is, the proportion, out of all people who are characterized by B and are hospitalized, that turns out to have A; and

$$p_2 = P(A|\bar{B} \text{ and } H),$$

that is, the proportion, out of all people who are not characterized by B and are hospitalized, that turns out to have A.

We make use of the following version of the definition of a conditional probability:

$$p_1 = P(A|B \text{ and } H) = \frac{P(B \text{ and } A|H)}{P(B|H)}. \tag{1.17}$$

Equation 1.17 differs from (1.2) only in that, once specified, the second condition H (for hospitalization) remains a condition qualifying all probabilities.

The numerator of (1.17) is, by Bayes' theorem [see (1.5)],

$$\begin{aligned} P(B \text{ and } A|H) &= \frac{P(H|B \text{ and } A)P(B \text{ and } A)}{P(H)} \\ &= \frac{P(H|B \text{ and } A)P(B)P(A)}{P(H)}, \end{aligned} \tag{1.18}$$

because of the assumed independence of A and B.

The denominator of (1.17) is, again by Bayes' theorem,

$$P(B|H) = \frac{P(H|B)P(B)}{P(H)}. \tag{1.19}$$

To find $P(H|B)$, we make use of the fact, represented by (1.11), that an overall rate is a weighted average of specific rates. This time, however, we apply the additional fact that a condition on a probability, in this case B, remains one in all subsequent probabilities. Thus

$$\begin{aligned} P(H|B) &= P(H|B \text{ and } A)P(A|B) + P(H|B \text{ and } \bar{A})P(\bar{A}|B) \\ &= P(H|B \text{ and } A)P(A) + P(H|B \text{ and } \bar{A})P(\bar{A}), \end{aligned}$$

because the assumed independence of A and B implies that $P(A|B)=P(A)$ and $P(\bar{A}|B)=P(\bar{A})$. Therefore, (1.19) becomes

$$P(B|H)=\frac{P(B)\left[P(H|B \text{ and } A)P(A)+P(H|B \text{ and } \bar{A})P(\bar{A})\right]}{P(H)}. \tag{1.20}$$

Substitution of (1.18) for the numerator of (1.17), and of (1.20) for the denominator, yields

$$p_1=\frac{P(H|B \text{ and } A)P(A)}{P(H|B \text{ and } A)P(A)+P(H|B \text{ and } \bar{A})P(\bar{A})}. \tag{1.21}$$

Similarly,

$$p_2=\frac{P(H|\bar{B} \text{ and } A)P(A)}{P(H|\bar{B} \text{ and } A)P(A)+P(H|\bar{B} \text{ and } \bar{A})P(\bar{A})}. \tag{1.22}$$

These two probabilities are not equal if the rates of hospitalization are not equal, even though the corresponding probabilities in the community, $P(A|B)$ and $P(A|\bar{B})$, are equal.

In our example, with A denoting locomotor disease, $P(A)=7\%$. The various hospitalization rates were

$$P(H|B \text{ and } A)=29\%,$$

$$P(H|B \text{ and } \bar{A})=7\%,$$

$$P(H|\bar{B} \text{ and } A)=10\%,$$

and

$$P(H|\bar{B} \text{ and } \bar{A})=9\%.$$

Substituting these values into (1.21) and (1.22), we find that

$$p_1=\frac{.29\times .07}{.29\times .07+.07\times .93}=\frac{.020}{.085}=.24$$

and that

$$p_2=\frac{.10\times .07}{.10\times .07+.09\times .93}=\frac{.007}{.091}=.08,$$

close to the values we found in Table 1.3.

The moral of this exercise is clear. Unless something is known about differential hospitalization rates or differential autopsy rates, a good amount of skepticism should be applied to any generalization from associations found for hospitalized patients or for autospy cases to associations for people at large. This caveat obviously applies also to associations obtained from reports by volunteers.

1.4. INFERENCES ABOUT A SINGLE PROPORTION

One is interested most frequently in comparing two or more proportions. On occasion, however, interest resides in a single proportion. This section presents a brief survey of inferential methods for this case.

Let n denote the number of people in the sample studied, p the proportion of the sample possessing the characteristic under study, and P the underlying (but unknown) proportion in the population who possess the characteristic. Inferences about P may be made using the *binomial probability distribution*, which is described in all good elementary statistics texts. When n is large (in the sense that $nP \geqq 5$ and $nQ \geqq 5$, where $Q = 1 - P$), the following procedures based on the normal distribution provide excellent approximations to the corresponding exact binomial procedures.

When P is the underlying proportion, the sample proportion p is approximately normally distributed with mean P and standard error

$$\text{s.e.}(p) = \sqrt{\frac{PQ}{n}}\,. \tag{1.23}$$

In order to test the hypothesis that P is equal to a prespecified P_0 against the alternative hypothesis that $P \neq P_0$, one may calculate the critical ratio

$$z = \frac{|p - P_0| - 1/(2n)}{\sqrt{\dfrac{P_0 Q_0}{n}}}, \tag{1.24}$$

where $Q_0 = 1 - P_0$, and reject the hypothesis if z exceeds the critical value of the normal curve for the desired two-tailed significance level. The quantity $1/(2n)$ subtracted in the numerator is a correction for continuity, bringing normal curve probabilities into closer agreement with binomial probabilities. It should be applied only when it is numerically smaller than $|p - P_0|$.

Consider the data in Table 1.3. The sample size is $n = 257$, and the proportion with locomotor disease is $p = 23/257 = .09$. Suppose (for illustrative purposes) that the rate of locomotor disease in comparable hospital samples is $P_0 = .05$. Is the present sample typical or atypical with respect to this condition? The value of the critical ratio given in (1.24) is

$$z = \frac{|.09 - .05| - \dfrac{1}{2 \times 257}}{\sqrt{\dfrac{.05 \times .95}{257}}} = 2.80,$$

which indicates a significant difference at the .01 level between the present sample and supposedly comparable ones.

A *confidence interval* may be desired for the underlying proportion. A point concerning confidence intervals that must be borne in mind is that they are more intimately connected with significance testing than with simple point estimation. In the present case a $100(1 - \alpha)\%$ confidence interval for the underlying proportion (usual values for α are .01 and .05) consists of all those values of P that would not be rejected by a two-tailed significance test at the α level of significance. If the test is based on the critical ratio given in (1.24), and if $c_{\alpha/2}$ denotes the value cutting off the area $\alpha/2$ in the upper tail of the standard normal distribution, an approximate $100(1 - \alpha)\%$ confidence interval consists of all those values of P satisfying

$$\frac{|p - P| - 1/(2n)}{\sqrt{\dfrac{PQ}{n}}} \leqq c_{\alpha/2}. \tag{1.25}$$

The limits of this interval are given by the two roots of the quadratic equation obtained by setting the square of the left-hand side of (1.25) equal to $c^2_{\alpha/2}$. Define $q = 1 - p$. The lower limit is given explicitly by

$$P_L = \frac{\left(2np + c^2_{\alpha/2} - 1\right) - c_{\alpha/2}\sqrt{c^2_{\alpha/2} - (2 + 1/n) + 4p(nq + 1)}}{2\left(n + c^2_{\alpha/2}\right)} \tag{1.26}$$

and the upper limit by

$$P_U = \frac{\left(2np + c^2_{\alpha/2} + 1\right) + c_{\alpha/2}\sqrt{c^2_{\alpha/2} + (2 - 1/n) + 4p(nq - 1)}}{2\left(n + c^2_{\alpha/2}\right)} \tag{1.27}$$

(see Problem 1.5).

For the data being analyzed, $n = 257$, $p = .09$, and $q = .91$. If a 95% confidence interval is desired for P, $c_{\alpha/2} = c_{.025} = 1.96$. It may be checked that $P_L = .059$ and $P_U = .133$, so

$$.059 \leqq P \leqq .133 \tag{1.28}$$

is an approximate 95% confidence interval for P. Furthermore, it may be checked that, if $P_0 = .059$ or $P_0 = .133$ is hypothesized as the value of P, the resulting value of z in (1.24) is exactly equal to 1.96.

The preceding somewhat complicated procedure for setting a confidence interval around P is preferred when p is near zero or near unity. When p is of moderate size (say, $.3 \leqq p \leqq .7$), the following more familiar and more simple procedure may be employed.

Because $\sqrt{x(1-x)}$ remains fairly constant for $.3 \leqq x \leqq .7$, $\sqrt{PQ}$ in the denominator of (1.25) may be replaced by $\sqrt{pq}$. This yields the interval

$$p - c_{\alpha/2}\sqrt{\frac{pq}{n}} - \frac{1}{2n} \leqq P \leqq p + c_{\alpha/2}\sqrt{\frac{pq}{n}} + \frac{1}{2n} \tag{1.29}$$

as an approximate $100(1-\alpha)\%$ confidence interval for P. For the data at hand, the resulting interval is

$$.053 \leqq P \leqq .126, \tag{1.30}$$

which is shifted to the left relative to the technically more correct interval given in (1.28).

When p is very small, the shift to the left of the interval in (1.29) relative to the one given by (1.26) and (1.27) may be extreme. Problem 1.4 illustrates this possibility.

Problem 1.1. Characteristics A and B are independent if $P(A \text{ and } B) = P(A)P(B)$. Show that, if this is so, then $P(A \text{ and } \bar{B}) = P(A)P(\bar{B})$; $P(\bar{A} \text{ and } B) = P(\bar{A})P(B)$; and $P(\bar{A} \text{ and } \bar{B}) = P(\bar{A})P(\bar{B})$. [*Hint.* $P(A) = P(A \text{ and } B) + P(A \text{ and } \bar{B})$, so that $P(A \text{ and } \bar{B}) = P(A) - P(A \text{ and } B)$. Use the given relation, $P(A \text{ and } B) = P(A)P(B)$, and use the fact that $P(\bar{B}) = 1 - P(B)$.]

Problem 1.2. Assume that the case rate for a specified disease, $P(B)$, is one case per 1000 population and that a screening test for the disease is being studied.

(*a*) Suppose that the test is administered a single time to a sample of people with the disease, of whom 99% respond positive. Suppose that it is also administered to a sample of people without the disease, of whom 1%

respond positive. What are the false positive and false negative rates? Do you think the test is a good one?

(*b*) Suppose now that the test is administered twice, with a positive overall result declared only if both tests are positive. Suppose, according to this revised definition, that 98% of the diseased sample respond positive, but that only one out of 10,000 nondiseased people responds positive. What are the false positive and false negative rates now? Would you be willing to employ the test for screening under these revised conditions?

(*c*) Note that not all people have to be tested twice. Only if the first test result is positive must a second test be administered; the final result will be declared positive only if the second test is positive as well. It is important to estimate the proportion of all people who will have to be tested again, that is, the proportion who are positive on their first test. What is this proportion? Out of every 100,000 people tested once, how many will not need to be tested again?

Problem 1.3. The opposite kind of bias from that considered in Section 1.3 can occur. That is, two characteristics may be associated in the community but may be independent when hospitalized samples are studied. Let A represent living alone, $\bar{A}$ living with family, B having a neurosis, and $\bar{B}$ not having a neurosis. Suppose that $P(A|B) = .40$ and $P(A|\bar{B}) = .20$. Thus 40% of neurotics live alone and 20% of nonneurotics live alone. Suppose that, in the population at large, 100,000 people are neurotic and one million are not neurotic.

(*a*) Consider first the 100,000 neurotics. (1) How many of them live alone? (2) If the annual hospitalization rate for neurotics living alone is 5/1000, how many such people will be hospitalized? Note that this is the number of hospitalized neurotics one would find who lived alone, that is, the numerator of p_1. (3) How many of the 100,000 neurotics live with their families? (4) If the annual hospitalization rate for neurotics living with their families is 6/1000, how many such people will be hospitalized? Note that the sum of the numbers found in (2) and (4) is the total number of hospitalized neurotics, that is, the denominator of p_1. (5) What is the value of p_1, the proportion of hospitalized neurotics who lived alone? (6) How does p_1 compare with $P(A|B)$, the proportion of neurotics in the community who live alone?

(*b*) Consider now the one million nonneurotics. (1) How many of them live alone? (2) If the annual hospitalization rate for nonneurotics living alone is 5/1000, how many such people will be hospitalized? Note that this is the number of hospitalized nonneurotics one would find who lived alone, that is, the numerator of p_2. (3) How many of the one million nonneurotics live with their families? (4) If the annual hospitalization rate for nonneurotics living with their families is 225/100,000, how many such people will be hospitalized? Note that the sum of the numbers found in ($b2$) and ($b4$) is the total number of hospitalized nonneurotics, that is, the denominator of p_2. (5) What is the value of p_2, the proportion of hospitalized nonneurotics who lived alone? (6) How does p_2 compare with $P(A|\bar{B})$, the proportion of nonneurotics in the community who live alone?

(*c*) What inference would you draw from the comparison of p_1 and p_2? How does this inference compare with the inference drawn from the comparison of $P(A|B)$ and $P(A|\bar{B})$?

Problem 1.4. Suppose that out of a sample of $n = 100$ subjects a proportion $p = .05$ had a specified characteristic.

(*a*) To two decimal places, find the lower and upper 99% confidence limits on P using (1.26) and (1.27). Use $c_{.005} = 2.576$.

(*b*) To two decimal places, find the lower and upper 99% confidence limits on P using (1.29).

(*c*) How do the two intervals compare? Would matters be improved any in (*b*) if the continuity correction were ignored?

Problem 1.5. Prove that (1.26) and (1.27) provide the two solutions to the equation

$$\frac{n(|p-P|-1/(2n))^2}{PQ} = c_{\alpha/2}^2.$$

[*Hint.* Bear in mind that the purpose of the continuity correction is to bring the difference between p and P closer to zero. In deriving the lower limit, therefore, work with $p - P - 1/(2n)$ in the numerator; in deriving the upper limit, work with $p - P + 1/(2n)$.]

Problem 1.6. Prove that P_L in (1.26) is never less than zero and that P_U in (1.27) is never greater than unity.

REFERENCES

Berkson, J. (1946). Limitations of the application of fourfold table analysis to hospital data. *Biom. Bull.* (now *Biometrics*), **2**, 47–53.

Berkson, J. (1955). The statistical study of association between smoking and lung cancer. *Proc. Staff Meet. Mayo Clin.*, **30**, 319–348.

Brown, G. W. (1976). Berkson fallacy revisited: Spurious conclusions from patient surveys. *Am. J. Dis. Child.*, **130**, 56–60.

Cochrane, A. L. and Holland, W. W. (1971). Validation of screening procedures. *Br. Med. Bull.*, **27**, 3–8.

Galen, R. S. and Gambino, S. R. (1975). *Beyond normality: The predictive value and efficiency of medical diagnoses*. New York: Wiley.

Greenhouse, S. W. and Mantel, N. (1950). The evaluation of diagnostic tests. *Biometrics*, **6**, 399–412.

Mainland, D. (1953). The risk of fallacious conclusions from autopsy data on the incidence of diseases with applications to heart disease. *Am. Heart J.*, **45**, 644–654.

Mainland, D. (1963). *Elementary medical statistics*, 2nd ed. Philadelphia: W. W. Saunders.

Mantel, N. and Haenszel, W. (1959). Statistical aspects of the analysis of data from retrospective studies of disease. *J. Natl. Cancer Inst.*, **22**, 719–748.

McNeil, B. J., Keeler, E., and Adelstein, S. J. (1975). Primer on certain elements of medical decision making. *New Engl. J. Med.*, **293**, 211–215.

Neyman, J. (1947). Outline of statistical treatment of the problem of diagnosis. *Public Health Rep.*, **62**, 1449–1456.

Nissen-Meyer, S. (1964). Evaluation of screening tests in medical diagnosis. *Biometrics*, **20**, 730–755.

Pearl, R. (1929). Cancer and tuberculosis. *Am. J. Hyg.* (now *Am. J. Epidemiol.*), **9**, 97–159.

Roberts, R. S., Spitzer, W. O., Delmore, T., and Sackett, D. L. (1978). An empirical demonstration of Berkson's bias. *J. Chronic Dis.*, **31**, 119–128.

Rogan, W. J. and Gladen, B. (1978). Estimating prevalence from the results of a screening test. *Am. J. Epidemiol.*, **107**, 71–76.

Sackett, D. L. (1979). Bias in analytic research. *J. Chronic Dis.*, **32**, 51–63.

Sandifer, M. G., Fleiss, J. L., and Green, L. M. (1968). Sample selection by diagnosis in clinical drug evaluations. *Psychopharmacologia*, **13**, 118–128.

White, C. (1953). Sampling in medical research. *Br. Med. J.*, **2**, 1284–1288.

Yerushalmy, J. (1947). Statistical problems in assessing methods of medical diagnosis, with special reference to X-ray techniques. *Public Health Rep.*, **62**, 1432–1449.

CHAPTER 2

Assessing Significance in a Fourfold Table

The fourfold table (see Table 2.1) has been and probably still is the most frequently employed means of presenting statistical evidence. The simplest and most frequently applied statistical test of the significance of the association indicated by the data is the classic chi square test. It is based on the magnitude of the statistic

$$\chi^2 = \frac{n_{..}\left(|n_{11}n_{22} - n_{12}n_{21}| - \frac{1}{2}n_{..}\right)^2}{n_{1.}n_{2.}n_{.1}n_{.2}}. \tag{2.1}$$

The value obtained for χ^2 is referred to tables of the chi square distribution with one degree of freedom (see Table A.1). If the value exceeds the entry tabulated for a specified significance level, the inference is made that *A* and *B* are associated. An interesting graphic assessment of significance is due to Zubin (1939).

Table 2.2 presents some hypothetical frequencies. The value of χ^2 is, by (2.1),

$$\chi^2 = \frac{200\left(|15 \times 40 - 135 \times 10| - \frac{1}{2}200\right)^2}{150 \times 50 \times 25 \times 175} = 2.58. \tag{2.2}$$

Since χ^2 would have to exceed 3.84 in order for significance at the .05 level

Table 2.1. Model fourfold table

	Characteristic *B*		
Characteristic *A*	Present	Absent	Total
Present	n_{11}	n_{12}	$n_{1.}$
Absent	n_{21}	n_{22}	$n_{2.}$
Total	$n_{.1}$	$n_{.2}$	$n_{..}$

Table 2.2. A hypothetical fourfold table

	Characteristic *B*		
Characteristic *A*	Present	Absent	Total
Present	15	135	150
Absent	10	40	50
Total	25	175	200

to be declared, the conclusion for these data would be that no significant association was demonstrated.

It is perhaps unfortunate that the chi square statistic (2.1) takes such a simple form, both because its calculation does not require the investigator to determine explicitly the proportions being contrasted—these representing the association being studied, not the raw frequencies—and because it invites the investigator to ignore the fact that the proper inference to be drawn from the magnitude of χ^2 depends on how the data were generated, even though the formula for χ^2 does not. These ideas are developed in Section 2.1. There, three methods for generating the frequencies of a fourfold table are presented and the statistical hypothesis appropriate to each is specified.

The "exact" test due to Fisher and Irwin is presented in Section 2.2, and the need for incorporating Yates' correction for continuity into the chi square and critical ratio statistics is considered in Section 2.3. Some criteria for choosing between a one-tailed and a two-tailed significance test are offered in Section 2.4. Section 2.5 is devoted to setting confidence limits around the difference between two independent proportions; Section 2.6, to a critical ratio test, different from the classic one, that is closely related to the construction of confidence intervals.

2.1. METHODS FOR GENERATING A FOURFOLD TABLE

There are, in practice, essentially three methods of sampling that can give rise to the frequencies set out in a fourfold table (see Barnard, 1947, for a more complete discussion).

Method I. The first method of sampling, termed naturalistic or cross-sectional sampling, calls for the selection of a total of $n_{..}$ subjects from a larger population and then for the determination for each subject of the presence or absence of characteristic *A* and the presence or absence of characteristic *B*. Only the total sample size, $n_{..}$, can be specified prior to the collection of the data.

Much of survey research is conducted along such a line. Examples of the use of method I sampling are the following. In a study of the quality of medical care delivered to patients, all new admissions to a specified service of a hospital might be cross-classified by sex and by whether or not each of a number of examinations was made. In a study of the variation of disease prevalence in a community, a random sample of subjects may be drawn and cross-classified by race and by the presence or absence of each of a number of symptoms. In a study of the association between birthweight and maternal age, all deliveries in a given maternity hospital might be cross-classified by the weight of the offspring and by the age of the mother.

With method I sampling, the issue is whether the presence or absence of characteristic A is associated with the presence or absence of characteristic B. In the population from which the sample was drawn, the proportions (of course unknown) are as in Table 2.3.

Table 2.3. Joint proportions of A and B in the population

	Characteristic B		
Characteristic A	Present	Absent	Total
Present	P_{11}	P_{12}	$P_{1.}$
Absent	P_{21}	P_{22}	$P_{2.}$
Total	$P_{.1}$	$P_{.2}$	1

By the definition of independence (see Section 1.1), characteristics A and B are independent if and only if each joint proportion (e.g., P_{12}) is the product of the two corresponding total or marginal proportions (in this example, $P_{1.}P_{.2}$). Whether the proportions actually have this property can only be determined by how close the joint proportions in the sample are to the corresponding products of marginal proportions. The cross-classification table in the sample should therefore be the analog of Table 2.3 and is obtained by dividing each frequency in Table 2.1 by $n_{..}$. Table 2.4 results.

Table 2.4. Joint proportions of A and B in the sample

	Characteristic B		
Characteristic A	Present	Absent	Total
Present	p_{11}	p_{12}	$p_{1.}$
Absent	p_{21}	p_{22}	$p_{2.}$
Total	$p_{.1}$	$p_{.2}$	1

The tenability of the hypothesis that A and B are independent depends on the magnitudes of the four differences $p_{ij} - p_{i.}p_{.j}$, where i and j equal 1 or 2. The smaller these differences are, the closer the data come to the standard of independence. The larger these differences are, the more questionable the hypothesis of independence becomes. (Actually, only a single one of these four differences needs to be examined, the other three being equal to it except possibly for a change in sign—see Problem 2.1.)

Pearson (1900) suggested a criterion for assessing the significance of these differences. His statistic, incorporating the continuity correction, is

$$\chi^2 = n_{..} \sum_{i=1}^{2} \sum_{j=1}^{2} \frac{\left(|p_{ij} - p_{i.}p_{.j}| - 1/(2n_{..})\right)^2}{p_{i.}p_{.j}}. \tag{2.3}$$

Problem 2.2 is devoted to the proof that (2.1) and (2.3) are equal. If χ^2 is found by reference to the table of chi square with one degree of freedom to be significantly large, the investigator would infer that A and B were associated and would proceed to describe their degree of association. Chapter 5 is devoted to methods of describing association following method I sampling.

Suppose that the data of Table 2.2 were obtained from a study employing method I sampling. The proper summarization of the data is illustrated by Table 2.5. Note, for example, that $p_{22} = .20$, whereas if A and B were independent, we would have expected the proportion to be $p_{2.}p_{.2} = .25 \times .875 = .21875$. Furthermore, note that each of the four differences entering into the formula in (2.3) is equal to $|\pm .01875| - .0025 = .01625$.

Table 2.5. Joint proportions for hypothetical data of Table 2.2

Characteristic A	Characteristic B		
	Present	Absent	Total
Present	.075	.675	.75
Absent	.050	.200	.25
Total	.125	.875	1

Applied to the data of Table 2.5, (2.3) yields the value

$$\chi^2 = 200\left(\frac{.01625^2}{.09375} + \frac{.01625^2}{.65625} + \frac{.01625^2}{.03125} + \frac{.01625^2}{.21875}\right) = 2.58, \tag{2.4}$$

equal to the value in (2.2).

Method II. The second method of sampling, sometimes termed purposive sampling, calls for the selection and study of a predetermined number, $n_{1.}$, of subjects who possess characteristic A and for the selection and study of a predetermined number, $n_{2.}$, of subjects for whom characteristic A is absent. This method of sampling forms the basis of comparative prospective and of comparative retrospective studies. In the former, $n_{1.}$ subjects with and $n_{2.}$ subjects without a suspected antecedent factor would be followed to determine how many develop disease. In the latter, $n_{1.}$ subjects with and $n_{2.}$ subjects without the disease would be traced back to determine how many possessed the suspected antecedent factor.

Of interest in method II sampling is whether the proportions in the two populations from which we have samples, say P_1 and P_2, are equal. It is therefore indicated that the sample data be so presented that information about these two proportions is afforded. The appropriate means of presentation is given in Table 2.6.

Table 2.6. Proportions with a specified characteristic in two independent samples

	Sample Size	Proportion
Sample 1	$n_{1.}$	$p_1(=n_{11}/n_{1.})$
Sample 2	$n_{2.}$	$p_2(=n_{21}/n_{2.})$
Combined	$n_{..}$	$\bar{p}(=n_{.1}/n_{..})$

The statistical significance of the difference between p_1 and p_2 is assessed by means of the statistic

$$z=\frac{|p_2-p_1|-\frac{1}{2}(1/n_{1.}+1/n_{2.})}{\sqrt{\bar{p}\bar{q}(1/n_{1.}+1/n_{2.})}}, \tag{2.5}$$

where $\bar{q}=1-\bar{p}$. To test the hypothesis that P_1 and P_2 are equal, z may be referred to the standard normal distribution. If z exceeds the normal curve value for a prespecified significance level—see Table A.2—P_1 and P_2 are inferred to be unequal. Since, by definition, the square of a quantity that has the standard normal distribution will be distributed as chi square with one degree of freedom, z^2 may be referred to tables of chi square with one degree of freedom. Problem 2.3 is devoted to the proof that z^2 is equal to the quantity in (2.1).

The analysis subsequent to the finding of statistical significance with method II sampling is shown in Chapter 6 to be quite different from the analysis appropriate to method I sampling.

Suppose, for illustration, that the data of Table 2.2 had been generated by deliberately studying 150 subjects with characteristic A and 50 subjects without it. The appropriate presentation of the data is illustrated in Table 2.7.

Table 2.7. ***Proportions with characteristic B for subjects with and subjects without characteristic A—from hypothetical data of Table 2.2***

	Sample Size	Proportion with B
A Present	150	$.10 = p_1$
A Absent	50	$.20 = p_2$
Total	200	$.125 = \bar{p}$

The value of z (2.5) is

$$z = \frac{|.20 - .10| - \frac{1}{2}\left(\frac{1}{150} + \frac{1}{50}\right)}{\sqrt{.125 \times .875\left(\frac{1}{150} + \frac{1}{50}\right)}} = 1.60, \tag{2.6}$$

which fails to reach the value 1.96 needed for significance at the .05 level. The square of the obtained value of z is 2.56, equal except for rounding errors to the value of χ^2 in (2.2).

Method III. The third method of sampling is like method II in that two samples of predetermined size are contrasted. Unlike method II, however, method III calls for the two samples to be constituted at random. This method lies at the basis of the controlled comparative clinical trial: of a total of $n_{..}$ subjects, $n_{1.}$ are selected at random to be treated with the control treatment and the remaining $n_{2.}$ to be treated with the test treatment.

Of importance are the proportions from the two groups experiencing the outcome under study (e.g., the remission of symptoms). The significance of their difference is assessed by the same statistic (2.5) appropriate to method II. The appropriate further description of the data, however, is shown in Chapter 7 to be different for the two methods.

2.2. "EXACT" ANALYSIS OF A FOURFOLD TABLE

A version of the χ^2 statistic almost as familiar as that given in (2.1) is

$$\chi^2 = \sum_{i=1}^{2} \sum_{j=1}^{2} \frac{\left(|n_{ij} - N_{ij}| - \frac{1}{2}\right)^2}{N_{ij}}, \tag{2.7}$$

where

$$N_{ij} = \frac{n_{i.}n_{.j}}{n_{..}} \tag{2.8}$$

is the frequency one would expect to find in the ith row and jth column under the hypothesis of independence (for sampling method I; see Problem 2.4) or under the hypothesis of equal underlying probabilities (for sampling methods II and III; see Problem 2.5). If the marginal frequencies are small, in the sense that one or more values of N_{ij} are less than 5, it may not be accurate to base the significance test on the chi square (or equivalent normal curve) distribution.

An alternative procedure, due to Fisher (1934) and Irwin (1935), proceeds from restricting attention to fourfold tables in which the marginal frequencies $n_{1.}$, $n_{2.}$, $n_{.1}$, and $n_{.2}$ are fixed at the observed values. Under this restriction, exact probabilities associated with the cell frequencies n_{11}, n_{12}, n_{21}, and n_{22} may be derived from the hypergeometric probability distribution:

$$\Pr\{n_{11}, n_{12}, n_{21}, n_{22}\} = \frac{n_{1.}!n_{2.}!n_{.1}!n_{.2}!}{n_{..}!n_{11}!n_{12}!n_{21}!n_{22}!}, \tag{2.9}$$

where $n! = n(n-1)\ldots3\cdot2\cdot1$. By convention, $0! = 1$.

The Fisher–Irwin "exact" test consists of evaluating the probability in (2.9) for the fourfold table actually observed, say P_{obs}, as well as for all the other tables having the same marginal frequencies. Attention is restricted to those probabilities that are less than or equal to P_{obs}. If the sum of all these probabilities is less than or equal to the prespecified significance level, the hypothesis is rejected; otherwise, it is not.

Consider the hypothetical data in Table 2.8. The exact probability

Table 2.8. Hypothetical data representing small marginal frequencies

	B	$\bar{B}$	Total
A	2	3	5
$\bar{A}$	4	0	4
Total	6	3	9

associated with the table is

$$P_{obs} = \frac{5!4!6!3!}{9!2!3!4!0!} = .1190. \tag{2.10}$$

The three other possible tables consistent with the marginal frequencies of

Table 2.8, together with their associated probabilities, are presented in Table 2.9.

Table 2.9. Remaining possible fourfold tables consistent with the marginal frequencies of Table 2.8

Table		Associated Probability
3	2	
3	1	.4762
4	1	
2	2	.3571
5	0	
1	3	.0476

Only the last of these tables has an associated probability less than or equal to $P_{obs} = .1190$, so the exact significance level associated with the observed table is $.1190 + .0476 = .1666$. The value of χ^2 for the data of Table 2.8 is 1.41. The probability of finding a value this large or larger is 0.23, which is somewhat different from the exact value of 0.17.

Fairly extensive tables exist of the hypergeometric probability distribution. One of the more accessible ones is Table 38 of the Biometrika Tables (Pearson and Hartley, 1970). More extensive tabulations have been compiled by Lieberman and Owen (1961) and by Finney, Latscha, Bennett, and Hsu (1963).

In the illustrative examples in the remainder of this text, the marginal frequencies will be sufficiently large for the chi square or critical ratio tests to be valid.

2.3. YATES' CORRECTION FOR CONTINUITY

Yates (1934) suggested that the correction

$$C_1 = -\frac{1}{2}n_{..} \tag{2.11}$$

be incorporated into expression (2.1) for χ^2 and that the correction

$$C_2 = -\frac{1}{2}\left(\frac{1}{n_{1.}} + \frac{1}{n_{2.}}\right) \tag{2.12}$$

be incorporated into expression (2.5) for z. These corrections take account of the fact that a continuous distribution (the chi square and normal, respectively) is being used to represent the discrete distribution of sample frequencies.

Studies of the effects of the continuity correction have been made by Pearson (1947), Mote, Pavate, and Anderson (1958), and Plackett (1964). On the basis of these and of their own analyses, Grizzle (1967) and Conover (1968, 1974) recommend that the correction for continuity not be applied. They give as their reason an apparent lowering of the actual significance level when the correction is used. A lowered significance level results in a reduction in *power*, that is, in a reduced probability of detecting a real association or real difference in rates.

Mantel and Greenhouse (1968) point out the inappropriateness of Grizzle's (and, by implication, of Conover's) analyses and refute their argument against the use of the correction. The details of Mantel and Greenhouse's refutation are beyond the scope of this book. An outline of their reasoning is provided instead.

In method I, the investigator hypothesizes no association between factors A and B, which means that all four cell probabilities are functions of the marginal proportions $P_{1.}$, $P_{2.}$, $P_{.1}$, and $P_{.2}$ (see Table 2.3). Because the investigator is almost never in the position to specify what the values of these proportions are, he or she must use the obtained marginal frequencies to estimate them.

In methods II and III, the investigator hypothesizes no difference between two independent proportions, P_1 and P_2. Because the investigator is almost never in the position to specify what the value of the hypothesized common proportion is, he must use the obtained marginal frequencies to estimate it.

For each of the three sampling methods, the investigator must therefore proceed to analyze the data with the restriction that his marginal proportions instead of the unknown population proportions characterize the factors under study. This restriction is equivalent to considering the four marginal frequencies $n_{1.}$, $n_{2.}$, $n_{.1}$, and $n_{.2}$ obtained (see Table 2.1) as fixed. As pointed out in Section 2.2, exact probabilities associated with the observed cell frequencies may, under the restriction of fixed marginal frequencies, be derived from the hypergeometric probability distribution. Because the incorporation of the correction for continuity brings probabilities associated with χ^2 and z into closer agreement with the exact probabilities than when it is not incorporated, the correction should always be used.

2.4. ONE-TAILED VERSUS TWO-TAILED TESTS

The chi square test and equivalent normal curve test so far considered are examples of *two-tailed* tests. Specifically, a significant difference is declared either if p_2 is sufficiently *greater* than p_1 or if p_2 is sufficiently *less* than p_1. Suppose that the investigator is interested in an alternative hypothesis specifying a difference in one direction only, say in P_2, the underlying proportion in group 2, being greater than P_1, the underlying proportion in group 1. The power of the comparison can be increased by performing a *one-tailed* test. The investigator can make one of two inferences after a one-tailed test, either that p_2 is significantly greater than p_1 or that it is not; the possible inference that p_1 is significantly greater than p_2 is ruled out as unimportant.

The one-tailed test begins with an inspection of the data to see if they are in the direction specified by the alternative hypothesis. If they are not (e.g., if $p_1 > p_2$ but the investigator was only interested in a difference in the reverse direction), no further calculations are performed and the inference is made that P_2 might not be greater than P_1. If the data are consistent with the alternative hypothesis, the investigator proceeds to calculate either the χ^2 statistic (2.1) or the z statistic (2.5).

The magnitude of χ^2 is assessed for significance as follows. If the investigator desires to have a significance level of α, he enters Table A.1 under the column for 2α. If the calculated value of χ^2 exceeds the tabulated critical value, the investigator infers that the underlying proportions differ in the direction predicted by the alternative hypothesis (e.g., that $P_2 > P_1$). If not, he or she infers that the underlying proportions might not differ in that direction. The magnitude of z is assessed similarly. When the desired significance level is α, Table A.2 is entered with 2α.

It is seen from Tables A.1 and A.2 that critical values for a significance level of 2α are less than those for a significance level of α. An obtained value for the test statistic (either χ^2 or z) that fails to exceed the critical value for a significance level of α may nevertheless exceed the critical value for a significance level of 2α. Because it is easier to reject a hypothesis of no difference with a one-tailed than with a two-tailed test when the proportions differ in the direction specified by the alternative hypothesis, the former test is more powerful than the latter.

As presented here, a one-tailed test is called for only when the investigator is not interested in a difference in the reverse direction from that hypothesized. For example, if the hypothesis is that $P_2 > P_1$, then it will make no difference if either $P_2 = P_1$ or $P_2 < P_1$. Such an instance is assuredly rare. One example where a one-tailed test is called for is when an investigator is comparing the response rate for a new treatment (p_2) with the response rate for a standard treatment (p_1), and when the new treatment will be substituted for the standard in practice only if p_2 is significantly greater than p_1. It will make no difference if the two treatments are equally effective or if the new treatment is actually worse than the standard; in either case, the investigator will stick with the standard.

If, however, the investigator intends to report the results to professional colleagues, he is ethically bound to perform a two-tailed test. For if the results indicate that the new treatment is actually worse than the standard —an inference possible only with a two-tailed test—the investigator is obliged to report this as a warning to others who might plan to study the new treatment.

In the vast majority of research undertakings, two-tailed tests are called for. Even if a theory or a large accumulation of published data suggests that the difference being studied should be in one direction and not the other, the investigator should nevertheless guard against the unexpected by performing a two-tailed test. Especially in such cases, the scientific importance of a difference in the unexpected direction may be greater than yet another confirmation of the difference being in the expected direction.

2.5. A SIMPLE CONFIDENCE INTERVAL FOR THE DIFFERENCE BETWEEN TWO INDEPENDENT PROPORTIONS

When the underlying proportions P_1 and P_2 are not hypothesized to be equal, a good estimate of the standard error of $p_2 - p_1$ is

$$\text{s.e.}(p_2 - p_1) = \sqrt{\frac{p_1 q_1}{n_{1.}} + \frac{p_2 q_2}{n_{2.}}}, \tag{2.13}$$

where $q_1 = 1 - p_1$ and $q_2 = 1 - p_2$. Suppose that both $n_{1.}$ and $n_{2.}$ are large in the sense that $n_{i.} p_i \geqq 5$ and $n_{i.} q_i \geqq 5$ for $i = 1, 2$, and that a $100(1 - \alpha)\%$ confidence interval is desired for the difference $P_2 - P_1$. Let $c_{\alpha/2}$ denote the value cutting off the proportion $\alpha/2$ in the upper tail of the standard normal curve. The interval

$$(p_2 - p_1) - c_{\alpha/2}\sqrt{\frac{p_1 q_1}{n_{1.}} + \frac{p_2 q_2}{n_{2.}}} - \frac{1}{2}\left(\frac{1}{n_{1.}} + \frac{1}{n_{2.}}\right) \leqq P_2 - P_1 \leqq$$
$$(p_2 - p_1) + c_{\alpha/2}\sqrt{\frac{p_1 q_1}{n_{1.}} + \frac{p_2 q_2}{n_{2.}}} + \frac{1}{2}\left(\frac{1}{n_{1.}} + \frac{1}{n_{2.}}\right) \tag{2.14}$$

is such that it will include the true difference approximately $100(1 - \alpha)\%$ of the time.

Consider, for example, the data of Table 2.7. The sample difference is $p_2 - p_1 = .10$, and its estimated standard error is

$$\text{s.e.}(p_2 - p_1) = \sqrt{\frac{.10 \times .90}{150} + \frac{.20 \times .80}{50}} = .062.$$

An approximate 95% confidence interval for the true difference is

$$.10 - 1.96 \times .062 - .013 \leqq P_2 - P_1 \leqq .10 + 1.96 \times .062 + .013,$$

or

$$-.035 \leqq P_2 - P_1 \leqq .235.$$

The interval includes the value 0, which is consistent with the failure above (see equation 2.6) to find a significant difference between p_2 and p_2.

2.6. AN ALTERNATIVE CRITICAL RATIO TEST

Occasionally, the consistency just found between the test for the significance of the difference between p_1 and p_2 and the confidence interval for $P_2 - P_1$ will not obtain. For example, the critical ratio test (2.5) may fail to reject the hypothesis that $P_1 = P_2$, but the confidence interval (2.14) may exclude the value zero. Partly in order to overcome such a possible inconsistency, Eberhardt and Fligner (1977) and Robbins (1977) considered an alternative critical ratio test in which the denominator of the statistic (2.5) is replaced by (2.13). The test statistic then becomes, say,

$$z' = \frac{|p_2 - p_1| - \frac{1}{2}(1/n_{1.} + 1/n_{2.})}{\sqrt{\dfrac{p_1 q_1}{n_{1.}} + \dfrac{p_2 q_2}{n_{2.}}}}. \tag{2.15}$$

Eberhardt and Fligner (1977) compared the performance of the test based on the critical ratio in (2.5) with that of the test based on the statistic in (2.15), although they did not include the continuity correction in their analysis. When $n_{1.} = n_{2.}$, they found that the test based on z' is always more powerful than the test based on z, but that it also tends to reject the hypothesis, when the hypothesis is true, more frequently than the nominal proportion of times, α. When $n_{1.} \neq n_{2.}$, there are some pairs of proportions P_1 and P_2 for which the test based on z' is more powerful and other pairs for which the test based on z is more powerful.

Consider again the data of Table 2.7. The value of z' is

$$z' = \frac{|.20 - .10| - \frac{1}{2}\left(\frac{1}{150} + \frac{1}{50}\right)}{\sqrt{\frac{.10 \times .90}{150} + \frac{.20 \times .80}{50}}} = 1.41,$$

which is less than the value of z in (2.6). For these data, it happens that the classic test based on z [see (2.5)] comes closer to rejecting the hypothesis than the test based on z' [see (2.15)].

Further analysis shows that when the test based on z' is more powerful than the one based on z, the increase in power is slight except when P_1 and P_2 are greatly different (measured by an odds ratio greater than 10; see equation 3.1 for a definition of the odds ratio). There do not, therefore, seem to be any overwhelming reasons for replacing the familiar test based on z (and the equivalent classic chi square test) with the test based on z'.

Problem 2.1. Consider the joint proportions of Table 2.4. Prove that $p_{12} - p_{1.}p_{.2} = -(p_{11} - p_{1.}p_{.1})$; that $p_{21} - p_{2.}p_{.1} = -(p_{11} - p_{1.}p_{.1})$; and that $p_{22} - p_{2.}p_{.2} = p_{11} - p_{1.}p_{.1}$. (*Hint.* Because $p_{11} + p_{12} = p_{1.}$, therefore $p_{12} = p_{1.} - p_{11}$. Use the fact that $1 - p_{.2} = p_{.1}$.)

Problem 2.2. Prove that formulas (2.3) and (2.1) for χ^2 are equal. [*Hint.* Begin by using the result of Problem 2.1 to factor $(|p_{11} - p_{1.}p_{.1}| - 1/(2n_{..}))^2$ out of the summation in formula (2.3). Bring the four remaining terms, $1/(p_{i.}p_{.j})$, over a common denominator and show, using the facts that $p_{1.} + p_{2.} = p_{.1} + p_{.2} = 1$, that the numerator of the resulting expression is unity. Finally, replace each proportion by its corresponding ratio of frequencies.]

Problem 2.3. Prove that the square of z—see (2.5)—is equal to the expression for χ^2 given in (2.1).

Problem 2.4. Show that the estimated expected entry in the ith row and jth column of a fourfold table generated by sampling method I is given by expression 2.8 under the hypothesis of independence. (*Hint.* The expected entry is equal to $n_{..}P_{ij}$. What is the estimate of P_{ij} under the hypothesis of independence?)

Problem 2.5. Show that the estimated expected entry in the ith row and jth column of a fourfold table generated by sampling methods II or III is given by expression 2.8 under the hypothesis of equal underlying probabilities. (*Hint.* Under the hypothesis, $P_1 = P_2 = P$, say. The expected entries are equal to $N_{i1} = n_{i.}P$ and $N_{i2} = n_{i.}Q$, where $Q = 1 - P$. What are the estimates of P and Q under the hypothesis?)

Problem 2.6. When the two sample sizes are equal, the denominator of (2.5) involves $2\bar{p}\bar{q}$, whereas the denominator of (2.15) involves $p_1q_1 + p_2q_2$. Prove that $p_1q_1 + p_2q_2 \leqq 2\bar{p}\bar{q}$ when $n_{1.} = n_{2.}$, with equality if and only if $p_1 = p_2$.

REFERENCES

Barnard, G. A. (1947). Significance tests for 2×2 tables. *Biometrika*, **34**, 123–138.

Conover, W. J. (1968). Uses and abuses of the continuity correction. *Biometrics*, **24**, 1028.

Conover, W. J. (1974). Some reasons for not using the Yates continuity correction on 2×2 contingency tables. (With comments). *J. Am. Stat. Assoc.*, **69**, 374–382.

Eberhardt, K. R. and Fligner, M. A. (1977). A comparison of two tests for equality of two proportions. *Am. Stat.*, **31**, 151–155.

Finney, D. J., Latscha, R., Bennett, B. M., and Hsu, P. (1963). *Tables for testing significance in a 2×2 contingency table*. Cambridge, England: Cambridge University Press.

Fisher, R. A. (1934). *Statistical methods for research workers*, 5th ed. Edinburgh: Oliver and Boyd.

Grizzle, J. E. (1967). Continuity correction in the χ^2-test for 2×2 tables. *Am. Stat.*, **21** (October), 28–32.

Irwin, J. O. (1935). Tests of significance for differences between percentages based on small numbers. *Metron*, **12**, 83–94.

Lieberman, G. J. and Owen, D. B. (1961). *Tables of the hypergeometric probability distribution*. Stanford: Stanford University Press.

Mantel, N. and Greenhouse, S. W. (1968). What is the continuity correction? *Am. Stat.*, **22** (December), 27–30.

Mote, V. L., Pavate, M. V., and Anderson, R. L. (1958). Some studies in the analysis of categorical data. *Biometrics*, **14**, 572–573.

Pearson, E. S. (1947). The choice of statistical tests illustrated on the interpretation of data classed in a 2×2 table. *Biometrika*, **34**, 139–167.

Pearson, E. S. and Hartley, H. O. (Eds.) (1970). *Biometrika tables for statisticians*, Vol. 1, 3rd ed. Cambridge, England: Cambridge University Press.

Pearson, K. (1900). On the criterion that a given system of deviations from the probable in the case of a correlated system of variables is such that it can be reasonably supposed to have arisen from random sampling. *Philos. Mag.*, **50**(5), 157–175.

Plackett, R. L. (1964). The continuity correction in 2×2 tables. *Biometrika*, **51**, 327–337.

Robbins, H. (1977). A fundamental question of practical statistics. *Am. Stat.*, **31**, 97.

Yates, F. (1934). Contingency tables involving small numbers and the χ^2 test. *J. R. Stat. Soc. Suppl.*, **1**, 217–235.

Zubin, J. (1939). Nomographs for determining the significance of the differences between the frequencies of events in two contrasted series or groups. *J. Am. Stat. Assoc.*, **34**, 539–544.

CHAPTER 3

Determining Sample Sizes Needed to Detect a Difference Between Two Proportions

There are two kinds of errors one must guard against in designing a comparative study. Even though these errors can occur in any statistical evaluation, their discussion here is restricted to the case where proportions from two independent samples are compared, that is, to sampling methods II and III. The reader is referred to Cohen (1977, Chapter 7) for a discussion of the two kinds of errors in sampling method I.

The first error, called the Type I error, consists in declaring that the difference in proportions being studied is real when in fact the difference is zero. This kind of error has been given the greater amount of attention in elementary statistics books, and hence in practice. It is typically guarded against simply by setting the significance level for the chosen statistical test, denoted α, at a suitably small probability such as .01 or .05.

This kind of control is not totally adequate, because a literal Type I error probably never occurs in practice. The reason is that the two populations giving rise to the observed samples will inevitably differ to some extent, albeit possibly by a trivially small amount. This is as true in the case of the improvement rates associated with any two treatments as in the case of the disease rates for people possessing and for people not possessing any suspected antecedent factor. It is shown in Problem 3.2 that no matter how small the difference is between the two underlying proportions—provided it is nonzero—samples of sufficiently large size can virtually guarantee statistical significance. Assuming that an investigator desires to detect only differences that are of practical importance, and not merely differences of any magnitude, he should impose the added safeguard of not employing sample sizes that are larger than he needs to guard against the second kind of error.

The second kind of error, called the Type II error, consists in failing to declare the two proportions significantly different when in fact they are different. As just pointed out, such an error is not serious when the proportions are only trivially different. It becomes serious only when the proportions differ to an important extent. The practical control over the Type II error must therefore begin with the investigator's specifying just what difference is of sufficient importance to be detected, and must continue with the investigator's specifying the desired probability of actually detecting it. This probability, denoted $1-\beta$, is called the *power* of the test; the quantity β is the probability of failing to find the specified difference to be statistically significant.

Some means of specifying an important difference between proportions are given in Section 3.1. Having specified the quantities α, $1-\beta$, and the minimum difference in proportions considered important, the investigator may use the mathematical results of Section 3.2 or the values in Table A.3 (described in Section 3.3) to find the sample sizes necessary to assure that (1) any smaller sample sizes will reduce the chances below $1-\beta$ of detecting the specified difference and (2) any appreciably larger sample sizes may increase the chances well above α of declaring a trivially small difference to be significant.

Frequently an investigator is restricted to working with sample sizes dictated by a prescribed budget or by a set time limit. He or she will still find the values in Table A.3 useful, for they can be used to find those differences that the investigator has a reasonable probability of detecting and thus to obtain a realistic appraisal of the chances for success of the study.

Section 3.4 is devoted to the case where unequal sample sizes are planned for beforehand. Some additional points are made in Section 3.5.

3.1. SPECIFYING A DIFFERENCE WORTH DETECTING

An investigator will often have some idea of the order of magnitude of the proportions he or she is studying. This knowledge might come from previous research, from an accumulation of clinical experience, from small-scale pilot work, or from vital statistics reports. Given at least some information, the investigator can, using his or her imagination and expertise, come up with an estimate of a difference between two proportions that is scientifically or clinically important. Given no information, the investigator has no basis for designing the study intelligently and would be hard put to justify designing it at all.

In this section only two of the many approaches to the specification of a difference are illustrated, each with two examples. Let P_1 denote the proportion of members of the first group who possess the attribute or experience the outcome being studied. In general, the designation of one of the groups as the first and the other as the second is arbitrary. Here, however, we designate the first group to be the one that might be viewed as a standard, typically because more information may be available for it than for the other group. Our problem is to determine that value of P_2, the proportion in the second group, which, if actually found, would be deemed on practical grounds to differ sufficiently from P_1 to warrant the conclusion that the two groups are different.

Example 1. In a comparative clinical trial, the first group might represent patients treated by a standard form of therapy. The proportion P_1 might then refer to their observed response (e.g., remission) within a specified period of time following the beginning of treatment. The second group might represent patients treated with an as yet untested alternative form of therapy. A clinically important proportion P_2 associated with the alternative treatment might be determined as follows.

Suppose that it can be assumed that all patients responding to the standard treatment would also respond to the new therapy. Suppose further that if at least an added fraction f, specified by the investigator, of nonresponders to the standard treatment respond to the new one, then the investigator would wish to identify the new treatment as superior to the old. Since the proportion of nonresponders to the standard treatment is $1 - P_1$, a clinically important value of P_2 is therefore $P_1 + f(1 - P_1)$.

For example, the remission rate associated with the standard treatment might be $P_1 = .60$. If the investigator will view the alternative treatment as superior to the standard only if it succeeds in remitting the symptoms of at least one quarter of those patients who would not otherwise show remission, so that $f = .25$, then he is in effect specifying a value $P_2 = .60 + .25 \times (1 - .60) = .70$ as one that is different to a practically important extent from $P_1 = .60$.

In this example, the proportions P_1 and P_2 refer to a favorable outcome, namely, a remission of symptoms. Similar reasoning can be applied to studies in which an untoward event (e.g., morbidity or mortality) is of especial interest.

Example 2. Suppose that the rate of premature births is P_1 among women of a certain age and race who attend the prenatal clinic in their community hospital. An intensive education program aimed at nonattenders is to be undertaken only if P_2, the rate of premature births among prospective mothers not attending the clinic but otherwise similar to the clinic attenders, is sufficiently greater than P_1.

It is reasonable to assume that a mother who delivered a premature offspring even after having attended the clinic would also have done so if she had not attended the clinic. The added risk of prematurity associated with nonattendance can thus only operate on mothers who do not deliver premature offspring after attending the clinic. If f denotes an added risk that is of practical importance, the hypothesized value of P_2 is then $P_1 + f(1 - P_1)$.

Suppose, for example, that the prematurity rate for clinic attenders is $P_1 = .25$. Suppose further that an education program is to be undertaken only if, of women who attend the clinic and who do not deliver a premature offspring, at least 20% would have delivered a premature offspring by not attending the clinic. The value of f is then .20, and the hypothesized value of P_2 is $.25 + .20 \times (1 - .25) = .40$.

The approach to the comparison of two proportions exemplified by these two examples has been recommended and applied by Sheps (1958, 1959, 1961). It is considered again in Chapter 7, where the *relative difference* $f = (P_2 - P_1)/(1 - P_1)$ is studied in greater detail.

Example 3. One often undertakes a study in order to replicate (or refute) another's research findings, or to see if one's own previous findings hold up in a new setting. One must be careful, however, to control for the possibility that the rates in the groups being compared are at levels in the new setting different from those in the old. This possibility effectively rules out attempting to recapture the simple difference between rates found previously.

For example, suppose that the rate of depression among women aged 20–49 was found in the mental hospitals of one community to be 40% higher than the rate among men aged 20–49. It will be impossible to find the same difference in the mental hospitals of a new community if, there, the rate of depression among males aged 20–49 is 70%, since a difference of 40% implies an impossible rate of 110% for women similarly aged.

A measure of the degree of inequality between two rates is therefore needed that may be expected to remain constant if the levels at which the rates apply vary across settings. A measure frequently found to have this property is the *odds ratio*, denoted ω. The odds ratio is discussed in greater detail in Chapters 5 and 6. Here we give only its definition.

If P_1 is the rate at which an event occurs in the first population, then the odds associated with that event in the first population are, say, $\Omega_1 = P_1/Q_1$, where $Q_1 = 1 - P_1$. Similarly, the odds associated with the event in the second population are $\Omega_2 = P_2/Q_2$. The odds ratio is simply the ratio of these two odds,

$$\omega = \frac{\Omega_2}{\Omega_1} = \frac{P_2 Q_1}{P_1 Q_2}. \tag{3.1}$$

The odds ratio is also termed the cross-product ratio (Fisher, 1962) and the approximate relative risk (Cornfield, 1951). If $P_2 = P_1$, then $\omega = 1$. If $P_2 < P_1$, then $\omega < 1$. If $P_2 > P_1$, then $\omega > 1$.

Suppose that a study is to be carried out in attempt to replicate the results of a previous study in which the odds ratio was found to be ω. If, in the community in which the new study is to be conducted, the rate of occurrence of the event in the first group is P_1, and if the same value ω for the odds ratio is hypothesized to apply in the new community, then the value hypothesized for P_2 is

$$P_2 = \frac{\omega P_1}{\omega P_1 + Q_1}. \tag{3.2}$$

For example, suppose that the value $\omega = 2.5$ had previously been found as the ratio of the odds for depression among female mental hospital patients aged 20–49 to the odds for male mental hospital patients similarly aged. If the same value for the odds ratio is hypothesized to obtain in the mental hospitals of a new community, and if in that community's mental hospitals the rate of depression among male patients aged 20–49 is approximately $P_1 = .70$, then the rate among female patients aged 20–49 is hypothesized to be approximately

$$P_2 = \frac{2.5 \times .70}{2.5 \times .70 + .30} = .85.$$

An important property of the odds ratio to be demonstrated in Chapters 5 and 6 is that the same value should be obtained whether the study is a prospective or retrospective one. This fact may be taken advantage of if an investigator wishes to replicate a previous study but alters the research design from, say, a retrospective to a prospective study.

Example 4. Suppose that a case-control (retrospective) study was conducted in a certain school district. School children with emotional disturbances requiring psychological care were compared with presumably normal children on a number of antecedent characteristics. Suppose it was found that one quarter of the emotionally disturbed children versus one tenth of the normal controls had lost (by death, divorce, or separation) at least one parent before age 5. The odds ratio is then, from (3.1),

$$\omega = \frac{.25 \times .90}{.10 \times .75} = 3.0.$$

Suppose that a study of this association is to be conducted prospectively in a new community by following through their school years a sample of children who begin school with both parents alive and at home (group 1) and a sample who begin with at least one parent absent from the home (group 2), with the proportions developing emotional problems being compared.

From a survey of available school records, the investigator in the new school district is able to estimate that P_1, the proportion of children beginning school with both parents at home who ultimately develop emotional problems, is $P_1 = .05$. If the value $\omega = 3.0$ found in the retrospective study is hypothesized to apply in the new school district, the investigator is effectively hypothesizing a value (see equation 3.2)

$$P_2 = \frac{3.0 \times .05}{3.0 \times .05 + .95} = .136,$$

or approximately 15%, as the rate of emotional disturbance during school years among children who have lost at least one parent before age 5.

The methods just illustrated may be of use in generating hypotheses for studies to be carried out within a relatively short time, but are likely to prove inadequate for long-term comparative studies. Halperin, Rogot, Gurian, and Ederer (1968) give a model and some numerical results when two long-term therapies are to be compared and when few or no dropouts are expected. When dropouts are likely to occur, the model of Schork and Remington (1967) may be useful for generating hypotheses. If the study calls for the comparison of more than two treatments, or if outcome is measured on a scale with more than two categories, the results of Lachin (1977) should be useful.

3.2. THE MATHEMATICS OF SAMPLE SIZE DETERMINATION

We assume in this section and the next that the sample sizes from the two populations being compared, n_1 and n_2, are equal to a common n. We find the value for the common sample size n so that (1) if in fact there is no difference between the two underlying proportions, then the chance is α of falsely declaring the two proportions to differ; and (2) if in fact the proportions are P_1 and $P_2 \neq P_1$, then the chance is $1-\beta$ of correctly declaring the two proportions to differ. Since this section only derives the mathematical results on which the values in Table A.3 (described in Section 3.3) are based, it is not essential to the sections that follow.

We begin by deriving the sample size, say n', required in both the groups if we ignore the continuity correction. With n' as a first approximation, we then obtain a formula for the desired sample size per group, n, that is appropriate when the test statistic incorporates the continuity correction.

Suppose that the proportions found in the two samples are p_1 and p_2. The statistic used for testing the significance of their difference is, temporarily ignoring the continuity correction,

$$z = \frac{p_2 - p_1}{\sqrt{2\bar{p}\bar{q}/n'}}, \tag{3.3}$$

where

$$\bar{p} = \tfrac{1}{2}(p_1 + p_2)$$

and

$$\bar{q} = 1 - \bar{p}.$$

To assure that the probability of a Type I error is α, the difference between p_1 and p_2 will be declared significant only if

$$|z| > c_{\alpha/2}, \tag{3.4}$$

where $c_{\alpha/2}$ denotes the value cutting off the proportion $\alpha/2$ in the upper tail of the standard normal curve and $|z|$ is the absolute value of z, always a nonnegative quantity. For example, if $\alpha = .05$, then $c_{.05/2} = c_{.025} = 1.96$, and the difference is declared significant if either $z > 1.96$ or $z < -1.96$.

If the difference between the underlying proportions is actually $P_2 - P_1$, we wish the chances to be $1 - \beta$ of rejecting the hypothesis, that is, of having the outcome represented in (3.4) actually occur. Thus we must find the value of n' such that, when $P_2 - P_1$ is the difference between the proportions,

$$\Pr\left\{ \frac{|p_2 - p_1|}{\sqrt{2\bar{p}\bar{q}/n'}} > c_{\alpha/2} \right\} = 1 - \beta. \tag{3.5}$$

The probability in (3.5) is the sum of two probabilities,

$$1 - \beta = \Pr\left\{ \frac{p_2 - p_1}{\sqrt{2\bar{p}\bar{q}/n'}} > c_{\alpha/2} \right\} + \Pr\left\{ \frac{p_2 - p_1}{\sqrt{2\bar{p}\bar{q}/n'}} < -c_{\alpha/2} \right\}. \tag{3.6}$$

If P_2 is hypothesized to be greater than P_1, then the second probability on the right-hand side of (3.6)—representing the event that p_2 is appreciably less than p_1—is near zero (see Problem 3.1). Thus we need only find the value of n' such that, when $P_2 - P_1$ is the actual difference,

$$1 - \beta = \Pr\left\{ \frac{p_2 - p_1}{\sqrt{2\bar{p}\bar{q}/n'}} > c_{\alpha/2} \right\}. \tag{3.7}$$

The probability in (3.7) cannot yet be evaluated because the mean and the standard error of $p_2 - p_1$ appropriate when $P_2 - P_1$ is the actual difference have not yet been taken into account. The mean of $p_2 - p_1$ is $P_2 - P_1$, and its standard error is

$$\text{s.e.}(p_2 - p_1) = \sqrt{(P_1 Q_1 + P_2 Q_2)/n'}\ , \tag{3.8}$$

where $Q_1 = 1 - P_1$ and $Q_2 = 1 - P_2$.

The following development of (3.7) can be traced using only simple algebra:

$$\begin{aligned}
1 - \beta &= \Pr\left\{ (p_2 - p_1) > c_{\alpha/2}\sqrt{2\bar{p}\bar{q}/n'} \right\} \\
&= \Pr\left\{ (p_2 - p_1) - (P_2 - P_1) > c_{\alpha/2}\sqrt{2\bar{p}\bar{q}/n'} - (P_2 - P_1) \right\} \\
&= \Pr\left\{ \frac{(p_2 - p_1) - (P_2 - P_1)}{\sqrt{(P_1 Q_1 + P_2 Q_2)/n'}} > \frac{c_{\alpha/2}\sqrt{2\bar{p}\bar{q}/n'} - (P_2 - P_1)}{\sqrt{(P_1 Q_1 + P_2 Q_2)/n'}} \right\}.
\end{aligned} \tag{3.9}$$

The final probability in (3.9) can be evaluated using tables of the normal distribution because, when the underlying proportions are P_2 and P_1, the quantity

$$Z = \frac{(p_2 - p_1) - (P_2 - P_1)}{\sqrt{(P_1 Q_1 + P_2 Q_2)/n'}} \tag{3.10}$$

has, to a good approximation if n' is large, the standard normal distribution.

Let $c_{1-\beta}$ denote the value cutting off the proportion $1 - \beta$ in the upper tail and β in the lower tail of the standard normal curve. Then, by definition,

$$1 - \beta = \Pr\{Z > c_{1-\beta}\}. \tag{3.11}$$

By matching (3.11) with the last probability of (3.9), we find that the value of n' we seek is the one that satisfies

$$c_{1-\beta} = \frac{c_{\alpha/2}\sqrt{2\bar{p}\bar{q}/n'} - (P_2 - P_1)}{\sqrt{(P_1Q_1 + P_2Q_2)/n'}} = \frac{c_{\alpha/2}\sqrt{2\bar{p}\bar{q}} - (P_2 - P_1)\sqrt{n'}}{\sqrt{P_1Q_1 + P_2Q_2}}. \tag{3.12}$$

Before presenting the final expression for n', we note that (3.12) is a function not only of P_1 and P_2, which may be hypothesized by the investigator, but also of $\bar{p}\bar{q}$, which is observable only after the study is complete. If n' is fairly large, however, $\bar{p}$ will be close to

$$\bar{P} = \frac{P_1 + P_2}{2}, \tag{3.13}$$

and, more importantly, $\bar{p}\bar{q}$ will be close to $\bar{P}\bar{Q}$, where $\bar{Q} = 1 - \bar{P}$. Therefore, replacing $\sqrt{2\bar{p}\bar{q}}$ in (3.12) by $\sqrt{2\bar{P}\bar{Q}}$ and solving for n', we find

$$n' = \frac{\left(c_{\alpha/2}\sqrt{2\bar{P}\bar{Q}} - c_{1-\beta}\sqrt{P_1Q_1 + P_2Q_2}\right)^2}{(P_2 - P_1)^2} \tag{3.14}$$

to be the required sample size from *each* of the two populations being compared when the continuity correction is not employed.

The minus sign associated with $c_{1-\beta}$ in the numerator of (3.14) is not a typographical error. Recall the definition of c_p given earlier, namely, the solution of the equation $\Pr\{z > c_p\} = p$, where z has the standard normal distribution. When p exceeds .5, c_p will be negative.

Haseman (1978) found that equation 3.14 gives values that are too low, in the sense that the power of the test based on sample sizes $n_1 = n_2 = n'$ is less than $1 - \beta$ when P_1 and P_2 are the underlying probabilities. Kramer and Greenhouse (1959) proposed an adjustment to (3.14) based on a double use of the continuity correction, once in the statistic (3.3) and again in the statistic (3.10). Their adjustment, which was tabulated in the first edition of this book, was found by Casagrande, Pike, and Smith (1978b) to result in an overcorrection.

By incorporating the continuity correction only in the test statistic (3.3), the latter authors derived

$$n = \frac{n'}{4}\left[1 + \sqrt{1 + \frac{4}{n'|P_2 - P_1|}}\right]^2 \tag{3.15}$$

as the sample size required in each group to provide, to an excellent degree of approximation, the desired significance level and power. The sample sizes tabulated in Table A.3 (which is different from Table A.3 of the first edition) are based on this formula. The values there agree very well with those tabulated by Casagrande, Pike, and Smith (1978a) and Haseman (1978). Ury and Fleiss (1980) present comparisons with some other formulas.

To a remarkable degree of accuracy (especially when n' and $|P_2 - P_1|$ are such that $n'|P_2 - P_1| \geqq 4$),

$$n = n' + \frac{2}{|P_2 - P_1|}. \tag{3.16}$$

This result, due to Fleiss, Tytun, and Ury (1980), is useful both in arriving quickly at an estimate of required sample sizes and in estimating the power associated with a study involving prespecified sample sizes. Suppose that one can study no more than a total of $2n$ subjects. If the significance level is α and if the two underlying proportions one is seeking to distinguish are P_1 and P_2, one can invert (3.16) and (3.14) to obtain

$$c_{1-\beta} = \frac{c_{\alpha/2}\sqrt{2\bar{P}\bar{Q}} - |P_2 - P_1|\sqrt{n - \dfrac{2}{|P_2 - P_1|}}}{\sqrt{P_1Q_1 + P_2Q_2}} \tag{3.17}$$

as the equation defining the normal curve deviate corresponding to the power associated with the proposed sample sizes. Table A.2 may then be used to find the power itself.

3.3. USING THE SAMPLE SIZE TABLES

Table A.3 gives the equal sample sizes necessary in each of the two groups being compared for varying values of the hypothesized proportions P_1 and P_2, for varying significance levels ($\alpha = .01$, .02, .05, .10, and .20) and for varying powers [$1 - \beta = .50$, .65(.05).95, .99]. The value $1 - \beta = .50$

is included not so much because an investigator will intentionally embark on a study for which the chances of success are only 50 : 50, but rather to help provide a baseline for the minimum sample sizes necessary.

The probability of a Type I error, α, is frequently specified first. If, on the basis of declaring the two proportions to differ significantly, the decision is made to conduct further (possibly expensive) research or to replace a standard form of treatment with a new one, then the Type I error is serious and α should be kept small (say, .01 or .02). If the study is aimed only at adding to the body of published knowledge concerning some theory, then the Type I error is less serious, and α may be increased to .05 or .10 (the more the published evidence points to a difference, the higher may α safely be set).

Having specified α, the investigator needs next to specify the chances $1 - \beta$ of detecting the proportions as different if, in the underlying populations, the proportions are P_1 and P_2. The criterion suggested by Cohen (1977, p. 56) seems reasonable. He supposes it to be the typical case that a Type I error is some four times as serious as a Type II error. This implies that one should set β, the probability of a Type II error, approximately equal to 4α, so that the power becomes, approximately, $1 - \beta = 1 - 4\alpha$. Thus when $\alpha = .01$, $1 - \beta$ may be set at .95; for $\alpha = .02$, set $1 - \beta = .90$; and for $\alpha = .05$, set $1 - \beta = .80$. When α is larger than .05, it seems safe to take $1 - \beta = .75$ or less. The use of Table A.3 will be illustrated for each of the examples of Section 3.1.

Example 1. The investigator hypothesizes a remission rate of $P_1 = .60$ for the standard treatment and one of $P_2 = .70$ for the new treatment. The significance level α is set at .01 and the power $1 - \beta$ at .95. It is necessary to study 827 patients under the standard treatment and 827 under the new one, the assignment of patients to treatment groups being at random, in order to guarantee the desired significance level and power.

To reduce the chances of detecting a difference to $1 - \beta = .75$ without increasing the significance level, it is necessary to study 499 patients with each treatment. If the investigator can afford to study no more than a total of 600 patients, so that each treatment would be applied to no more than 300 subjects, the chances of detecting the hypothesized difference become less than 50 : 50.

Example 2. The investigator hypothesizes a prematurity rate of $P_1 = .25$ for clinic attenders and one of $P_2 = .40$ for nonattenders. The significance level α is set at .01 and the power $1 - \beta$ at .95. It is necessary to study 357 mothers from each group, all women being within a specified age range. If the significance level is increased to $\alpha = .02$ and the power lowered to $1 - \beta = .90$, 265 mothers from each group are needed.

Example 3. The investigator hypothesizes the rate of depression among male mental hospital patients aged 20–49 to be $P_1 = .70$, and the rate among similarly aged female patients to be $P_2 = .85$. The significance level α is set at .05 and the power $1 - \beta$ at .80. It is necessary to study 134 patients of each sex. If the investigator had planned to study 250 patients of each sex, the chances of picking up the hypothesized difference would be over 95%, a value that might be larger than necessary.

Example 4. The investigator hypothesizes that the proportion developing emotional problems among children beginning school with both parents at home is $P_1 = .05$, and that the proportion among children beginning school with at least one parent absent is approximately $P_2 = .15$. The significance level α is set at .05 and the power $1 - \beta$ at .80. It is necessary to follow up 160 of each kind of child, making certain that the two cohorts are similar with respect to sex and race. If the investigator can afford to study no more than 120 of each kind of child, and if he or she is willing to increase the chance of making a Type I error to $\alpha = .10$, there will still be over a 75% chance of finding the groups to be different if the hypothesized values of P_1 and P_2 are correct.

3.4. UNEQUAL SAMPLE SIZES

Suppose that considerations of relative cost, the desire for more precise estimates for one group than for the other, or other factors (Walter, 1977) lead to the selection of samples of unequal size from the two populations. Let the required sample size from the first population be denoted m and that from the second, rm $(0 < r < \infty)$, with r specified in advance. The total sample size is, say, $N = (r + 1)m$.

If p_1 and p_2 are the two resulting sample proportions, the test statistic is

$$z = \frac{|p_2 - p_1| - \dfrac{1}{2m}\left(\dfrac{r+1}{r}\right)}{\sqrt{\dfrac{\bar{p}\bar{q}(r+1)}{mr}}},$$

where $\bar{p} = (p_1 + rp_2)/(r + 1)$ and $\bar{q} = 1 - \bar{p}$. If the desired significance level is α, and if a power of $1 - \beta$ is desired against the alternative hypothesis that $P_1 \neq P_2$, with P_1 and P_2 specified, the same kind of development as in

Section 3.2 leads to the value

$$m = \frac{m'}{4}\left\{1 + \sqrt{1 + \frac{2(r+1)}{m'r|P_2 - P_1|}}\right\}^2 \tag{3.18}$$

as the required sample size from the first population and rm as that from the second. In (3.18),

$$m' = \frac{\left[c_{\alpha/2}\sqrt{(r+1)\bar{P}\bar{Q}} - c_{1-\beta}\sqrt{rP_1Q_1 + P_2Q_2}\right]^2}{r(P_2 - P_1)^2}, \tag{3.19}$$

where $\bar{P} = (P_1 + rP_2)/(r+1)$ and $\bar{Q} = 1 - \bar{P}$.

As found by Fleiss, Tytun, and Ury (1980), m is approximately equal to

$$m = m' + \frac{r+1}{r|P_2 - P_1|}. \tag{3.20}$$

Note that equations 3.14–3.16 are special cases of those presented above when the two sample sizes are equal (i.e., when $r = 1$).

Consider again the example of comparing the rates of prematurity in the offspring of clinic attenders (P_1 is hypothesized to be .25) and in the offspring of nonattenders (P_2 is hypothesized to be .40). The significance level is again set at $\alpha = .01$ and the power at $1 - \beta = .95$. Suppose that recruitment of clinic attenders is easier than recruitment of nonattenders and that the investigator decides to study half as many nonattenders as attenders, so that $r = .5$. The value of m' (3.19) is

$$m' = \frac{\left[2.576\sqrt{1.5 \times .30 \times .70} - (-1.645)\sqrt{.5 \times .25 \times .75 + .40 \times .60}\right]^2}{.5(.15)^2}$$

$$= 510.34$$

so that, by (3.18), the required sample size from the population of clinic attenders is

$$m = \frac{510.34}{4}\left(1 + \sqrt{1 + \frac{2(1.5)}{510.34 \times .5 \times .15}}\right)^2$$

$$= 530.$$

Equation (3.20) yields the same value to the nearest integer.

The two required sample sizes are therefore $n_1 = m = 530$ and $n_2 = .5m = 265$. The total number of women required is $N = 795$, some 80 more than were required for the case of equal sample sizes (see Example 2 of Section 3.3).

Problem 3.5 is devoted to further applications of the results of this section.

3.5. SOME ADDITIONAL COMMENTS

Cohen (1977, Chapter 6) gives a set of tables and Feigl (1978) gives a set of graphs for determining sample sizes when the same parameters as those preceding are specified. Since the significance test they consider is different from the standard one and does not incorporate the continuity correction, their sample sizes are slightly different from those in Table A.3. In general, Cohen's tables and Feigl's charts have to be used if the investigator can hypothesize the order of magnitude of the difference between P_1 and P_2, but not their separate magnitudes. If the investigator can hypothesize the separate values of P_1 and P_2, the current table is preferable.

It has so far been assumed that a two-tailed test (see Section 2.4) would be used in comparing the two proportions. If the investigator chooses to perform a one-tailed test, he can still use Table A.3, but should enter it with twice the significance level. Thus for a one-tailed significance level of .01, one uses $\alpha = .02$; for a one-tailed significance level of .05, one uses $\alpha = .10$. No change in the value of $1 - \beta$ is necessary.

Problem 3.1. Suppose that $P_2 > P_1$ and that n' is the sample size studied in each group. Let Z represent a random variable having the standard normal distribution.

(*a*) Show that the probability that p_2 is significantly *less* than p_1,

$$\Pr\left\{\frac{p_2 - p_1}{\sqrt{2\bar{p}\bar{q}/n'}} < -c_{\alpha/2}\right\},$$

is approximately equal to

$$\Pi = \Pr\left\{Z < \frac{-c_{\alpha/2}\sqrt{2\bar{P}\bar{Q}} - (P_2 - P_1)\sqrt{n'}}{\sqrt{P_1Q_1 + P_2Q_2}}\right\}.$$

(*b*) If $P_2 = P_1$, then $\Pi = \alpha/2$. Thus if $P_2 > P_1$, show why $\Pi < \alpha/2$. (*Hint*. Prove that $\sqrt{P_1Q_1 + P_2Q_2} < \sqrt{2\bar{P}\bar{Q}}$ whenever $P_2 \neq P_1$. Therefore, if $P_2 > P_1$,

$$\frac{-c_{\alpha/2}\sqrt{2\bar{P}\bar{Q}} - (P_2 - P_1)\sqrt{n'}}{\sqrt{P_1Q_1 + P_2Q_2}} < -c_{\alpha/2} - \frac{(P_2 - P_1)\sqrt{n'}}{\sqrt{P_1Q_1 + P_2Q_2}} < -c_{\alpha/2}.)$$

(c) Π is small even if P_2 is only slightly larger than P_1 and even if n' is small. Find the value of Π when $P_1 = .10$, $P_2 = .11$, $n' = 9$, and $\alpha = .05$. Note that the probability found in Table A.2 must be *halved*.

Problem 3.2. Let the notation and assumptions of Problem 3.1 be used again. The power of the test for comparing p_1 and p_2 is approximately

$$1 - \beta = \Pr\left\{Z > \frac{c_{\alpha/2}\sqrt{2\bar{P}\bar{Q}} - (P_2 - P_1)\sqrt{n'}}{\sqrt{P_1Q_1 + P_2Q_2}}\right\}.$$

(a) Show that $1 - \beta$ approaches unity as n' becomes large but α remains fixed. (*Hint*. What is the probability that a standard normal random variable exceeds -1? -2? -3? What value does the expression to the right of the inequality sign above approach as n' increases?)

(b) Show that $1 - \beta$ decreases as α becomes small but n' remains fixed.

Problem 3.3. Show that, when the test statistic incorporates the continuity correction, equation 3.15 gives the sample size needed in each group to achieve a power of $1 - \beta$ when the underlying probabilities are P_1 and P_2. (*Hint*. The hypothesis that $P_1 = P_2$ is rejected if

$$\frac{|p_2 - p_1| - \frac{1}{n}}{\sqrt{2\bar{p}\bar{q}/n}} > c_{\alpha/2}.$$

Assume that $P_2 > P_1$, and apply the same algebraic development as in Section 3.2 to arrive at

$$\sqrt{n} - \frac{1}{\sqrt{n}\,(P_2 - P_1)} = \sqrt{n'}$$

as the equation defining n, where n' is defined in (3.14). Finally, solve the above quadratic equation for $\sqrt{n}$ and then for n, noting that one of the two roots is negative and therefore inadmissible.)

Problem 3.4. An investigator hypothesizes that the improvement rate associated with a placebo is $P_1 = .45$, and that the improvement rate associated with an active drug is $P_2 = .65$. The plan is to perform a *one-tailed* test.

(*a*) If a significance level of $\alpha = .01$ and a power of $1 - \beta = .95$ are desired, how large a sample per treatment must he study?

(*b*) How large must the sample sizes be if the significance level is relaxed to $\alpha = .05$ and the power to $1 - \beta = .80$?

(*c*) What is the power of his one-tailed test if he can study only 52 patients per group, and if his significance level is $\alpha = .05$? (*Hint*. Because we are considering a one-tailed test, be sure to replace $c_{\alpha/2}$ in equation 3.17 by 1.645.)

Problem 3.5. The comparison of two populations with hypothesized probabilities $P_1 = .25$ and $P_2 = .40$ was considered several times in the text.

(*a*) Continuing to take $\alpha = .01$ and $1 - \beta = .95$, fill in the remaining values in the second, third, and fourth columns of the following table.

Ratio of Sample Sizes ($n_2/n_1 = r$)	n_1	n_2	Required Total Sample Size	Total Cost
.5	530	265	795	$8,480
.6	—	—	—	—
.7	—	—	—	—
.8	—	—	—	—
.9	—	—	—	—
1	357	357	714	$7,854

(*b*) Suppose that the average cost of studying a member of group 1 is $10, and that the average cost for group 2 is $12. Find the total cost associated with each tabulated value of *r*. For which of these ratios of sample sizes is the total cost minimized?

(*c*) Suppose that the investigator can afford to spend only $6,240 on the study and decides to employ the value of *r* found in part (*b*). What is the value of *m*? What is the corresponding value of m' when $P_1 = .25$ and $P_2 = .40$? (Solve equation 3.20 for m'.) What is the corresponding value of $c_{1-\beta}$ when $\alpha = .01$? (Solve equation 3.19 for $c_{1-\beta}$.) What, finally, is the power of the test?

REFERENCES

Casagrande, J. T., Pike, M. C., and Smith, P. G. (1978a). The power function of the "exact" test for comparing two binomial distributions. *Appl. Stat.*, **27**, 176–180.

Casagrande, J. T., Pike, M. C., and Smith, P. G. (1978b). An improved approximate formula for calculating sample sizes for comparing two binomial distributions. *Biometrics*, **34**, 483–486.

Cohen, J. (1977). *Statistical power analysis for the behavioral sciences*, revised ed. New York: Academic Press.

Cornfield, J. (1951). A method of estimating comparative rates from clinical data. Applications to cancer of the lung, breast and cervix. *J. Natl. Cancer Inst.*, **11**, 1269–1275.

Feigl, P. (1978). A graphical aid for determining sample size when comparing two independent proportions. *Biometrics*, **34**, 111–122.

Fisher, R. A. (1962). Confidence limits for a cross-product ratio. *Aust. J. Stat.*, **4**, 41.

Fleiss, J. L., Tytun, A., and Ury, H. K. (1980). A simple approximation for calculating sample sizes for comparing independent proportions. *Biometrics*, **36**, 343–346.

Halperin, M., Rogot, E., Gurian, J., and Ederer, F. (1968). Sample sizes for medical trials with special reference to long-term therapy. *J. Chronic Dis.*, **21**, 13–24.

Haseman, J. K. (1978). Exact sample sizes for use with the Fisher-Irwin test for 2×2 tables. *Biometrics*, **34**, 106–109.

Kramer, M. and Greenhouse, S. W. (1959). Determination of sample size and selection of cases. Pp. 356–371 in National Academy of Sciences—National Research Council Publication 583, *Psychopharmacology: Problems in evaluation*. Washington, D.C.

Lachin, J. M. (1977). Sample size determinations for $r \times c$ comparative trials. *Biometrics*, **33**, 315–324.

Schork, M. A. and Remington, R. D. (1967). The determination of sample size in treatment-control comparisons for chronic disease studies in which drop-out or non-adherence is a problem. *J. Chronic Dis.*, **20**, 233–239.

Sheps, M. C. (1958). Shall we count the living or the dead? *New Engl. J. Med.*, **259**, 1210–1214.

Sheps, M. C. (1959). An examination of some methods of comparing several rates or proportions. *Biometrics*, **15**, 87–97.

Sheps, M. C. (1961). Marriage and mortality. *Am. J. Public Health*, **51**, 547–555.

Ury, H. K. and Fleiss, J. L. (1980). On approximate sample sizes for comparing two independent proportions with the use of Yates' correction. *Biometrics*, **36**, 347–351.

Walter, S. D. (1977). Determination of significant relative risks and optimal sampling procedures in prospective and retrospective comparative studies of various sizes. *Am. J. Epidemiol.*, **105**, 387–397.

CHAPTER 4

How to Randomize

A number of references have been made so far to randomization. Subjects were described as being "randomly selected" from a larger group of subjects, as being "randomly assigned" to one or another treatment group, and so on. This chapter gives some methods for achieving randomness of selection (needed in sampling methods I and II; see Section 4.1) or of assignment (needed in sampling method III; see Sections 4.2 and 4.3).

It is important to bear in mind that randomness inheres not in the samples one ends up with, but in the method used to generate those samples. When we say that a group of a given size is a *simple random sample* from a larger group, we mean that each possible sample of that size has the same chance of being selected. When we say that treatments are assigned to subjects *at random*, we mean that each subject is equally likely to receive each of the treatments.

The necessity for randomization in controlled experiments was first pointed out by Fisher (1935). In the context of comparative trials, Hill (1962) describes what is accomplished by the random assignment of treatments to subjects:

> [Randomization] ensures three things: it ensures that neither our personal idiosyncracies... nor our lack of balanced judgement has entered into the construction of the different treatment groups...; it removes the danger, inherent in an allocation based upon personal judgement, that believing we may be biased in our judgements we endeavour to allow for that bias, to exclude it, and that in so doing we may overcompensate and by thus "leaning over backward" introduce a lack of balance from the other direction; and, having used a random allocation, the sternest critic is unable to say when we eventually dash into print that quite probably the groups were differentially biased through our predilections or through our stupidity [p. 35].

In spite of the widespread use of randomization in comparative trials, debate continues as to its universal appropriateness. This issue will be addressed in Section 7.3.

Table A.4 presents 20,000 random digits, arrayed on each page in ten columns of 50 numbers, with five digits to each number. Some illustrations of the use of the table follow.

4.1. SELECTING A SIMPLE RANDOM SAMPLE

Suppose that a firm has 250 employees, of whom 100 are to be selected for a thorough physical examination and an interview to determine health habits. A simple random sample of 100 out of the larger group of 250 may be selected as follows.

Examine consecutive three digit numbers, ignoring any that are 000 or between 251 and 999. Of the numbers between 001 and 250, list the first 100 distinct ones that are encountered. When a column is completed, proceed to the next one. The 100 numbers listed designate the employees to be selected. If a card file exists containing the names of all employees—the order in which the names appear is immaterial—the employees may be numbered from 1 to 250 and those whose numbers appear on the list of random numbers would be selected.

To illustrate, let us begin in the second column of the second page of Table A.4. Each number in the column contains five digits, of which only the first three will be examined. The first five numbers in the column, after deleting their last two digits, are 670, 716, 367, 988, and 283—all greater than 250. The sixth number, 142, is between 001 and 250 and thus designates one of the employees to be selected. The other numbers selected from the second column are seen to be 021, 166, 127, 060, 098, 219, 161, 042, 043, 157, 113, 234, 024, 028, and 128.

Having exhausted the second column, and still requiring 84 additional numbers, we proceed to examine the third column of the second page. At the end of the third column, 29 distinct numbers between 001 and 250 are available. In numerical order, they are

001	028	052	107	142	166
014	034	059	113	146	219
021	042	060	121	157	234
024	043	080	127	160	244
026	047	098	128	161	

Subsequent columns are examined similarly until an additional 71 distinct numbers between 001 and 250 are found. If a previously selected number is encountered (e.g., 244 is encountered again in column 4), it is ignored.

4.2. RANDOMIZATION IN A CLINICAL TRIAL

Suppose that a clinical trial is to be carried out to compare the effectiveness of a drug with that of an inert placebo, and suppose that 50 patients are to be studied under each drug, requiring a total of 100 patients. Suppose, finally, that patients enter the study serially over time, and so are not all available at once.

Two randomization methods exist. The first calls for selecting 50 distinct numbers between 001 and 100, as described in Section 4.1, and for letting these numbers designate the patients who will receive the active drug. The remaining 50 numbers designate the patients who will receive the placebo.

This method has two drawbacks. First, if the study must be terminated prematurely, there exists a strong likelihood that the total number of patients who had been assigned the active drug up to the termination date will not equal the total number who had been assigned the placebo. Statistical comparisons lose sensitivity if the sample sizes differ. Second, if the clinical characteristics of the patients entering the trial during one interval of time differ from those of patients entering during another, or if the standards of assessment change over time, then the two treatment groups might well end up being different, in spite of randomization, either in the kinds of patients they contain or in the standards of assessment applied to them (see Cutler, Greenhouse, Cornfield, and Schneiderman, 1966, p. 865).

The second possible method of randomization guards against these potential weaknesses in the first method. It calls for independently randomizing, to one treatment group or the other, patients who enter the trial within each successive short interval of time.

Suppose, for example, that ten patients are expected to enter the trial each month. A reasonable strategy is to assign at random five of the first ten patients to one treatment group and the other five to the second treatment group, and to repeat the random assignment of five patients to one group and five to the other for each successive group of ten.

The procedure is implemented as follows, beginning at the top of the fifth column of the first page of Table A.4. Because selection is from ten subjects, only single digits need be examined, with 0 designating the tenth subject. The first five distinct digits are found to be 2, 5, 4, 8, and 6. Therefore, the second, fourth, fifth, sixth, and eighth patients out of the first ten will be assigned the active drug, and the others—the first, third, seventh, ninth, and tenth—will be assigned the placebo.

Examination of the column continues for the second series of ten patients. The next five distinct digits are found to be 3, 1, 8, 0, and 5, implying that, of the second group of ten patients, the first, third, fifth, eighth, and tenth are assigned the active drug and the second, fourth, sixth, seventh, and ninth are assigned the placebo. As soon as the leading digits in a column are exhausted, the second digits may be examined.

It is important that a new set of random numbers be selected for each successive group, lest an unsuspected periodicity in the kind of patient entering the trial, or a pattern soon apparent to personnel who should be kept ignorant of which drugs the patients are receiving, introduce a bias.

A special case of the method just illustrated is the pairing of subjects, with one member of the pair randomly assigned the active drug and the other assigned the placebo. Random assignment becomes especially simple when subjects are paired. To begin with, one member of the pair must be designated the first and the other, the second. The designation might be on the basis of time of entry into the study, of alphabetical order of the surname, or of any other criterion, provided that the designation is made before the randomization is performed.

Table A.4 is entered at any convenient point, and successive single digits are examined, one digit for each pair. If the digit is odd—1, 3, 5, 7, or 9—the first member of the pair is assigned the active drug and the second member, the placebo. If the digit is even—2, 4, 6, 8, or 0—the second member of the pair is assigned the active drug and the first member the placebo.

To illustrate, let us begin at the top of the first column on the third page of Table A.4. The first digit encountered, 2, is even, indicating that, in the first pair, the active drug is given to the second member and the placebo to the first. The second digit encountered, 8, is also even, so that, in the second pair, too, the active drug is given to the second member and the placebo to the first. The sixth, seventh, and eighth digits—3, 9, and 1—are all odd. Therefore, in the sixth through eighth pairs, the active drug is given to the first member and the placebo to the second.

Investigators who require more than the 20,000 random digits of Table A.4 are referred to the Rand Corporation's extensive table (1955). Those whose research designs call for applying each of a number (more than two) of treatments to each of a sample of subjects will find Moses and Oakford's tables of random permutations (1963) indispensable. Their tables also facilitate each of the uses of randomization illustrated above.

4.3. VARIATIONS ON SIMPLE RANDOMIZATION

The randomization procedures discussed in the preceding section are all such that each new patient has a 50 : 50 chance of being assigned one or the other of the two treatments being compared. With the exception of studies in which patients are matched on characteristics associated with the outcome variable, these procedures run the risk of producing a lack of balance between the treatment groups in the distributions of age, sex, initial severity, or other prognostic factors. A solution other than matching on these characteristics is to stratify patients into predetermined strata (e.g., males aged 20–29, females aged 20–29, males aged 30–39, etc.), and to apply one of the simple randomization methods described above separately and independently within each stratum.

The risk of imbalance is reduced but not totally removed by simple stratified randomization, especially if patients enter the trial serially over time so that the total number of patients ending up in each stratum is not known until the recruitment of patients is completed. As a means of reducing even further the risk of imbalance, Efron (1971) introduced and Pocock and Simon (1975) elaborated on the concept of the "biased coin." Suppose that a new patient is in a stratum for which, to date, more patients have been assigned one treatment than the other. Biased coin allocation calls for this new patient to be assigned, with prespecified probability $p > .5$, the treatment that is currently underrepresented in his stratum. His chances of being assigned the currently overrepresented treatment are $1 - p < .5$. If the two treatments have so far been assigned to equal numbers of patients in his stratum, the new patient is assigned to one or the other with probability .5.

We illustrate Efron's biased coin scheme with the value of p he proposed as generally being good, that is, $p = \frac{2}{3}$. Suppose a new patient is in a stratum where unequal numbers of patients have so far been assigned the two treatments. A single digit from Table A.4 is examined for this patient, with the digit 0 ignored. If the selected digit is not divisible by 3 (1, 2, 4, 5, 7, or 8), the patient is assigned the currently underrepresented treatment. If the selected digit is divisible by 3 (3, 6, or 9), he is assigned the currently overrepresented treatment.

Probabilities of assignment other than 50 : 50 are also called for in so-called adaptive clinical trials, where the intent is to have, by the time the trial is concluded, more patients treated with the superior than with the inferior treatment (see Chapter 7 for discussion and references). Let $p > .5$ be the prespecified probability that a new patient is assigned the treatment that, to date, seems to be superior (if the two treatments appear to be equally effective, the new patient is assigned one or the other treatment with probability .5). Two-digit numbers in Table A.4 may be examined. If the selected number for a new patient lies between 01 and $100p$, inclusive, he is assigned the currently superior treatment; if it exceeds $100p$ (00 is taken to be 100), he is assigned the currently inferior treatment. For example, $p = .60$ leads to any number in the interval from 01 to 60 assigning the patient the currently superior treatment, and any number in the interval from 61 to 00 assigning him the currently inferior treatment.

REFERENCES

Cutler, S. J., Greenhouse, S. W., Cornfield, J., and Schneiderman, M. A. (1966). The role of hypothesis testing in clinical trials: Biometrics seminar. *J. Chronic Dis.*, **19**, 857–882.

Efron, B. (1971). Forcing a sequential experiment to be balanced. *Biometrika*, **58**, 403–417.

Fisher, R. A. (1935). *The design of experiments*. Edinburgh: Oliver and Boyd.

Hill, A. B. (1962). *Statistical methods in clinical and preventive medicine*. New York: Oxford University Press.

Moses, L. E. and Oakford, R. V. (1963). *Tables of random permutations*. Stanford: Stanford University Press.

Pocock, S. J. and Simon, R. (1975). Sequential treatment assignment with balancing for prognostic factors in the controlled clinical trial. *Biometrics*, **31**, 103–115.

Rand Corporation (1955). *A million random digits with 100,000 normal deviates.* New York: The Free Press.

CHAPTER 5

Sampling Method I: Naturalistic or Cross-Sectional Studies

In this chapter we study what had been identified in Section 2.1 as method I sampling. This method of sampling, referred to as *cross-sectional*, *naturalistic*, or *multinomial* sampling, does not attempt to prespecify any frequencies except the overall total.

We consider only the case where the resulting data are arrayed in a 2×2 table. Most statistics texts describe the chi square test for association when there are more than two rows or more than two columns in the resulting cross-classification table (e.g., Dixon and Massey, 1969, Section 13.3; and Everitt, 1977, Chapter 3). The accuracy of the chi square test for general contingency tables when the total sample size is small has been studied by Craddock and Flood (1970). Methods for estimating association in such tables are given by Goodman and Kruskal (1954, 1959), Goodman (1964), and Altham (1970a, 1970b).

In Section 5.1 we present some hypothetical data that are referred to repeatedly in this and the next chapter. In section 5.2 we examine the estimation by means of measures based on χ^2 of the degree of association between the two characteristics studied; we examine other measures in Section 5.3. Some properties of the odds ratio and of its logarithm are presented in Section 5.4.

Methods for testing hypotheses about the odds ratio are discussed in Section 5.5, and methods for constructing confidence intervals for the odds ratio in Section 5.6. Section 5.7 defines and presents methods for making inferences about the attributable risk.

5.1. SOME HYPOTHETICAL DATA

Suppose that we are studying the association, if any, between the age of the mother (A represents a maternal age less than or equal to 20 years; $\bar{A}$, a maternal age over 20 years) and the birthweight of her offspring (B represents a birthweight less than or equal to 2500 grams; $\bar{B}$, a birthweight over 2500 grams). Since the association might vary as a function of race and socioeconomic status, let us agree to study only black women from social classes IV and V who deliver in a single specified hospital.

Suppose that all the data we need are on file in the hospital's record room and that, for each delivery, a card is punched recording the birthweight of the infant and the age, color, and social class of the mother. After sorting out the cards for the mothers who do not fit the study criteria, we are left with a file of cards for deliveries to black mothers of social classes IV and V. Suppose that the number of cards remaining is quite large and that the determination of A versus $\bar{A}$ or of B versus $\bar{B}$ is not too simple. For example, the mother's date of birth might be recorded instead of her age, making card sorting complicated. The decision might then be made to examine only a sample, say 200, of all the records.

The sample should ideally be a simple random sample (see Chapter 4), but alternatives exist. Suppose that there are a total of 1000 records. A *systematic random sample* of 200 may be selected by drawing every fifth record, with the starting record (the first, second, third, fourth, or fifth) chosen at random. Another alternative is to base the selection on the last digit of the identification number, choosing only those records whose last digit is, for example, a 3 or a 7.

Table 5.1. Association between birthweight and maternal age: cross-sectional study

	Birthweight		
Maternal Age	B	$\bar{B}$	Total
A	10	40	50
$\bar{A}$	15	135	150
Total	25	175	200

Let us suppose that the sample of 200 records has been selected, and that the data are as in Table 5.1. As pointed out in Chapter 2, a more appropriate means of presenting the data resulting from method I sampling is as in Table 5.2.

Table 5.2. Joint proportions derived from Table 5.1

	Birthweight		
Maternal Age	B	$\bar{B}$	Total
A	.050$(=p_{11})$	.200$(=p_{12})$	.25$(=p_{1.})$
$\bar{A}$	.075$(=p_{21})$	.675$(=p_{22})$	.75$(=p_{2.})$
Total	.125$(=p_{.1})$	.875$(=p_{.2})$	1.

A consequence of sampling method I is that all probabilities may be estimated. Thus the proportion of all deliveries in which the mother was aged 20 years or less and in which the offspring weighed 2500 grams or less is estimated as $p(A \text{ and } B) = p_{11} = .05$. The proportion of all deliveries in which the mother was aged 20 years or less is estimated as $p(A) = p_{1.} = .25$, and the proportion of all deliveries in which the offspring weighed 2500 grams or less is $p(B) = p_{.1} = .125$.

The significance of the association between maternal age and birthweight (the first, but by no means the most important, issue) may be assessed by means of the standard chi square test. The value of the test statistic is

$$\chi^2 = \frac{200\left(|10 \times 135 - 40 \times 15| - \frac{1}{2}200\right)^2}{50 \times 150 \times 25 \times 175} = 2.58, \tag{5.1}$$

indicating an association that is not statistically significant.

5.2. MEASURES OF ASSOCIATION DERIVED FROM χ^2

The failure to find statistical significance would presumably signal the completion of the analysis. For later comparative purposes, however, we proceed to consider the estimation of the *degree* of association between the two characteristics, beginning with estimates based on the magnitude of χ^2.

A common mistake is to use the value of χ^2 itself as the measure of association. Even though χ^2 is excellent as a measure of the *significance* of the association, it is not at all useful as a measure of the *degree* of association. The reason is that χ^2 is a function both of the proportions in the various cells and of the total number of subjects studied. The degree of association present is really only a function of the cell proportions. The number of subjects studied plays a role in the chances of finding significance if association exists, but should play no role in determining the extent of association (see, e.g., Fisher, 1954, pp. 89–90).

Suppose, for example, that another investigator studied the characteristics of 400 births from the same hospital, and suppose that the resulting data were as in Table 5.3. The value of χ^2 for these data is

$$\chi^2 = \frac{400(|20 \times 270 - 80 \times 30| - \frac{1}{2}400)^2}{100 \times 300 \times 50 \times 350} = 5.97,$$

which indicates an association significant at the .05 level.

The inferences are different for the data of Tables 5.1 and 5.3: nonsignificance for the first but significance for the second. The only reason for the difference, however, is that twice as many births were used in Table 5.3 as in Table 5.1. The joint proportions (see Table 5.2) are obviously identical

Table 5.3. Association between birthweight and maternal age: sampling method I with 400 births

	Birthweight		
Maternal Age	B	$\bar{B}$	Total
A	20	80	100
$\bar{A}$	30	270	300
Total	50	350	400

for Tables 5.1 and 5.3, so that the associations between maternal age and birthweight implied by both are also identical. The larger value of χ^2 for the data of Table 5.3 is thus a reflection only of the larger total sample size (the doubling of all frequencies has in fact more than doubled the value of chi square), and not of a greater degree of association.

A measure of the degree of association between characteristics A and B which is derived from χ^2 but is free of the influence of the total sample size, $n_{..}$, is the *phi coefficient*:

$$\varphi = \sqrt{\frac{\chi_u^2}{n_{..}}}, \tag{5.2}$$

where χ_u^2 is the uncorrected chi square statistic,

$$\chi_u^2 = \frac{n_{..}(n_{11}n_{22} - n_{12}n_{21})^2}{n_{1.}n_{2.}n_{.1}n_{.2}}. \tag{5.3}$$

The phi coefficient is especially popular as a measure of association in the behavioral sciences, and is interpretable as a correlation coefficient. In fact, φ may be computed by assigning any two distinct numbers (0 and 1, for simplicity) to A and $\bar{A}$, assigning any two to B and $\bar{B}$, and calculating the usual product-moment correlation coefficient between the resulting values.

Values of φ close to zero indicate little if any association, whereas values close to unity indicate almost perfect predictability: if φ is near 1, then knowing whether a subject is A or $\bar{A}$ permits an accurate prediction of whether the subject is B or $\bar{B}$. The maximum value of φ is unity (if the marginal distributions are not equal, the maximum is less than unity) and, as a rule of thumb, any value less than .30 or .35 may be taken to indicate no more than trivial association.

For the data of Table 5.1,

$$\chi_u^2 = \frac{200(10 \times 135 - 40 \times 15)^2}{50 \times 150 \times 25 \times 175} = 3.43,$$

so that

$$\varphi = \sqrt{\frac{3.43}{200}} = .13, \qquad (5.4)$$

hardly of appreciable magnitude. The value of φ for the data of Table 5.3 is obviously also equal to .13.

The phi coefficient finds its greatest usefulness in the study of items contributing to psychological and educational tests (Lord and Novick, 1968) and in the *factor analysis* of a number of yes-no items. See Harman (1960) and Nunnally (1978) for a general description of factor analysis and Nunnally (1978) and Lord and Novick (1968) for the validity of factor analysis when applied to phi coefficients. Berger (1961) has presented a method for comparing phi coefficients from two independent studies.

The phi coefficient has a number of serious deficiencies, however. As shown in Chapter 6, the values of φ obtained when the association between characteristics A and B is studied prospectively and retrospectively are not comparable, nor is either value comparable to that obtained when the association is studied naturalistically. Carroll (1961) has shown that if either or both characteristics are dichotomized by cutting a continuous distribution into two parts, then the value of φ depends strongly on where the cutting point is set.

This lack of invariance of the phi coefficient and of other measures derived from χ^2 (see Goodman and Kruskal, 1954, pp. 739–740), plus presumably other reasons, led Goodman and Kruskal to assert that they "have been unable to find any convincing published defense of χ^2-like statistics as measures of association" (1954, p. 740). Whereas this assertion ignores the usefulness of φ in psychometrics, it does point to the avoidance of φ and of other statistics based on χ^2 as measures of association in those areas of research where comparability of findings but not necessarily of methods is essential.

5.3. OTHER MEASURES OF ASSOCIATION: THE ODDS RATIO

Goodman and Kruskal (1954, 1959) present a great many measures of association for 2×2 tables that are not functions of χ^2 and give their statistical properties in two later papers (1963, 1972). Here we concentrate on one such measure, the *odds ratio*.

Frequently, one of the two characteristics being studied is antecedent to the other. In the example we have been considering, maternal age is antecedent to birthweight. A measure of the risk of experiencing the outcome under study when the antecedent factor is present is

$$\Omega_A = \frac{P(B|A)}{P(\bar{B}|A)} \tag{5.5}$$

(see Section 1.1 for a definition of conditional probabilities). Ω_A is the *odds* that B will occur when A is present. Since $P(B|A)$ may be estimated by

$$p(B|A) = \frac{p_{11}}{p_{1.}}$$

and $P(\bar{B}|A)$ by

$$p(\bar{B}|A) = \frac{p_{12}}{p_{1.}},$$

therefore Ω_A may be estimated by

$$O_A = \frac{p_{11}/p_{1.}}{p_{12}/p_{1.}} = \frac{p_{11}}{p_{12}}. \tag{5.6}$$

For our example, the estimated odds that a mother aged 20 years or less will deliver an offspring weighing 2500 grams or less are, from Table 5.2,

$$O_A = \frac{.05}{.20} = \frac{1}{4} = .25. \tag{5.7}$$

Thus, for every four births weighing over 2500 grams to mothers aged 20 years or less, there is one birth weighing 2500 grams or less.

The information conveyed by these odds is exactly the same as that conveyed by the rate of low birthweight specific to young mothers,

$$p(B|A) = \frac{.05}{.25} = \frac{1}{5} = .20,$$

but the emphases differ. One can imagine attempting to educate prospective mothers aged 20 years or less. The impact of the statement, "One out of every five of you is expected to deliver an infant with a low birthweight," may well be different from the impact of "For every four of you who deliver infants of fairly high weight, one is expected to deliver an infant of low birthweight."

When A is absent, the odds of B's occurrence are defined as

$$\Omega_{\bar{A}} = \frac{P(B|\bar{A})}{P(\bar{B}|\bar{A})}, \tag{5.8}$$

which may be estimated as

$$O_{\bar{A}} = \frac{p_{21}/p_{2.}}{p_{22}/p_{2.}} = \frac{p_{21}}{p_{22}}. \tag{5.9}$$

For our example, the estimated odds that a mother aged more than 20 years will deliver an offspring weighing 2500 grams or less are

$$O_{\bar{A}} = \frac{.075}{.675} = \frac{1}{9} = .11. \tag{5.10}$$

Thus for every nine births weighing over 2500 grams to mothers aged more than 20 years (as opposed to every four to younger mothers), there is one birth weighing 2500 grams or less.

The two odds, Ω_A (5.5) and $\Omega_{\bar{A}}$ (5.8), may be contrasted in a number of ways in order to provide a measure of association. One such measure, due to Yule (1900), is

$$Q = \frac{\Omega_A - \Omega_{\bar{A}}}{\Omega_A + \Omega_{\bar{A}}}. \tag{5.11}$$

Another, also due to Yule (1912), is

$$Y = \frac{\sqrt{\Omega_A} - \sqrt{\Omega_{\bar{A}}}}{\sqrt{\Omega_A} + \sqrt{\Omega_{\bar{A}}}}. \tag{5.12}$$

The measure of association based on Ω_A and $\Omega_{\bar{A}}$ that is currently in greatest use is simply their ratio,

$$\omega = \frac{\Omega_A}{\Omega_{\bar{A}}}, \tag{5.13}$$

which may be estimated by the sample odds ratio,

$$o = \frac{O_A}{O_{\bar{A}}} = \frac{p_{11}/p_{12}}{p_{21}/p_{22}} = \frac{p_{11}p_{22}}{p_{12}p_{21}}. \tag{5.14}$$

If the two rates $P(B|A)$ and $P(B|\bar{A})$ are equal, indicating the independence or lack of association between the two characteristics, then the two odds Ω_A and $\Omega_{\bar{A}}$ are also equal (see Problem 5.1) so that the odds ratio $\omega = 1$. If $P(B|A) > P(B|\bar{A})$, then $\Omega_A > \Omega_{\bar{A}}$ and $\omega > 1$ (see Problem 5.2). If $P(B|A) < P(B|\bar{A})$, then $\Omega_A < \Omega_{\bar{A}}$ and $\omega < 1$.

For our data, the estimated odds ratio is

$$o = \frac{.05 \times .675}{.20 \times .075} = 2.25, \tag{5.15}$$

indicating that the odds of a young mother's delivering an offspring with low birthweight are $2\frac{1}{4}$ times those for an older mother. Because the odds ratio may also be estimated as

$$o = \frac{n_{11}n_{22}}{n_{12}n_{21}}, \tag{5.16}$$

it is sometimes also referred to as the *cross-product ratio*.

The standard error of the estimated odds ratio is estimated by

$$\text{s.e.}(o) = \frac{o}{\sqrt{n_{..}}}\sqrt{\frac{1}{p_{11}} + \frac{1}{p_{12}} + \frac{1}{p_{21}} + \frac{1}{p_{22}}}. \tag{5.17}$$

For the data of Table 5.2, the standard error is found to be

$$\text{s.e.}(o) = \frac{2.25}{\sqrt{200}}\sqrt{\frac{1}{.05} + \frac{1}{.20} + \frac{1}{.075} + \frac{1}{.675}} = 1.00. \tag{5.18}$$

An equivalent formula in terms of the original frequencies is

$$\text{s.e.}(o) = o\sqrt{\frac{1}{n_{11}} + \frac{1}{n_{12}} + \frac{1}{n_{21}} + \frac{1}{n_{22}}}. \tag{5.19}$$

The standard error is useful in gauging the precision of the estimated odds ratio, but not in testing its significance or in constructing confidence intervals. The classic chi square test should be used as a test of the hypothesis that the odds ratio in the population is equal to 1; the methods of Section 5.6 should be used for constructing confidence intervals.

Anscombe (1956), Gart (1966), and Gart and Zweifel (1967) have studied the sampling properties of o and its standard error. Note that if either n_{12} or n_{21} is equal to zero, then o in (5.16) is undefined. If any one of the four cell frequencies is equal to zero, then s.e.(o) in (5.19) is undefined. Suggested improved estimates are

$$o' = \frac{(n_{11} + .5)(n_{22} + .5)}{(n_{12} + .5)(n_{21} + .5)} \tag{5.20}$$

for the odds ratio and

$$\text{s.e.}(o') = o' \sqrt{\frac{1}{n_{11} + .5} + \frac{1}{n_{12} + .5} + \frac{1}{n_{21} + .5} + \frac{1}{n_{22} + .5}} \tag{5.21}$$

for its standard error.

A number of important properties of the odds ratio as a measure of association will be demonstrated in the sequel. Advantages of using the odds ratio instead of other measures have been illustrated by Mosteller (1968). Edwards (1963) considered the advantages to be so great that he recommended that only the odds ratio or functions of it be used to measure association in 2×2 tables.

5.4. SOME PROPERTIES OF THE ODDS RATIO AND ITS LOGARITHM

The odds ratio was originally proposed by Cornfield (1951) as a measure of the degree of association between an antecedent factor and an outcome event such as morbidity or mortality, but only because it provided a good approximation to another measure he proposed, the *relative risk*. If the risk of the occurrence of event B when A is present is taken simply as the rate of B's occurrence specific to the presence of A, $P(B|A)$, and similarly for the risk of B when A is absent, then the relative risk is simply the ratio of the two risks,

$$R = \frac{P(B|A)}{P(B|\bar{A})}. \tag{5.22}$$

R may be estimated by

$$r=\frac{p_{11}/p_{1.}}{p_{21}/p_{2.}}=\frac{p_{11}p_{2.}}{p_{21}p_{1.}}. \tag{5.23}$$

If the occurrence of event B is unlikely, whether or not characteristic A is present, then, as shown in Problem 5.3, r is approximately equal to o [see (5.14)]. For the data of Table 5.2,

$$r=\frac{.05\times .75}{.075\times .25}=2.0, \tag{5.24}$$

only slightly less than the value found for the odds ratio, $o=2.25$ [see (5.15)].

There is more to the odds ratio, however, than merely an approximation to the relative risk. There exists a mathematical model, the so-called *logistic model*, that naturally gives rise to the odds ratio as a measure of association. Consider, for specificity, the association between cigarette smoking and lung cancer. Mortality from lung cancer is a function not only of whether one smokes but also, as but one example, of the amount of air pollution in the environment of the community where he works or lives.

Let us agree to study the association between smoking and lung cancer in one community only, and let x represent the mean amount of a specified pollutant in the atmosphere surrounding that community. A possible representation of the mortality rate from lung cancer for cigarette smokers is

$$P_S=\frac{1}{1+e^{-(ax+b_S)}}, \tag{5.25}$$

and of the mortality rate for nonsmokers,

$$P_N=\frac{1}{1+e^{-(ax+b_N)}}, \tag{5.26}$$

where $e=2.718$, the base of natural logarithms. The parameter a measures the dependence of mortality on the specified air pollutant. The use of the same parameter, a, in (5.25) and (5.26) is equivalent to the assumption of no synergistic effect of smoking and air pollution on mortality. If a is positive, then both P_S and P_N approach unity as x, the mean amount of the pollutant, becomes large.

According to the model represented by (5.25) and (5.26), the effect of smoking on mortality is reflected only in the possible difference between the parameters b_S and b_N. When $x=0$, that is, when a community is completely free of the specified pollutant, then b_S is directly related to the mortality rate for smokers and b_N is directly related to the mortality rate for nonsmokers.

Consider, now, the odds that a smoker from the selected community will die of lung cancer. These odds are

$$\Omega_S = \frac{P_S}{1 - P_S}.$$

Since

$$1 - P_S = \frac{e^{-(ax+b_S)}}{1 + e^{-(ax+b_S)}},$$

therefore

$$\Omega_S = \frac{1}{e^{-(ax+b_S)}} = e^{ax+b_S}. \tag{5.27}$$

Similarly, the odds that a nonsmoker from that community will die of lung cancer are

$$\Omega_N = e^{ax+b_N}. \tag{5.28}$$

Thus, if the logistic model is correct, the odds ratio, that is, the ratio of the odds in (5.27) to the odds in (5.28), becomes simply

$$\omega = \frac{\Omega_S}{\Omega_N} = \frac{e^{ax+b_S}}{e^{ax+b_N}} = e^{(b_S - b_N)}, \tag{5.29}$$

independent of x. The natural logarithm of the odds ratio is then simply

$$\ln(\omega) = b_S - b_N, \tag{5.30}$$

which is also independent of x, and, moreover, is the simple difference between the two parameters assumed to distinguish smokers from nonsmokers.

The importance of this result is that, if the odds ratio or its logarithm is found to be stable across many different kinds of populations, then one may reasonably infer that the logistic model is a fair representation of the phenomenon under study. Given this inference, one may predict the value of the odds ratio in a new population and test the difference between the observed and predicted values; one may predict the effects on mortality of controlling the factor represented by x (in our example, an air pollutant); and one may of course predict the effects on mortality of controlling smoking habits.

The representation (5.30) of the logarithm of ω suggests that the logarithm of the sample odds ratio,

$$L = \ln(o), \tag{5.31}$$

is an important measure of association. Natural logarithms are tabulated in Table A.5. The standard error of L has been studied by Woolf (1955), Haldane (1956), and Gart (1966). A better estimate of $\ln(\omega)$ was found to be

$$L' = \ln(o'), \tag{5.32}$$

where o' is defined in (5.20), and a good estimate of its standard error was found to be

$$\text{s.e.}(L') = \sqrt{\frac{1}{n_{11}+.5} + \frac{1}{n_{12}+.5} + \frac{1}{n_{21}+.5} + \frac{1}{n_{22}+.5}}\ . \tag{5.33}$$

When the logistic model of (5.25) and (5.26) obtains, $\ln(\omega)$ is seen by (5.30) to be completely independent of x. Even if, instead, a model specified by a cumulative normal distribution is assumed, $\ln(\omega)$ is nearly independent of x (Edwards, 1966; Fleiss, 1970). The logistic model is far more manageable for representing rates and proportions than the cumulative normal model, however, and has been so used by Bartlett (1935), Winsor (1948), Dyke and Patterson (1952), Cox (1958, 1970), Grizzle (1961, 1963), Maxwell and Everitt (1970), and Fienberg (1977).

5.5. TESTING HYPOTHESES ABOUT THE ODDS RATIO

Theoretical result. Let attention be restricted to fourfold tables with marginal frequencies $n_{1.}$, $n_{2.}$, $n_{.1}$, and $n_{.2}$ fixed at the values actually observed, and suppose that the value of the underlying odds ratio is equal to ω. The expected cell frequencies N_{ij} associated with ω are such that (a) they are consistent with the original data in the sense that they recapture the marginal frequencies:

Table 5.4. Expected frequencies in a fourfold table

	Factor B		
Factor A	Present	Absent	Total
Present	N_{11}	N_{12}	$n_{1.}$
Absent	N_{21}	N_{22}	$n_{2.}$
Total	$n_{.1}$	$n_{.2}$	$n_{..}$

and (b) they are consistent with the value ω in the sense that

$$\frac{N_{11}N_{22}}{N_{12}N_{21}} = \omega. \tag{5.34}$$

The hypothesis that the value of the underlying odds ratio is equal to ω may be tested by referring the value of

$$\chi^2 = \sum_{i=1}^{2} \sum_{j=1}^{2} \frac{\left(|n_{ij} - N_{ij}| - \frac{1}{2}\right)^2}{N_{ij}} \tag{5.35}$$

to the chi square distribution with one degree of freedom. The form of the test statistic in (5.35) is identical to the form of the classic statistic presented in (2.7), and the interpretations of the N_{ij}'s as expected cell frequencies associated with hypothesized values of the odds ratio [$\omega = 1$ in (2.7), ω arbitrary in (5.35)] are also identical.

When ω is the hypothesized value of the odds ratio, and when $\omega \neq 1$, the expected cell frequencies may be found as follows. Define

$$X = \omega(n_{1.} + n_{.1}) + (n_{2.} - n_{.1}) \tag{5.36}$$

and

$$Y = \sqrt{X^2 - 4n_{1.}n_{.1}\omega(\omega - 1)}\ ; \tag{5.37}$$

then

$$N_{11} = \frac{X - Y}{2(\omega - 1)}, \tag{5.38}$$

$$N_{12} = n_{1.} - N_{11}, \tag{5.39}$$

$$N_{21} = n_{.1} - N_{11}, \tag{5.40}$$

and

$$N_{22} = n_{.2} - n_{1.} + N_{11}. \tag{5.41}$$

The following result was proved by Stevens (1951) and by Cornfield (1956). When the marginal frequencies are held fixed and when ω is the value of the odds ratio, n_{ij} (for any one of the four cells) is approximately normally distributed with mean N_{ij} and standard error $1/\sqrt{W}$, where

$$W = \sum_{i=1}^{2} \sum_{j=1}^{2} \frac{1}{N_{ij}} \tag{5.42}$$

and the N_{ij}'s are defined by (5.38)–(5.41).

Application. Consider testing the hypothesis that the value of the odds ratio underlying the data of Table 5.1 is $\omega = 5$. The value of X in (5.36) is

$$X = 5(50 + 25) + (150 - 25) = 500, \tag{5.43}$$

and that of Y in (5.37) is

$$Y = \sqrt{500^2 - 4 \times 50 \times 25 \times 5 \times 4} = 387.30. \tag{5.44}$$

The four expected cell frequencies are presented in Table 5.5. Note that

Table 5.5. Expected frequencies for data in Table 5.1 when the odds ratio is 5

	Birthweight		
Maternal Age	B	$\bar{B}$	Total
A	14.1	35.9	50
$\bar{A}$	10.9	139.1	150
Total	25	175	200

the odds ratio for the expected frequencies is

$$\frac{14.1 \times 139.1}{35.9 \times 10.9} = 5.0. \tag{5.45}$$

The value of chi square in (5.35) is

$$\chi^2 = \frac{(|10 - 14.1| - 0.5)^2}{14.1} + \frac{(|40 - 35.9| - 0.5)^2}{35.9} + \frac{(|15 - 10.9| - 0.5)^2}{10.9} + \frac{(|135 - 139.1| - 0.5)^2}{139.1} = 2.56, \tag{5.46}$$

and so the hypothesis that $\omega = 5$ is not rejected.

Another test of this hypothesis may be based on results given in the preceding section. If $\lambda = \ln(\omega)$, the quantity

$$\chi^2 = \frac{(L' - \lambda)^2}{(\text{s.e.}(L'))^2} \tag{5.47}$$

may be referred to the chi square distribution with one degree of freedom. For the data at hand,

$$\lambda = \ln(5) = 1.61, \tag{5.48}$$

$$L' = \ln \frac{10.5 \times 135.5}{40.5 \times 15.5} = .82, \tag{5.49}$$

and

$$\text{s.e.}(L') = \sqrt{\frac{1}{10.5} + \frac{1}{40.5} + \frac{1}{15.5} + \frac{1}{135.5}} = .438. \tag{5.50}$$

The value of the chi square statistic in (5.47) is then

$$\chi^2 = \frac{(.82 - 1.61)^2}{.438^2} = 3.25, \tag{5.51}$$

which is larger than the value of the statistic in (5.46) but still indicates a nonsignificant difference from $\omega = 5$.

The value of the chi square statistic in (5.47) based on the log odds ratio usually exceeds that of the statistic in (5.35) based on a comparison of the n_{ij}'s with the N_{ij}'s, but the difference is small when the marginal frequencies are large. If the statistic in (5.35) were defined without the continuity correction, its value would be close to that of the statistic in (5.47) even for moderate sample sizes (see Problem 5.5).

The procedure described in this section is more complicated than the one based on the log odds ratio, but is more accurate. It should be used whenever a hypothesized value of ω is tested.

5.6. CONFIDENCE INTERVALS FOR THE ODDS RATIO

A $100(1-\alpha)\%$ confidence interval for ω may be constructed as follows. The interval consists of all those values of ω for which, when the N_{ij}'s are the associated expected cell frequencies from (5.38)–(5.41),

$$\chi^2 = \sum_{i=1}^{2} \sum_{j=1}^{2} \frac{\left(|n_{ij} - N_{ij}| - \frac{1}{2}\right)^2}{N_{ij}} \leq c^2_{\alpha/2}. \tag{5.52}$$

The upper and lower limits are those for which the value of χ^2 equals $c^2_{\alpha/2}$.

The statistic in (5.52) depends on ω not explicitly but only implicitly through (5.36)–(5.41). The criterion for finding the upper and lower limits is therefore not simple. However, it is not overly complicated and can be implemented as follows.

The lower confidence limit, say ω_L, is associated with values of N_{11} and N_{22} smaller than n_{11} and n_{22} and with values of N_{12} and N_{21} larger than n_{12} and n_{21}. The continuity correction in (5.52) is then such that the χ^2 criterion simplifies to, say,

$$\chi^2_L = \left(n_{11} - N_{11} - \tfrac{1}{2}\right)^2 W = c^2_{\alpha/2}, \tag{5.53}$$

where W is defined in (5.42).

The lower limit, ω_L, will have been found when

$$F = \left(n_{11} - N_{11} - \tfrac{1}{2}\right)^2 W - c^2_{\alpha/2} \tag{5.54}$$

is equal to zero. Define

$$T = \frac{1}{2(\omega-1)^2}\left(Y - n_{..} - \frac{\omega - 1}{Y}\left[X(n_{1.} + n_{.1}) - 2n_{1.}n_{.1}(2\omega - 1)\right]\right), \tag{5.55}$$

$$U = \frac{1}{N_{12}^2} + \frac{1}{N_{21}^2} - \frac{1}{N_{11}^2} - \frac{1}{N_{22}^2}, \tag{5.56}$$

and

$$V = T\left[\left(n_{11} - N_{11} - \tfrac{1}{2}\right)^2 U - 2W\left(n_{11} - N_{11} - \tfrac{1}{2}\right)\right]. \tag{5.57}$$

Let $\omega_L^{(1)}$ be a first approximation to ω_L, let X and Y be the corresponding solutions of (5.36) and (5.37), and let N_{11}, N_{12}, N_{21}, and N_{22} be the corresponding solutions of (5.38)–(5.41). If the value of F in (5.54) is not equal to zero, a second, better approximation to ω_L is

$$\omega_L^{(2)} = \omega_L^{(1)} - \frac{F}{V}. \tag{5.58}$$

If the value of F associated with the second approximation is still not zero (say, if its absolute value exceeds .05), the process has to be repeated.

Convergence to ω_U, the upper confidence limit, proceeds by exactly the same process, except that the continuity correction is taken as $+\frac{1}{2}$ in (5.53), (5.54), and (5.57). Good first approximations to ω_L and ω_U are provided by the limits of the interval based on the log odds ratio,

$$\omega_L^{(1)} = \text{antilog}\left[L' - c_{\alpha/2}\text{s.e.}(L')\right] \tag{5.59}$$

and

$$\omega_U^{(1)} = \text{antilog}\left[L' + c_{\alpha/2}\text{s.e.}(L')\right]. \tag{5.60}$$

Consider again the data of Table 5.1, and suppose that a 95% confidence interval is desired for ω. From (5.49) and (5.50),

$$\begin{aligned}\omega_L^{(1)} &= \text{antilog}(.82 - 1.96 \times .438)\\ &= \text{antilog}(-.04) = .96\end{aligned} \tag{5.61}$$

and

$$\begin{aligned}\omega_U^{(1)} &= \text{antilog}(.82 + 1.96 \times .438)\\ &= \text{antilog}(1.68) = 5.37.\end{aligned} \tag{5.62}$$

Consider first the lower confidence limit. The value of X in (5.36) associated with $\omega_L^{(1)} = .96$ is

$$X = .96(50 + 25) + (150 - 25) = 197.00 \tag{5.63}$$

and that of Y in (5.37) is

$$Y = \sqrt{197^2 - 4 \times 50 \times 25 \times .96 \times (-.04)} = 197.49. \tag{5.64}$$

The value of N_{11} in (5.38) is therefore

$$N_{11} = \frac{197.00 - 197.49}{2(-.04)} = 6.13, \tag{5.65}$$

and the values of the other cell frequencies to be expected if the odds ratio is .96 are $N_{12} = 43.87$, $N_{21} = 18.87$, and $N_{22} = 131.13$. The value of W in (5.42) is

$$W = \frac{1}{6.13} + \frac{1}{43.87} + \frac{1}{18.87} + \frac{1}{131.13} = .2465 \tag{5.66}$$

and that of F in (5.54) is

$$F = (10 - 6.13 - .5)^2 \times .2465 - 3.84 = -1.04. \tag{5.67}$$

The value of the chi square criterion is 1.04 units below the desired value of 3.84, and the iterative process therefore has to be initiated.

The required values of the quantities in (5.55)–(5.57) are

$$\begin{aligned} T &= \frac{1}{2(-.04)^2}\Big(197.49 - 200 - \frac{-.04}{197.49} \times \\ &\quad \big[197(50 + 25) - 2 \times 50 \times 25 \times (1.92 - 1)\big]\Big) \\ &= 5.22, \end{aligned} \tag{5.68}$$

$$U = \frac{1}{43.87^2} + \frac{1}{18.87^2} - \frac{1}{6.13^2} - \frac{1}{131.13^2} = -.0233, \tag{5.69}$$

and

$$\begin{aligned} V &= 5.22\big[(10 - 6.13 - .5)^2 \times (-.0233) - 2 \times .2465 \times (10 - 6.13 - .5)\big] \\ &= -10.05. \end{aligned} \tag{5.70}$$

The second approximation to ω_L is then, from (5.58),

$$\omega_L^{(2)} = .96 - \frac{-1.04}{-10.05} = .86. \tag{5.71}$$

This turns out to be, to two decimal places, the lower 95% confidence limit on ω. The table of expected cell frequencies is given in Table 5.6, and

Table 5.6. Expected frequencies for data in Table 5.1 when the odds ratio is .86

	Birthweight		
Maternal Age	B	$\bar{B}$	Total
A	5.66	44.34	50
$\bar{A}$	19.34	130.66	150
Total	25	175	200

the value of the chi square criterion in (5.53) is

$$\chi_L^2 = (10 - 5.66 - .5)^2 \times .2586 = 3.81, \tag{5.72}$$

which is close to the desired value of 3.84.

Problem 5.6 is devoted to applying the same kind of iterative procedure to the upper 95% confidence limit, which turns out to be 5.84. The desired 95% confidence interval for the odds ratio underlying the data of Table 5.1 is therefore

$$.86 \leqq \omega \leqq 5.84. \tag{5.73}$$

Note that it is somewhat wider than the interval based on the log odds ratio, (.96, 5.37). This phenomenon has been found in other analyses by Gart (1962), Gart and Thomas (1972), and Fleiss (1979a); that is, the interval defined by end points satisfying equality in (5.52) is wider (but more accurate) than the interval defined by (5.59) and (5.60), and in fact wider (but more accurate) than a number of other suggested approximate confidence intervals.

For a limited number of tables, exact upper and lower confidence bounds on the odds ratio have been tabulated by Thomas and Gart (1977). For general fourfold tables, the procedure just described is the method of choice. It is more complicated than its competitors, but it is more accurate and is easily programmed for analysis on a programmable pocket or desktop calculator.

The tables of expected frequencies associated with the lower and upper confidence limits on the odds ratio can be used for more than making inferences about the odds ratio alone. The values of any function (phi coefficient, relative risk, etc.) evaluated for the entries in these two tables provide lower and upper confidence limits on the corresponding parameter. The expected frequencies associated with $\omega_L = .86$ were tabulated in Table 5.6. The lower 95% confidence limit on the phi coefficient, for example, is

$$\varphi_L = \frac{N_{11}N_{22} - N_{12}N_{21}}{\sqrt{n_{1.}n_{2.}n_{.1}n_{.2}}} = \frac{5.66 \times 130.66 - 44.34 \times 19.34}{\sqrt{50 \times 150 \times 25 \times 175}}$$

$$= -.02, \tag{5.74}$$

and the lower 95% confidence limit on the relative risk is

$$R_L = \frac{N_{11}n_{2.}}{N_{21}n_{1.}} = \frac{5.66 \times 150}{19.34 \times 50} = .88. \tag{5.75}$$

Problem 5.7 is devoted to finding the upper 95% confidence limits on these two parameters.

Although presented in the context of a fourfold table generated by sampling method I, the procedures described in Sections 5.5 and 5.6 are also applicable to tables generated by sampling methods II and III. The reason is that, once attention is restricted to fourfold tables having a specified set of marginal frequencies, the probability structure of the internal cell frequencies is independent of the way the data were generated.

5.7. ATTRIBUTABLE RISK

Let A denote the presence of the risk factor under study and B the presence of the outcome condition. The overall rate of occurrence of the condition is $P(B) = P(B|A)P(A) + P(B|\bar{A})P(\bar{A})$. Unfortunately there are several definitions available for the *attributable risk*; of these, the one due to Levin (1953) makes the most substantive sense. It is that fraction of $P(B)$ that can uniquely be attributed to the presence of the risk factor.

Exposure to the risk factor is not necessary for the occurrence of the outcome condition; a proportion, $P(B|\bar{A})$, of people without the risk factor will develop it. If the risk factor were without any effect, we would expect this same proportion to apply to the exposed group, so that its contribution to $P(B)$ would be $P(B|\bar{A})P(A)$. Its actual contribution is $P(B|A)P(A)$, and the difference between these two quantities relative to

$P(B)$ is Levin's attributable risk, say,

$$R_A = \frac{P(B|A)P(A) - P(B|\bar{A})P(A)}{P(B)} = \frac{P(B|A)P(A) - P(B|\bar{A})P(A)}{P(B|A)P(A) + P(B|\bar{A})P(\bar{A})}$$

$$= \frac{P(A)\left[P(B|A) - P(B|\bar{A})\right]}{P(B|A)P(A) + P(B|\bar{A})[1 - P(A)]}$$

$$= \frac{P(A)\left[P(B|A) - P(B|\bar{A})\right]}{P(B|\bar{A}) + P(A)\left[P(B|A) - P(B|\bar{A})\right]}$$

$$= \frac{P(A)(R-1)}{1 + P(A)(R-1)}, \tag{5.76}$$

where R is the relative risk,

$$R = \frac{P(B|A)}{P(B|\bar{A})}. \tag{5.77}$$

The attributable risk is interpreted as follows. If the risk factor could be eliminated, R_A is the proportion by which $P(B)$, the rate of occurrence of the outcome characteristic in the population, would be reduced. Therefore it has important uses in educational programs and in health planning, but its dependence on $P(A)$, the rate of exposure to the risk factor, limits its use in making comparisons between populations in which this rate varies. (There are some striking exceptions, however; see Problem 5.9.)

When, as pointed out by Markush (1977), data are collected in cross-sectional surveys (such as those conducted by the National Center for Health Statistics) or as part of routine registration (such as the recording of vital events by local health departments), the attributable risk may be estimated by, say,

$$r_A = \frac{p_{11}p_{22} - p_{12}p_{21}}{p_{.1}p_{2.}} \tag{5.78}$$

(see Problem 5.8). Walter (1976) derived a complicated expression for the standard error of r_A. A related but much simpler expression was derived by Fleiss (1979b) for the standard error of $\ln(1 - r_A)$,

$$\text{s.e.}(\ln(1 - r_A)) = \sqrt{\frac{p_{12} + r_A(p_{11} + p_{22})}{n_{..}p_{21}}}. \tag{5.79}$$

Its use is illustrated on the data in Table 5.7, which cross-classifies birthweight and infant mortality among whites in 1974 in New York City.

Table 5.7. Infant mortality by birthweight for 72,730 live white births in 1974 in New York City

Birthweight	Outcome at One Year: Dead	Alive	Total
$\leqq$ 2500 gm	.0085	.0632	.0717
> 2500 gm	.0058	.9225	.9283
Total	.0143	.9857	1

The estimated risk of infant death attributable to low birthweight is

$$r_A = \frac{.0085 \times .9225 - .0632 \times .0058}{.0143 \times .9283} = .563, \tag{5.80}$$

with an estimated standard error of $\ln(1 - r_A)$ of

$$\text{s.e.}[\ln(1 - r_A)] = \sqrt{\frac{.0632 + .563(.0085 + .9225)}{72{,}730 \times .0058}} = .037. \tag{5.81}$$

The natural logarithm of $1 - r_A$ is

$$\ln(1 - r_A) = \ln(1 - .563) = -.828, \tag{5.82}$$

and an approximate 95% confidence interval for $\ln(1 - R_A)$ is

$$-.828 - 1.96 \times .037 \leqq \ln(1 - R_A) \leqq -.828 + 1.96 \times .037, \tag{5.83}$$

or

$$-.901 \leqq \ln(1 - R_A) \leqq -.755. \tag{5.84}$$

By taking antilogarithms of these limits, and then complements from unity, we obtain

$$.530 \leqq R_A \leqq .594 \tag{5.85}$$

as an approximate 95% confidence interval for the attributable risk itself. With 95% confidence, therefore, between 53% and 59% of all white infant deaths in New York City in 1974 could have been prevented if prematurity (birthweights less than or equal to 2500 grams) had been eliminated.

Problem 5.1. The odds Ω_A and $\Omega_{\bar{A}}$ are defined by (5.5) and (5.8). Prove that $\Omega_A = \Omega_{\bar{A}}$ if and only if $P(B|A) = P(B|\bar{A})$.

Problem 5.2. The odds ratio ω is defined by (5.13). Prove that $\omega > 1$ if and only if $P(B|A) > P(B|\bar{A})$.

Problem 5.3. The relative risk r is defined by (5.23) and the odds ratio o by (5.14). Prove that r is approximately equal to o if p_{21} is small relative to p_{22} and if p_{11} is small relative to p_{12}. [*Hint.* $p_{2.} = p_{22}(1 + p_{21}/p_{22})$ and $p_{1.} = p_{12}(1 + p_{11}/p_{12})$.]

Problem 5.4. It has long been known that, among first admissions to American public mental hospitals, schizophrenia as diagnosed by the hospital psychiatrists is more prevalent than the affective disorders, whereas the converse is true for British public mental hospitals. A cooperative study between New York and London psychiatrists was designed to determine the extent to which the difference was a function of differences in diagnostic habits. The following data are from a study reported by Cooper et al. (1972).

(*a*) One hundred and forty-five patients in a New York hospital and 145 in a London hospital were selected for study. The New York hospital diagnosed 82 patients as schizophrenic and 24 as affectively ill, whereas the London hospital diagnosed 51 as schizophrenic and 67 as affectively ill. Ignoring the patients given other diagnoses, set up the resulting fourfold table.

The project psychiatrists made diagnoses using a standard set of criteria after conducting standardized interviews with the patients. In New York, the project diagnosed 43 patients as schizophrenic and 53 as affectively ill. In London, the project diagnosed 33 patients as schizophrenic and 85 as affectively ill. Ignoring the patients given other diagnoses, set up the resulting fourfold table.

The results of the standardized interview served as input to a computer program that yields psychiatric diagnoses. In New York, the computer diagnosed 67 patients as schizophrenic and 27 as affectively ill. In London, the computer diagnosed 56 patients as schizophrenic and 37 as affectively ill. Ignoring the patients given other diagnoses, set up the resulting fourfold table.

(*b*) Three diagnostic contrasts between New York and London are possible: by the hospitals' diagnoses; by the project's diagnoses; and by the computer's diagnoses. Compute, for each of the three sources of diagnoses, the ratio of the odds that a New York patient will be diagnosed schizophrenic rather than affective to the corresponding odds for London.

How do the odds ratios for the project's and computer's diagnoses compare? How do these two compare with the odds ratio for the hospital's diagnoses?

(*c*) For each source of diagnosis, all four cell frequencies are large, indicating that the improved estimate (5.20) may not be necessary. Check that, for each source of diagnosis, the estimate of the odds ratio given by (5.20) is only slightly less than the estimate given by (5.14).

Problem 5.5. When the continuity correction is not incorporated into the test statistic in (5.35), the resulting value usually agrees very well with that of the statistic in (5.47). Find the value of

$$\chi^2 = \sum_{i=1}^{2} \sum_{j=1}^{2} (n_{ij} - N_{ij})^2 / N_{ij}$$

for the data in Tables 5.1 and 5.5, and compare with the value found in (5.51).

Problem 5.6. Apply the iterative procedure of Section 5.6 to find the upper 95% confidence limit on the odds ratio underlying the data of Table 5.1. Use as the initial approximation the value $\omega_U^{(1)} = 5.37$ from (5.62).

Problem 5.7. Find the expected cell frequencies associated with the upper 95% confidence limit found in Problem 5.6. From them, calculate upper 95% confidence limits on the phi coefficient and the relative risk underlying the data of Table 5.1.

Problem 5.8. Show that, when the components of the population attributable risk defined in (5.76) are replaced by their sample estimators [$P(A)$ by $p_{1.}$ and R by the expression in (5.23)], and when the resulting expression is simplified, equation (5.78) results.

Problem 5.9. Data on infant mortality by birthweight for whites were presented in Table 5.7. Data for 37,840 nonwhite live births in New York City in 1974 are

	Outcome at One Year		
Birthweight	Dead	Alive	Total
≦ 2500 gm	.0140	.1147	.1287
> 2500 gm	.0088	.8625	.8713
Total	.0228	.9772	1

(*a*) What is the estimated attributable risk for nonwhite live births? How does this compare with the value found in (5.80) for white live births?

(*b*) What is the standard error of $\ln(1 - r_A)$ for the estimate found in (*a*)? What is an approximate 95% confidence interval for R_A in nonwhite live births? How does this compare with the interval found in (5.85) for white live births?

REFERENCES

Altham, P. M. E. (1970a). The measurement of association of rows and columns for an $r \times s$ contingency table. *J. R. Stat. Soc., Ser. B*, **32**, 63–73.

Altham, P. M. E. (1970b). The measurement of association in a contingency table: Three extensions of the cross-ratios and metric methods. *J. R. Stat. Soc., Ser. B.*, **32**, 395–407.

Anscombe, F. J. (1956). On estimating binomial response relations. *Biometrika*, **43**, 461–464.

Bartlett, M. S. (1935). Contingency table interactions. *J. R. Stat. Soc. Suppl.*, **2**, 248–252.

Berger, A. (1961). On comparing intensities of association between two binary characteristics in two different populations. *J. Am. Stat. Assoc.*, **56**, 889–908.

Carroll, J. B. (1961). The nature of the data, or how to choose a correlation coefficient. *Psychometrika*, **26**, 347–372.

Cooper, J. E., Kendell, R. E., Gurland, B. J., Sharpe, L., Copeland, J. R. M., and Simon, R. (1972). *Psychiatric diagnosis in New York and London*. London: Oxford University Press.

Cornfield, J. (1951). A method of estimating comparative rates from clinical data. Applications to cancer of the lung, breast and cervix. *J. Natl. Cancer Inst.*, **11**, 1269–1275.

Cornfield, J. (1956). A statistical problem arising from retrospective studies. Pp. 135–148 in J. Neyman (Ed.). *Proceedings of the third Berkeley symposium on mathematical statistics and probability*, Vol. 4. Berkeley: University of California Press.

Cox, D. R. (1958). The regression analysis of binary sequences. *J. R. Stat. Soc., Ser. B*, **20**, 215–242.

Cox, D. R. (1970). *Analysis of binary data*. London: Methuen.

Craddock, J. M. and Flood, C. R. (1970). The distribution of the χ^2 statistic in small contingency tables. *Appl. Stat.*, **19**, 173–181.

Dixon, W. J. and Massey, F. J. (1969). *Introduction to statistical analysis*, 3rd ed. New York: McGraw-Hill.

Dyke, G. V. and Patterson, H. D. (1952). Analysis of factorial arrangements when the data are proportions. *Biometrics*, **8**, 1–12.

Edwards, A. W. F. (1963). The measure of association in a 2×2 table. *J. R. Stat. Soc., Ser. A*, **126**, 109–114.

Edwards, J. H. (1966). Some taxonomic implications of a curious feature of the bivariate normal surface. *Brit. J. Prev. Soc. Med.*, **20**, 42–43.

Everitt, B. S. (1977). *The analysis of contingency tables*. London: Chapman and Hall.

Fienberg, S. E. (1977). *The analysis of cross-classified categorical data*. Cambridge, Mass.: M.I.T. Press.

Fisher, R. A. (1954). *Statistical methods for research workers*, 12th ed. Edinburgh: Oliver and Boyd.

Fleiss, J. L. (1970). On the asserted invariance of the odds ratio. *Brit. J. Prev. Soc. Med.*, **24**, 45–46.

Fleiss, J. L. (1979a). Confidence intervals for the odds ratio in case-control studies: The state of the art. *J. Chronic Dis.*, **32**, 69–77.

Fleiss, J. L. (1979b). Inference about population attributable risk from cross-sectional studies. *Am. J. Epidemiol.*, **110**, 103–104.

Gart, J. J. (1962). Approximate confidence limits for the relative risk. *J. R. Stat. Soc., Ser. B*, **24**, 454–463.

Gart, J. J. (1966). Alternative analyses of contingency tables. *J. R. Stat. Soc., Ser. B*, **28**, 164–179.

Gart, J. J. and Thomas, D. G. (1972). Numerical results on approximate confidence limits for the odds ratio. *J. R. Stat. Soc., Ser. B*, **34**, 441–447.

Gart, J. J. and Zweifel, J. R. (1967). On the bias of various estimators of the logit and its variance, with application to quantal bioassay. *Biometrika*, **54**, 181–187.

Goodman, L. A. (1964). Simultaneous confidence limits for cross-product ratios in contingency tables. *J. R. Stat. Soc., Ser. B*, **26**, 86–102.

Goodman, L. A. and Kruskal, W. H. (1954). Measures of association for cross classifications. *J. Am. Stat. Assoc.*, **49**, 732–764.

Goodman, L. A. and Kruskal, W. H. (1959). Measures of association for cross classifications. II: Further discussion and references. *J. Am. Stat. Assoc.*, **54**, 123–163.

Goodman, L. A. and Kruskal, W. H. (1963). Measures of association for cross classifications. III: Approximate sampling theory. *J. Am. Stat. Assoc.*, **58**, 310–364.

Goodman, L. A. and Kruskal, W. H. (1972). Measures of association for cross classifications. IV. Simplification of asymptotic variances. *J. Am. Stat. Assoc.*, **67**, 415–421.

Grizzle, J. E. (1961). A new method of testing hypotheses and estimating parameters for the logistic model. *Biometrics*, **17**, 372–385.

Grizzle, J. E. (1963). Tests of linear hypotheses when the data are proportions. *Am. J. Public Health*, **53**, 970–976.

Haldane, J. B. S. (1956). The estimation and significance of the logarithm of a ratio of frequencies. *Ann. Hum. Genet.*, **20**, 309–311.

Harman, H. H. (1960). *Modern factor analysis*. Chicago: University of Chicago Press.

Levin, M. L. (1953). The occurrence of lung cancer in man. *Acta Unio Int. Contra Cancrum*, **19**, 531–541.

Lord, F. M. and Novick, M. R. (1968). *Statistical theories of mental test scores*. Reading, Mass.: Addison-Wesley.

Markush, R. E. (1977). Levin's attributable risk statistic for analytic studies and vital statistics. *Am. J. Epidemiol.*, **105**, 401–406.

Maxwell, A. E. and Everitt, B. S. (1970). The analysis of categorical data using a transformation. *Brit. J. Math. Stat. Psychol.*, **23**, 177–187.

Mosteller, F. (1968). Association and estimation in contingency tables. *J. Am. Stat. Assoc.*, **63**, 1–28.

Nunnally, J. (1978). *Psychometric theory*, 2nd ed. New York: McGraw-Hill.

Stevens, W. L. (1951). Mean and variance of an entry in a contingency table. *Biometrika*, **38**, 468–470.

Thomas, D. G. and Gart, J. J. (1977). A table of exact confidence limits for differences and ratios of two proportions and their odds ratios. *J. Am. Stat. Assoc.*, **72**, 73–76.

Walter, S. D. (1976). The estimation and interpretation of attributable risk in health research. *Biometrics*, **32**, 829–849.

Winsor, C. P. (1948). Factorial analysis of a multiple dichotomy. *Hum. Biol.*, **20**, 195–204.

Woolf, B. (1955). On estimating the relation between blood group and disease. *Ann Hum. Genet.*, **19**, 251–253.

Yule, G. U. (1900). On the association of attributes in statistics. *Philos. Trans. R. Soc. Ser. A*, **194**, 257–319.

Yule, G. U. (1912). On the methods of measuring the association between two attributes. *J. R. Stat. Soc.*, **75**, 579–642.

CHAPTER 6

Sampling Method II: Prospective and Retrospective Studies

Sampling method II was defined in Section 2.1 as the selection of a sample from each of two populations, a predetermined number n_1 from the first and a predetermined number n_2 from the second. Method II sampling is used in comparative prospective studies—in which one of the two populations is defined by the presence and the second by the absence of a suspected antecedent factor (MacMahon and Pugh, 1970, Chapter 11)—and is used in comparative retrospective studies—in which one of the two populations is defined by the presence and the second by the absence of the outcome under study (MacMahon and Pugh, 1970, Chapter 12).

The analysis of data from a comparative prospective study is discussed in Section 6.1, and the analysis of data from a comparative retrospective study in Section 6.2. Berkson's, Sheps's, and Feinstein's criticisms of the odds ratio are presented in Section 6.3. Inferences about the attributable risk when the data are from retrospective studies are considered in Section 6.4. A comparison of the prospective and retrospective approaches is made in Section 6.5.

6.1. PROSPECTIVE STUDIES

The comparative prospective study (also termed the *cohort*, or *forward-going*, or *follow-up* study) is characterized by the identification of the two study samples on the basis of the presence or absence of the antecedent factor and by the estimation for both samples of the proportions developing the disease or condition under study.

Consider again the hypothetical association between maternal age—the antecedent factor—and birthweight—the outcome—introduced in Chapter 5. A design fitting the paradigm of a comparative prospective study would be indicated if, for example, the file containing records for mothers aged 20 years or less were kept separate from the file containing records for mothers aged more than 20 years. Suppose that 100 of both kinds of mothers are sampled from the respective lists, and the weights of their offspring ascertained.

The precise outcome is of course subject to chance variation, but let us suppose that the data turn out to be perfectly consistent with those obtained with sampling method I (see Section 5.1). From Table 5.2, the rate of low birthweight specific to mothers aged 20 years or less is estimated to be

$$p(B|A) = \frac{p_{11}}{p_{1.}} = \frac{.05}{.25} = .20. \tag{6.1}$$

Thus we would expect to have 20% of the offspring of mothers aged 20 years or less, or 20 infants, weighing 2500 grams or less, and the remaining 80 weighing over 2500 grams.

The rate of low birthweight specific to mothers aged over 20 years is estimated from Table 5.2 to be

$$p(B|\bar{A}) = \frac{p_{21}}{p_{2.}} = \frac{.075}{.75} = .10. \tag{6.2}$$

We would therefore expect to have ten of the offspring of mothers aged over 20 years weighing 2500 grams or less, and the remaining 90 weighing over 2500 grams. The expected table is therefore as shown in Table 6.1.

Table 6.1. Association between birthweight and maternal age: prospective study

Maternal Age	Birthweight		Total	Proportion with Low Birthweight
	B	$\bar{B}$		
A	20	80	$100(=N_A)$	$.20[=p(B\|A)]$
$\bar{A}$	10	90	$100(=N_{\bar{A}})$	$.10[=p(B\|\bar{A})]$
Total	30	170	200	

The value of χ^2 for these data is

$$\chi^2 = 3.18, \tag{6.3}$$

so that the association fails to reach significance at the .05 level. What is noteworthy is that the total sample sizes of Tables 5.1 and 6.1 are equal and that the frequencies of the two tables are consistent. Nevertheless, the chi square value for the latter table is greater than that for the former. The inference from this comparison holds in general: a prospective study with equal sample sizes yields a more *powerful* chi square test than a cross-sectional study with the same total sample size (see Lehmann, 1959, p. 146).

The odds ratio ω was introduced in Section 5.3 as a measure of association between characteristics A and B. Because of the way the separate odds Ω_A in (5.5) and $\Omega_{\bar{A}}$ in (5.8) were defined, it is clear that the odds ratio may be estimated from a comparative prospective study as well as from a cross-sectional study. The estimate is

$$o = \frac{p(B|A)p(\bar{B}|\bar{A})}{p(\bar{B}|A)p(B|\bar{A})}. \tag{6.4}$$

For the data of Table 6.1, the estimated odds ratio is

$$o = \frac{.20 \times .90}{.80 \times .10} = 2.25, \tag{6.5}$$

which is precisely equal to the value found in (5.15) for the data from the cross-sectional study.

Formula 5.16 for o as a function of the cross-products of cell frequencies applies to the data from comparative prospective studies as well. For the data of Table 6.1, obviously,

$$o = \frac{20 \times 90}{80 \times 10} = 2.25.$$

The standard error of the odds ratio estimated from a comparative prospective study is estimated as

$$\text{s.e.}(o) = o\sqrt{\frac{1}{N_A p(B|A)p(\bar{B}|A)} + \frac{1}{N_{\bar{A}} p(B|\bar{A})p(\bar{B}|\bar{A})}}. \tag{6.6}$$

For the data of Table 6.1, the estimated standard error is

$$\begin{aligned}\text{s.e.}(o) &= 2.25\sqrt{\frac{1}{100 \times .20 \times .80} + \frac{1}{100 \times .10 \times .90}}\\ &= 0.94.\end{aligned} \tag{6.7}$$

An equivalent expression for the standard error in terms of the original frequencies (see Problem 6.1) is

$$\text{s.e.}(o) = o\sqrt{\frac{1}{n_{11}} + \frac{1}{n_{12}} + \frac{1}{n_{21}} + \frac{1}{n_{22}}}, \tag{6.8}$$

which is identical to expression (5.19).

We found above that the chi square test applied to data from a comparative prospective study with equal sample sizes is more powerful than the chi square test applied to data from a cross-sectional study. A similar phenomenon holds for the *precision* of the estimated odds ratio. Even though the total sample sizes in Tables 5.1 and 6.1 are equal, and even though the association between the two characteristics is the same, the odds ratio from the latter table is estimated more precisely [s.e.(o) = 0.94, from (6.7)] than the odds ratio from the former [s.e.(o) = 1.00, from (5.18)]. A prospective study with equal sample sizes is thus superior in terms of both power and precision to a cross-sectional study with the same total sample size.

Using the methods described in Section 5.6, an approximate 95% confidence interval for the odds ratio based on the data of Table 6.1 may be shown to be

$$.93 \leqq \omega \leqq 5.51, \tag{6.9}$$

which is narrower than (and thus superior to) the interval found in (5.73) for data from the corresponding cross-sectional study.

The value of the uncorrected chi square statistic (see equation 5.3) for the data of Table 6.1 is

$$\chi_u^2 = \frac{200(20 \times 90 - 80 \times 10)^2}{100 \times 100 \times 30 \times 170} = 3.92.$$

It yields a phi coefficient (see equation 5.2) of

$$\varphi = \sqrt{\frac{3.92}{200}} = .14, \tag{6.10}$$

which is only slightly greater than the value of .13 found in (5.4) for the data of Table 5.1. Recall, however, that the data of Tables 5.1 and 6.1 are perfectly consistent.

6.2. RETROSPECTIVE STUDIES

The comparative retrospective study (also termed the *case-control* study) is characterized by the identification of the two study samples on the basis of the presence or absence of the outcome factor and by the estimation for both samples of the proportions possessing the antecedent factor under study.

This method might be easily applied to the study of the association between maternal age and birthweight if, for example, the file containing records for infants of low birthweight (less than or equal to 2500 grams) were kept separate from the file containing records for infants of higher birthweight. Suppose that 100 of both kinds of infants are sampled from the respective lists and that the ages of their mothers are ascertained.

Let us suppose, as we did in Section 6.1, that the data turn out to be perfectly consistent with those already given. We need the rates $p(A|B)$ and $p(A|\bar{B})$, that is, the proportions of mothers aged 20 years or less among infants of low and among infants of high birthweight. From Table 5.2 we find that

$$p(A|B) = \frac{p_{11}}{p_{.1}} = \frac{.05}{.125} = .40, \tag{6.11}$$

implying that 40% of the mothers of the 100 low-birthweight infants, or 40, should be aged 20 years or less, and the remaining 60, over 20 years. We also find that

$$p(A|\bar{B}) = \frac{p_{12}}{p_{.2}} = \frac{.20}{.875} = .23, \tag{6.12}$$

implying that 23% of the mothers of the 100 higher-birthweight infants, or 23, should be aged 20 years or less, and the remaining 77, over 20 years.

Standard practice is to set out the data resulting from sampling method II as in Table 2.6, so that the two study samples appear one above the other and the characteristic determined for each subject is located across the top. This causes something of an anomaly when applied to data from a comparative retrospective study because the characteristic determined for each subject is the suspected antecedent factor and not the outcome characteristic that is hypothesized to follow from it. Miettinen (1970) stressed the necessary shift in thinking required for analyzing data from a comparative retrospective study: that the outcome characteristic follows the antecedent factor in a logical sequence, but precedes it in a retrospective study.

Following, then, the format of Table 2.6, we present the expected data as

Table 6.2. Association between birthweight and maternal age: retrospective study

	Maternal Age			
Birthweight	A	$\bar{A}$	Total	Proportion with Low Age
B	40	60	$100(=N_B)$	$.40[=p(A\|B)]$
$\bar{B}$	23	77	$100(=N_{\bar{B}})$	$.23[=p(A\|\bar{B})]$
Total	63	137	200	

shown in Table 6.2. The value of χ^2 for these data is

$$\chi^2 = 5.93, \tag{6.13}$$

which indicates that the association is significant at the .05 level.

The gradient in the magnitude of χ^2 from the cross-sectional study ($\chi^2 = 2.58$) to the prospective study ($\chi^2 = 3.18$) to the retrospective study ($\chi^2 = 5.93$) is noteworthy because the three sets of data were all generated according to the same set of underlying rates and because the three total sample sizes were equal. It is true in general that a retrospective study with equal sample sizes yields a more powerful chi square test than a cross-sectional study with the same total sample size. If, in addition, the outcome characteristic is rarer than the antecedent factor (more precisely, if

$$|P(B) - .5| > |P(A) - .5|), \tag{6.14}$$

then the chi square test on the data of a retrospective study with equal sample sizes is more powerful than the chi square test on the data of a prospective study with equal sample sizes (Lehmann, 1959, p. 146).

For the data of Table 6.2, the value of the uncorrected chi square statistic is

$$\chi_u^2 = \frac{200(40 \times 77 - 60 \times 23)^2}{100 \times 100 \times 63 \times 137} = 6.70.$$

The value of the associated phi coefficient is

$$\varphi = \sqrt{\frac{6.70}{200}} = .18, \tag{6.15}$$

nearly 40% higher than the value, $\varphi = .13$, associated with the data of

Table 5.1 and nearly 30% higher than the value, $\varphi = .14$, associated with the data of Table 6.1. Now, if a given measure is to be more than a mere uninterpretable index, it should have the property that different investigators studying the same phenomenon should emerge with at least similar estimates even though they studied the phenomenon differently. Since the phi coefficient obviously lacks the property of *invariance*, it should not be used as a measure of association for data from comparative prospective or retrospective studies.

The odds ratio ω, on the other hand, is invariant across the three kinds of studies we are considering. It is defined by

$$\omega = \frac{P(B|A)P(\bar{B}|\bar{A})}{P(\bar{B}|A)P(B|\bar{A})}. \qquad (6.16)$$

When expressed in this form, ω seems to be estimable only from cross-sectional and comparative prospective studies, because only these two kinds of studies provide estimates of the rates $P(B|A)$ and $P(B|\bar{A})$. An equivalent expression for ω (see Problem 6.2), however, is

$$\omega = \frac{P(A|B)P(\bar{A}|\bar{B})}{P(\bar{A}|B)P(A|\bar{B})}, \qquad (6.17)$$

and, when expressed in this form, ω is clearly estimable from comparative retrospective studies, too (see Cornfield, 1956).

The estimate is

$$o = \frac{p(A|B)p(\bar{A}|\bar{B})}{p(\bar{A}|B)p(A|\bar{B})}. \qquad (6.18)$$

For the data of Table 6.2, the estimated odds ratio is

$$o = \frac{.40 \times .77}{.60 \times .23} = 2.23, \qquad (6.19)$$

equal except for rounding errors to the value $o = 2.25$ found previously. Formula 5.16, which involves the cell frequencies, continues to apply as well.

The standard error of the odds ratio estimated from a comparative retrospective study is estimated as

$$\text{s.e.}(o) = o\sqrt{\frac{1}{N_B p(A|B)p(\bar{A}|B)} + \frac{1}{N_{\bar{B}} p(A|\bar{B})p(\bar{A}|\bar{B})}}. \qquad (6.20)$$

Expressions 5.19 and 6.8 for the standard error as a function of the original frequencies are also valid for the data of a comparative retrospective study.

For the data of Table 6.2,

$$\begin{aligned}\text{s.e.}(o) &= 2.23\sqrt{\frac{1}{100 \times .40 \times .60} + \frac{1}{100 \times .23 \times .77}} \\ &= 2.23\sqrt{\frac{1}{40} + \frac{1}{60} + \frac{1}{23} + \frac{1}{77}} \\ &= 0.70. \end{aligned} \tag{6.21}$$

The gradient noted above in the magnitude of χ^2 is matched by a gradient in the precision with which the odds ratio is estimated. For the cross-sectional study, s.e.$(o) = 1.00$ [see (5.18)]; for the comparative prospective study, s.e.$(o) = 0.94$ [see (6.7)]; and for the comparative retrospective study, s.e.$(o) = 0.70$ [see (6.21)].

An approximate 95% confidence interval for the odds ratio based on the data of Table 6.2, using the methods described in Section 5.6, is

$$1.16 \leqq \omega \leqq 4.33. \tag{6.22}$$

Exactly the same gradient across the three kinds of studies is therefore found for the length of the 95% confidence interval for the odds ratio. The length of the interval for the cross-sectional study [see (5.73)] is largest; that for the prospective study [see (6.9)], smaller; and that for the retrospective study [see (6.22)], smallest of all.

According to the criteria of precision, power, and length of confidence interval, therefore, when the total sample sizes are equal, a comparative retrospective study with equal sample sizes is superior to both a cross-sectional study and a comparative prospective study with equal sample sizes.

6.3. CRITICISMS OF THE ODDS RATIO

It was pointed out in the paragraph following (6.16) that the two rates $P(B|A)$ and $P(B|\bar{A})$ are estimable only from cross-sectional and prospective studies (we continue to let A denote the presence of the antecedent factor and B the undesirable outcome). Only when association is measured by the odds ratio, which is a function of the *ratio* of these rates, does the retrospective study provide data comparable to the data from the other two kinds of studies.

However, Berkson (1958) and, more recently, Feinstein (1973) have strongly criticized taking the ratio of rates as a measure of association, pointing out that the level of the rates is lost. Thus a tenfold increase over a rate of one per million would be considered equivalent to a tenfold increase over a rate of one per thousand, even though the latter increase is far more serious than the former. Berkson and Feinstein maintain that the simple difference between two rates is the proper measure of the practical magnitude, in terms of public health importance, of an association.

Data from Table 26 of *Smoking and Health* (1964, p. 110) will illustrate their point. Table 6.3 gives the approximate death rates per 100,000 person-years for smokers and for nonsmokers of cigarettes.

Table 6.3. Mortality rates per 100,000 person-years from lung cancer and coronary artery disease for smokers and nonsmokers of cigarettes

	Smokers	Nonsmokers	o	Difference
Cancer of the Lung	48.33	4.49	10.8	43.84
Coronary Artery Disease	294.67	169.54	1.7	125.13

If we compared only the odds ratios, o, we would conclude that cigarette smoking has a greater effect on lung cancer than on coronary artery disease. It was this conclusion, from a number of studies, that Berkson felt was unwarranted. He contended, quite correctly, that the odds ratio throws away all information on the number of deaths due to either cause. Berkson went farther, however, and maintained that it is *only* the excess in mortality that permits a valid assessment of the effect of smoking on a cause of death: "... of course, from a strictly practical viewpoint, it is only the total number of increased deaths that matters" [1958, p. 30].

He maintained that the effect of smoking is greater on that cause of death with the greater number of excess deaths. Thus since smoking is associated with an excess mortality of over 120 per 100,000 person-years from coronary artery disease and with an excess mortality of under 50 per 100,000 person-years from lung cancer, Berkson concluded that the association is stronger for coronary artery disease than for lung cancer.

Sheps (1958, 1961) has proposed a simple and elegant modification of Berkson's index. Let p_c denote the mortality rate (in general, the rate at which an untoward event occurs) in a control sample and p_s the corresponding rate in a study sample presumed to be at higher risk. Thus by assumption, $p_s > p_c$.

Sheps contends that the excess risk associated with being in the study group, say p_e, can operate only on those individuals who would not have had the event occurring to them otherwise. Thus she postulates the model

$$p_s = p_c + p_e(1 - p_c); \tag{6.23}$$

that is, the rate in the study group, p_s, is the sum of the rate in the control group, p_c, and of the excess risk, p_e, applied to those who would not otherwise have had the event, $1 - p_c$. Sheps suggests that p_e be used as a measure of added or excess risk. Because, clearly,

$$p_e = \frac{p_s - p_c}{1 - p_c}, \tag{6.24}$$

p_e may also be called the *relative difference*.

This differs from Berkson's index, $p_s - p_c$, only in that the difference is divided by $1 - p_c$, the proportion of the people actually at added risk. If p_c is small, then Sheps's and Berkson's indexes are close. Thus for the data of Table 6.3, the excess mortality due to cancer of the lung is $p_e = 43.84/(100{,}000 - 4.49) = 43.84$ additional deaths attributable to cigarette smoking per 100,000 person-years saved by not having smoked. For coronary artery disease, the excess mortality is $p_e = 125.13/(100{,}000 - 169.54) = 125.34$ deaths per 100,000 person-years saved by not having smoked.

If research into the etiology of disease were concerned *solely* with public health issues, then Berkson's simple difference or Sheps's relative difference would be the only valid measures of the association between the antecedent factor and the outcome event. Because retrospective studies are incapable of providing estimates of either measure, such studies are necessarily useless from the point of view of public health. As Cornfield et al. (1959) and Greenberg (1969) have pointed out, however, etiological research is also concerned with the search for regularities in many sets of data, with the development of models of disease causation and distribution, and with the generation of hypotheses by one set of data that can be tested on other sets.

Given these concerns, that measure of association is best which is suggested by a mathematical model, which remains valid under alternative models, which is capable of assuming predicted values for certain kinds of populations and can thus serve as the basis of a test of a hypothesis, and which is invariant under different methods of studying association. Because the odds ratio, or such a function of it as its logarithm, comes closest to providing such a measure (Cox, 1970, pp. 20–21), and because the odds ratio is estimable from a retrospective study, retrospective studies are eminently valid from the more general point of view of the advancement of knowledge. Peacock (1971) warns, however, against the uncritical assumption that the odds ratio should be constant across different kinds of populations.

That Sheps's measure, as does Berkson's, lacks the regularity observed with the odds ratio is seen from the data of Table 6.4. It gives overall death rates for smokers and nonsmokers of varying ages, taken from the graph on page 88 of *Smoking and Health* (1964). In that graph two rates appear for the nonsmokers aged 75–79. The value that seemed the more reasonable was used.

Table 6.4. Death rates from all causes per 100,000 by age and smoking

Age Interval	Smokers	Nonsmokers	o	p_e per 100,000
45–49	580	270	2.2	310
50–54	1050	440	2.4	610
55–59	1600	850	1.9	750
60–64	2500	1500	1.7	1000
65–69	3700	2000	1.9	1700
70–74	5300	3000	1.8	2400
75–79	9200	4800	2.0	4600

As is so often the case, one comes to different conclusions depending on which measure is chosen. Looking at p_e, the conclusion is that the effect of smoking steadily increases with increasing age. In terms of lives lost among those not otherwise expected to die, this conclusion is correct. The increase in p_e, however, is so erratic that no precise mathematical extrapolation or even interpolation seems possible.

Looking at o, on the other hand, the conclusion is that the effect of smoking on overall mortality is essentially constant across all the tabulated ages. The epidemiological importance of this conclusion is that one may validly predict that the odds ratio would be approximately 2.0 for specific ages between 45 and 79, for ages below 45, and for ages above 79. If the observed odds ratio for any such interpolated or extrapolated age departed appreciably from 2.0, further research would clearly be indicated.

6.4. ESTIMATING ATTRIBUTABLE RISK FROM RETROSPECTIVE STUDIES

Levin's (1953) attributable risk, R_A, was presented in Section 5.7 as the fraction of all occurrences of a condition due to exposure to a specified risk factor. As motivated there, and as defined in (5.76), R_A appears to be estimable only from a cross-sectional study, which simultaneously permits the estimation of $P(A)$, the proportion of the population exposed to the risk factor, and of R, the relative risk. As pointed out by Levin (1953), Walter (1975, 1976) and Taylor (1977), however, R_A is estimable from a retrospective study under certain assumptions. If $P(B)$, the rate of occurrence of the outcome characteristic in the population, is low, then the odds ratio, o, estimates R. If, in addition, the control group ($\bar{B}$) in the study is a random sample of the corresponding group in the population, then $p(A|\bar{B})$ estimates $P(A)$.

Under these assumptions,

$$r_A = \frac{p(A|\bar{B})(o-1)}{1+p(A|\bar{B})(o-1)} = \frac{p(A|B)-p(A|\bar{B})}{1-p(A|\bar{B})} \tag{6.25}$$

is a good estimate of the population attributable risk (Levin and Bertell, 1978). For the hypothetical data of Table 6.2, the estimated risk for low birthweight attributable to low maternal age is

$$r_A = \frac{.40 - .23}{1 - .23} = .22. \tag{6.26}$$

For the corresponding data from a cross-sectional study (Table 5.2), $p(A) = .25$ and the relative risk is $r = 2.0$; thus

$$r_A = \frac{.25(2.0-1)}{1+.25(2.0-1)} = .20, \tag{6.27}$$

which is reasonably close to the estimate in (6.26) from a retrospective study. In either case, the hypothetical data suggest that the rate of low birthweight could be reduced about a fifth if pregnancy among women aged 20 years or less could be avoided.

Walter (1975) has shown that when the attributable risk is estimated from a retrospective study using formula (6.25), the estimated standard error of the natural logarithm of its complement, $1 - r_A$, is given by

$$\text{s.e.}[\ln(1-r_A)] = \sqrt{\frac{p(A|B)}{N_B p(\bar{A}|B)} + \frac{p(A|\bar{B})}{N_{\bar{B}} p(\bar{A}|\bar{B})}}, \tag{6.28}$$

where N_B is the number of cases and $N_{\bar{B}}$ the number of controls. For the data of Table 6.2,

$$\ln(1-r_A) = \ln(1-.22) = -.25 \tag{6.29}$$

and

$$\text{s.e.}[\ln(1-r_A)] = \sqrt{\frac{.40}{100 \times .60} + \frac{.23}{100 \times .77}} = .10. \qquad (6.30)$$

An approximate 95% confidence interval for $\ln(1-R_A)$ is

$$-.25 - 1.96 \times .10 \leqq \ln(1-R_A) \leqq -.25 + 1.96 \times .10, \qquad (6.31)$$

or

$$-.45 \leqq \ln(1-R_A) \leqq -.05. \qquad (6.32)$$

By taking antilogarithms of these limits, and then complements from unity, we obtain an approximate 95% confidence interval for the attributable risk itself:

$$.05 \leqq R_A \leqq .36. \qquad (6.33)$$

6.5. THE RETROSPECTIVE APPROACH VERSUS THE PROSPECTIVE APPROACH

If a scientist accepts the argument of Section 6.3 that the odds ratio and thus the retrospective study are inherently valid, he must still bear in mind that retrospective studies are subject to more sources of error than prospective studies. Hammond (1958), for example, has cited the bias that may arise because historical data are obtained only after subjects become ill, and frequently only after they are diagnosed. A patient's knowledge that he has a certain disease might easily affect his recollection, intentionally or unintentionally, of which factors preceded his illness.

Another difficultly pointed out by Hammond is in finding an adequate control series for the sample of patients: one must, after all, find groups of subjects who are like the cases in all respects save for having the disease. Mantel and Haenszel (1959) cite these and other deficiencies in the retrospective approach. When, for example, the subjects having the disease are found in hospitals or clinics, inferences from retrospective studies may be subject to the kind of bias illustrated in Section 1.3. Specifically, the antecedent factors may appear to be associated with the disease, but might in reality be more associated with admission to a treatment facility. Feinstein (1973) describes the closely related bias due to differential surveillance of individuals with and individuals without the suspected risk factor.

This does not imply that only the retrospective approach, and not the prospective, is open to bias. Similar biases have been shown to operate in prospective studies as well (Mainland and Herrera, 1956; Yerushalmy and Palmer, 1959; Mantel and Haenszel, 1959; Greenland, 1977). For example, the bias possible in those retrospective studies that require hospitalized patients to be evaluated is matched by the potential bias in those prospective studies that require volunteers to be followed up. Errors of diagnosis and insufficiently frequent screening are also problems associated with prospective studies (Schlesselman, 1977).

What does seem to be true, however, is that a greater degree of ingenuity is needed for the proper design of a retrospective study than for the design of a prospective study. Thus Levin (1954) controlled for the first bias cited above—that due to a patient's report possibly being influenced by his knowledge that he has the disease being studied—by questioning all subjects prior to the final diagnosis. By this means, the possible bias due to the examiner's applying different standards to the responses of cases and controls is also controlled. Rimmer and Chambers (1969) suggested another means of control. They found greater accuracy in the recollections of relatives than in those of the patients themselves.

As a means of reducing the bias possible in contrasting cases with only one control series, Doll and Hill (1952) studied two control groups. One was a sample of hospitalized patients with other diseases than the one studied, and the second was a sample from the community. Matching (see Chapter 8) is another device for reducing bias. The validity of the retrospective approach can only increase as investigators learn which kinds of information can be accurately recalled by a subject and which cannot. For example, Gray (1955) and Klemetti and Saxén (1967) have shown that the occurrence or nonoccurrence of a past event can be recalled accurately, but not the precise time when the event occurred.

Jick and Vessey (1978) have reviewed the major sources of error in those retrospective studies in particular that seek to elucidate the role of prescribed drug use in the development of illness, and they indicate methods for their control. The January 1979 issue of the *Journal of Chronic Diseases* (Ibrahim, 1979) is devoted to the state of the art of retrospective studies; the contribution by Sackett (1979) is especially useful for its cataloguing of several sources of bias and for its prescriptions for their control or measurement.

The reduction of bias is discussed again in Chapter 12. To the extent that bias can be controlled, the following points made by Mantel and Haenszel argue strongly for the retrospective approach:

> Among the desirable attributes of the retrospective study is the ability to yield results from presently collectible data.... The retrospective approach is also adapted to the limited resources of an individual investigator.... For especially

rare diseases a retrospective study may be the only feasible approach.... In the absence of important biases in the study setting, the retrospective method could be regarded, according to sound statistical theory, as the study method of choice [1959, p. 720].

Problem 6.1. Prove the equality of expressions 6.6 and 6.8 for the standard error of the odds ratio. [*Hint.* $p(B|A)=n_{11}/N_A$; $p(\bar{B}|A)=n_{12}/N_A$; $p(B|\bar{A})=n_{21}/N_{\bar{A}}$; and $p(\bar{B}|\bar{A})=n_{22}/N_{\bar{A}}$. Also, $N_A=n_{11}+n_{12}$ and $N_{\bar{A}}=n_{21}+n_{22}$.]

Problem 6.2. Prove the equality of expressions 6.16 and 6.17 for the odds ratio. [*Hint.* Use the definitions of Section 1.1 to replace all conditional probabilities in (6.16) by joint probabilities. The probabilities $P(A)$ and $P(\bar{A})$ are seen to cancel out. Multiply and divide by $P(B)$ and by $P(\bar{B})$, and use the definition of conditional probabilities to arrive at (6.17).]

Problem 6.3. The phi coefficient [see (5.2)] is a valid measure of association only for method I (naturalistic or cross-sectional) sampling. Phi coefficients applied to data from method II (prospective or retrospective) studies are not at all comparable to those applied to data from method I studies.

Even more is true. When two studies are both conducted according to either the prospective or retrospective approaches, but with proportionately different allocations of the total sample, the phi coefficient for one will not in general be comparable to that for the other.

(*a*) In a retrospective study of factors associated with cancer of the oral cavity, Wynder, Navarrette, Arostegui, and Llambes (1958) studied 34 women with cancer of the oral cavity and 214 women, matched by age, with nonmalignant conditions. Twenty-four percent of the cancer cases, as opposed to 66% of the controls, were nonsmokers. Set up the resulting fourfold table and calculate uncorrected chi square and the associated phi coefficient.

(*b*) Suppose that Wynder et al. had studied, instead, 214 cancer cases and 34 controls. Assuming the same proportions of nonsmokers as above, set up the expected fourfold table and calculate uncorrected chi square and the associated phi. How do the phi coefficients compare?

(*c*) Suppose, now, that 124 of both kinds of women had been studied, and assume the same proportions of nonsmokers as above. Set up the resulting expected fourfold table and calculate uncorrected chi square and the associated phi. How does this phi coefficient compare with those in (*a*) and (*b*)? What is the percentage difference between the phi coefficient of (*a*) and that of (*c*)? What would you conclude about the comparability of phi coefficients in retrospective studies with varying allocations of a total sample?

Problem 6.4. Three criteria were suggested in this chapter for comparing the cross-sectional, prospective, and retrospective approaches. Another criterion is the total sample size necessary for the standard error of the odds ratio to assume some specified value. The data for the following questions are those employed throughout Chapters 5 and 6. Suppose that the correct values of $o = 2.25$ and of all the necessary proportions are known.

(*a*) The approximate standard error of o with cross-sectional sampling is given by (5.17). What value of $n_{..}$ is needed to give a standard error of 0.50?

(*b*) The approximate standard error of o with prospective sampling is given by (6.6). Let $N_A + N_{\bar{A}}$, the total sample size, be denoted N_P, and suppose for simplicity that $N_A = N_{\bar{A}} = N_P/2$. What value of N_P is needed to give a standard error of 0.50? What is the percentage reduction from $n_{..}$ to N_P?

(*c*) The approximate standard error of o with retrospective sampling is given by (6.20). Let $N_B + N_{\bar{B}}$, the total sample size, be denoted N_R, and suppose for simplicity that $N_B = N_{\bar{B}} = N_R/2$. What value of N_R is needed to give a standard error of 0.50? What is the percentage reduction from $n_{..}$ to N_R? From N_P to N_R?

(*d*) Is the reduction in (*b*) of much practical (e.g., monetary) importance? How about the reductions in (*c*)?

REFERENCES

Berkson, J. (1958). Smoking and lung cancer: Some observations on two recent reports. *J. Am. Stat. Assoc.*, **53**, 28–38.

Cornfield, J. (1956). A statistical problem arising from retrospective studies. Pp. 135–148 in J. Neyman (Ed.). *Proceedings of the third Berkeley symposium on mathematical statistics and probability*, Vol. 4. Berkeley: University of California Press.

Cornfield, J., Haenszel, W., Hammond, E. C., Lilienfeld, A. M., Shimkin, M. B., and Wynder, E. L. (1959). Smoking and lung cancer: Recent evidence and a discussion of some questions. *J. Natl. Cancer Inst.*, **22**, 173–203.

Cox, D. R. (1970). *Analysis of binary data*. London: Methuen.

Doll, R. and Hill, A. B. (1952). A study of the etiology of carcinoma of the lung. *Brit. Med. J.*, **2**, 1271–1286.

Feinstein, A. R. (1973). Clinical biostatistics XX. The epidemiologic trohoc, the ablative risk ratio, and retrospective research. *Clin. Pharmacol. Ther.*, **14**, 291–307.

Gray, P. G. (1955). The memory factor in social surveys. *J. Am. Stat. Assoc.*, **50**, 344–363.

Greenberg, B. G. (1969). Problems of statistical inference in health with special reference to the cigarette smoking and lung cancer controversy. *J. Am. Stat. Assoc.*, **64**, 739–758.

Greenland, S. (1977). Response and follow-up bias in cohort studies. *Am. J. Epidemiol.*, **106**, 184–187.

Hammond, E. C. (1958). Smoking and death rates: A riddle in cause and effect. *Am. Sci.*, **46**, 331–354.

Ibrahim, M. A. (Ed.) (1979). The case-control study: Consensus and controversy. *J. Chronic Dis.*, **32**, 1–144.

Jick, H. and Vessey, M. P. (1978). Case-control studies in the evaluation of drug-induced illness. *Am. J. Epidemiol.*, **107**, 1–7.

Klemetti, A. and Saxén, L. (1967). Prospective versus retrospective approach in the search for environmental causes of malformations. *Am. J. Public Health*, **57**, 2071–2075.

Lehmann, E. L. (1959). *Testing statistical hypotheses*. New York: Wiley.

Levin, M. L. (1953). The occurrence of lung cancer in man. *Acta Unio Int. Contra Cancrum*, **19**, 531–541.

Levin, M. L. (1954). Etiology of lung cancer: Present status. *N. Y. State J. Med.*, **54**, 769–777.

Levin, M. L. and Bertell, R. (1978). Re: "Simple estimation of population attributable risk from case-control studies." *Am. J. Epidemiol.*, **108**, 78–79.

MacMahon, B. and Pugh, T. F. (1970). *Epidemiology*: *Principles and methods*. Boston: Little, Brown.

Mainland, D. and Herrera, L. (1956). The risk of biased selection in forward-going surveys with nonprofessional interviewers. *J. Chronic Dis.*, **4**, 240–244.

Mantel, N. and Haenszel, W. (1959). Statistical aspects of the analysis of data from retrospective studies of disease. *J. Natl. Cancer Inst.*, **22**, 719–748.

Miettinen, O. S. (1970). Matching and design efficiency in retrospective studies. *Am. J. Epidemiol.*, **91**, 111–118.

Peacock, P. B. (1971). The noncomparability of relative risks from different studies. *Biometrics*, **27**, 903–907.

Rimmer, J. and Chambers, D. S. (1969). Alcoholism: Methodological considerations in the study of family illness. *Am. J. Orthopsychiatry*, **39**, 760–768.

Sackett, D. L. (1979). Bias in analytic research. *J. Chronic Dis.*, **32**, 51–63.

Schlesselman, J. J. (1977). The effect of errors of diagnosis and frequency of examination on reported rates of disease. *Biometrics*, **33**, 635–642.

Sheps, M. C. (1958). Shall we count the living or the dead? *New Engl. J. Med.*, **259**, 1210–1214.

Sheps, M. C. (1961). Marriage and mortality. *Am. J. Public Health*, **51**, 547–555.

Smoking and Health (1964). Report of the Advisory Committee to the Surgeon General of the Public Health Service. Princeton: Van Nostrand.

Taylor, J. W. (1977). Simple estimation of population attributable risk from case-control studies. *Am. J. Epidemiol.*, **106**, 260.

Walter, S. D. (1975). The distribution of Levin's measure of attributable risk. *Biometrika*, **62**, 371–374.

Walter, S. D. (1976). The estimation and interpretation of attributable risk in health research. *Biometrics*, **32**, 829–849.

Wynder, E. L., Navarrette, A., Arostegui, G. E., and Llambes, J. L. (1958). Study of environmental factors in cancer of the respiratory tract in Cuba. *J. Natl. Cancer Inst.*, **20**, 665–673.

Yerushalmy, J. and Palmer, C. E. (1959). On the methodology of investigations of etiologic factors in chronic diseases. *J. Chronic Dis.*, **10**, 27–40.

CHAPTER 7

Sampling Method III: Controlled Comparative Trials

Sampling method III is exemplified by the comparative clinical trial in which treatments are assigned to subjects at random. The philosophy of the controlled clinical trial is discussed by Hill (1962, Chapters 1–3), and solutions to some practical problems arising in the execution of a clinical trial are offered by Mainland (1960) and in an entire issue of *Clinical Pharmacology and Therapeutics* (Roth and Gordon, 1979). Ethical issues are discussed by Fox (1959) and Meier (1975). Considerations needed in deciding how many patients to study and some unfortunate consequences of studying too few are discussed by Frieman et al. (1978).

A problematic feature of clinical trials is the opportunity to terminate the study prematurely if one of the groups being compared is found to be experiencing an alarmingly high rate of serious adverse reactions or if the therapeutic difference between the treatments is so overwhelming that to withhold the superior treatment from all future eligible patients is deemed unethical. The University Group Diabetes Program (Report of the Committee for the Assessment of Biometric Aspects of Controlled Trials of Hypoglycemic Agents, 1975) provides a noteworthy example of the former reason to terminate a trial, and the Anturane Trial (Anturane Reinfarction Trial Research Group, 1978) provides a noteworthy example of the latter. Meier (1979) suggests a framework for deciding whether to continue or to terminate a trial.

In this chapter we consider only the comparison of two treatments. Section 7.1 describes the analysis of data from a simple comparative trial and Section 7.2 discusses the crossover design, both for the case where the outcome is a yes-no variable, such as recovery-no recovery. Some proposed alternatives to simple randomization in comparative trials are discussed in Section 7.3.

7.1. THE SIMPLE COMPARATIVE TRIAL

Suppose that the data of Table 7.1 resulted from a trial in which one treatment was applied to a sample of $n_1 = 80$ subjects randomly selected from a total of $n = 150$ and the other treatment was applied to the remaining $n_2 = 70$ subjects.

Table 7.1. Hypothetical data from a comparative clinical trial

	Number of Patients	Proportion Improved
Treatment 1	$80 (= n_1)$	$.60 (= p_1)$
Treatment 2	$70 (= n_2)$	$.80 (= p_2)$
Overall	$150 (= n)$	$.69 (= \bar{p})$

The statistical significance of the difference between the two improvement rates is tested using the statistic given in (2.5). For the data of Table 7.1, the value is

$$z = \frac{|.80 - .60| - \frac{1}{2}\left(\frac{1}{80} + \frac{1}{70}\right)}{\sqrt{.69 \times .31\left(\frac{1}{80} + \frac{1}{70}\right)}} = 2.47, \tag{7.1}$$

indicating that the difference is significant at the .05 level.

The simple difference between the two improvement rates,

$$d = p_2 - p_1, \tag{7.2}$$

is the measure most frequently used to describe the differential effectiveness of the second treatment over the first. The approximate standard error of d is

$$\text{s.e.}(d) = \sqrt{\frac{p_1 q_1}{n_1} + \frac{p_2 q_2}{n_2}}\ . \tag{7.3}$$

For the data of Table 7.1, the simple difference is

$$d = .80 - .60 = .20, \tag{7.4}$$

implying that among every 100 patients given the first treatment an additional 20 would have been expected to improve had they been given the second treatment. The estimated standard error of d is

$$\text{s.e.}(d) = \sqrt{\frac{.60 \times .40}{80} + \frac{.80 \times .20}{70}} = .07. \tag{7.5}$$

An approximate 95% confidence interval for the difference between the two underlying rates of improvement is

$$.20 - 1.96 \times .07 \leqq P_2 - P_1 \leqq .20 + 1.96 \times .07, \tag{7.6}$$

or

$$.06 \leqq P_2 - P_1 \leqq .34. \tag{7.7}$$

One can occasionally assume that the two treatments are such that any patient who responds to the first treatment is also expected to respond to the second. This assumption may be tenable if the first treatment is an inert placebo, or if the first treatment is an active drug and the second is that drug plus another compound or that drug at a greater dosage level. A consequence of the assumption is that any greater effectiveness of the second treatment can only be manifested on subjects who were refractory to the first (see Sheps, 1958, for further examples and discussion).

Let P_1 denote the proportion improving in the population of patients given the first treatment and P_2 the proportion improving in the population given the second. Let f denote the proportion of patients, among those failing to respond to the first treatment, who would be expected to respond to the second. It is then assumed that

$$P_2 = P_1 + f(1 - P_1), \tag{7.8}$$

that is, the improvement rate under the second treatment is equal to that under the first plus an added improvement rate which applies only to patients who fail to improve under the first treatment. The value of f is clearly

$$f = \frac{P_2 - P_1}{1 - P_1}, \tag{7.9}$$

so that f may be termed the *relative difference*.

Because the sample proportions p_1 and p_2 are estimates of the corresponding population proportions, an estimate of the relative difference is

$$p_e = \frac{p_2 - p_1}{1 - p_1}. \tag{7.10}$$

Its standard error (see Sheps, 1959) is approximately

$$\text{s.e.}(p_e) = \frac{1}{q_1}\sqrt{\frac{p_2 q_2}{n_2} + (1-p_e)^2 \frac{p_1 q_1}{n_1}} \; . \tag{7.11}$$

Walter (1975) showed, however, that more accurate inferences about f could be made by taking $\ln(1-p_e)$ as normally distributed with a mean of $\ln(1-f)$ and a standard error of

$$\text{s.e.}[\ln(1-p_e)] = \sqrt{\frac{p_2}{n_2 q_2} + \frac{p_1}{n_1 q_1}} \; . \tag{7.12}$$

For the data of Table 7.1, the relative difference is

$$p_e = \frac{.80 - .60}{1 - .60} = .50, \tag{7.13}$$

implying that, of every 100 patients who fail to improve under the first treatment, 50 would be expected to improve under the second. The value of $\ln(1-p_e)$ is $-.69$, and an estimate of its standard error is

$$\text{s.e.}[\ln(1-p_e)] = \sqrt{\frac{.80}{70 \times .20} + \frac{.60}{80 \times .40}} = .28. \tag{7.14}$$

An approximate 95% confidence interval for $\ln(1-f)$ is

$$-.69 - 1.96 \times .28 \leqq \ln(1-f) \leqq -.69 + 1.96 \times .28,$$

or

$$-1.24 \leqq \ln(1-f) \leqq -.14. \tag{7.15}$$

By taking antilogarithms of the limits of this interval and then their complements from unity, one obtains an approximate 95% confidence interval for the relative difference itself:

$$.13 \leqq f \leqq .71. \tag{7.16}$$

A perspective different from the usual one is required when the aim of the study is to demonstrate that two treatments are therapeutically equivalent or, at least, that they differ by an amount that is clinically unimportant. Examples of such studies and a method of analysis are given by Dunnett and Gent (1977).

7.2. THE TWO-PERIOD CROSSOVER DESIGN

The next chapter presents means for analyzing data from, *inter alia*, a controlled trial in which patients are first matched on characteristics associated with the outcome and then randomly assigned the treatments. An extreme example of matching is when, as in a *crossover design*, each patient serves as his own control, that is, when each patient receives each treatment.

Half of the sample of patients is randomly selected to be given the two treatments in one order, and the other half to be given the treatments in the reverse order. A number of factors must be guarded against in analyzing the data from such studies, however.

Meier, Free, and Jackson (1958) have shown that the order in which the treatments are given may affect the response. The following test, due to Gart (1969), is valid when order effects are present and when the outcome is measured as good or poor.

Let the data be arrayed as in Table 7.2, where, for example, n_{12} denotes the number of patients, among those receiving the treatments in the order

Table 7.2. Layout of data from a two-period crossover design

Order AB			Order BA		
Response to A	Response to B Good	Poor	Response to A	Response to B Good	Poor
Good	n_{11}	n_{12}	Good	m_{11}	m_{12}
Poor	n_{21}	n_{22}	Poor	m_{21}	m_{22}

AB, who had a good response to treatment A but a poor response to treatment B. For the sample of patients receiving the two treatments in the order AB, those with a good response to both treatments (n_{11} in number) and those with a poor response to both (n_{22} in number) provide no information about a difference between the two treatments and may be ignored in the analysis. Similarly, the $m_{11}+m_{22}$ patients with similar responses to the two treatments, among those receiving them in the order BA, are uninformative about a difference between A and B and may also be ignored.

The resulting data should be arrayed as in Table 7.3, where $n=n_{12}+n_{21}$ and $m=m_{12}+m_{21}$. If treatments A and B are equally effective, the two proportions $p_1=n_{12}/n$ and $p_2=m_{21}/m$ should be close; if A and B are

Table 7.3. Layout of data from Table 7.2 to test hypothesis of equal treatment effectiveness

Order of Treatment	Outcome: First Treatment Better	Outcome: Second Treatment Better	Total
AB	n_{12}	n_{21}	n
BA	m_{21}	m_{12}	m
Total	$n_{12}+m_{21}$	$n_{21}+m_{12}$	$n+m$

different, p_1 and p_2 should be different. The hypothesis of equal effectiveness of the two treatments may be tested by comparing p_1 with p_2 in the standard manner (see Problem 7.2).

A factor to be guarded against in crossover studies is the possibility that a treatment's effectiveness is long-lasting and hence may affect the response to the treatment given after it. When this so-called *carry-over effect* operates, and when it is unequal for the two treatments, Grizzle (1965) has shown that, for comparing their effectiveness, only the data from the first period may be used. Specifically, the responses by the subjects given one of the treatments first must be compared with the responses by the subjects given the other treatment first. The responses to the treatments given second shed light on the carry-over effects, but might just as well not have been determined if the simple effectiveness of the treatments is all that is of interest.

Differential carry-over effects may be eliminated by interposing a long dry-out period between the termination of the treatment given first and the beginning of the treatment given second. The longer the dry-out period, however, the greater the chances that patients drop out of the trial.

Crossover designs are safe when the treatments are short-acting. When the possibility exists that they are long-acting, the crossover design is to be avoided.

7.3. ALTERNATIVES TO SIMPLE RANDOMIZATION

The need always to randomize the assignment of treatments to patients in clinical trials continues to be debated. Gehan and Freireich (1974) and Weinstein (1974), for example, offer criticisms of and propose some alternatives to the strictly randomized clinical trial, whereas Byar et al. (1976) and Peto et al. (1976) come to its defense. There appear to be some instances in which competitors to randomization should be seriously considered (e.g., when only a small number of patients is available for study, so that recent historical controls may have to be relied on), but these are small in number.

Debate continues, too, with respect to the design of trials in which patients treated with the experimental treatment are compared to concurrent controls. The need to stratify or to match patients on prognostic factors (e.g., age, sex, and initial severity) and just how to assure comparable distributions across the strata in the two treatment groups are two areas of debate and research. On one side of the debate are Peto et al. (1976), who recommend not bothering to stratify the eligible sample of patients before randomizing them to one or the other treatment group, especially when the number of patients is large.

The consensus of opinion among those responsible for the design of clinical trials, however, seems to be that some degree of control, preferably stratification, is desirable to guard against unlikely but devastating imbalances between the two groups (see, e.g., Simon, 1979). Separate and independent randomizations of the patients falling into the several strata will usually suffice, but some recently proposed modifications in simple stratified randomization should be borne in mind.

These modifications are intended especially for studies in which patients enter the trial serially over time, so that it is impossible to determine at the start of the trial exactly how many patients will end up in each stratum. Suppose, for example, that a patient enters the trial at a point at which, within his stratum, more patients have been assigned treatment A than treatment B. Instead of assigning this new patient to one treatment group or the other independently of the current imbalance in his stratum, one may adopt an allocation method that seeks to bring the numbers in the two groups closer to each other.

At one extreme is Efron's (1971) "biased coin" scheme, which, in the current example, would assign this patient treatment B (the underrepresented one) with a fixed, prespecified probability that is greater than .5 but less than 1. At the other extreme, in effect, is Taves's (1974) "minimization" scheme, which would assign this patient treatment B with certainty. Randomization plays no role in "minimization" except for the first group of patients to enter the trial, unless the two treatment groups are balanced at the time a new patient enters the trial. In between these two extremes is the scheme proposed by Pocock and Simon (1975) and simplified by Freedman and White (1976).

As Pocock (1979) has pointed out, however, the ability to execute a design is at least as important as the design's theoretical optimality. Complexity of execution plus the likely sufficiency in most instances of simple stratified randomization appear to rule out the large-scale adoption of these proposed schemes for establishing balance between the two treatment groups.

Other alternatives to the simple comparative study, termed adaptive trials, have been proposed because of the uniqueness of comparative clinical trials as studies aimed as much at the testing of scientific hypotheses as at the alleviation of symptoms of the patients being studied. Whereas the classic design and its modifications summarized in the preceding call for assigning the different treatments to approximately equal numbers of patients, adaptive designs call for assigning to an increasing proportion of patients the treatment that, on the basis of the accumulated data, appears to be superior.

Adaptive designs are still in the process of development and apparently have not yet been adopted in practice in spite of their intuitive appeal. The interested reader is referred to Anscombe (1963), Colton (1963), Cornfield, Halperin, and Greenhouse (1969), Zelen (1969), Canner (1970), and Robbins (1974) for some suggested means of balancing the statistical requirement of equal sample sizes with the ethical requirement of applying the superior treatment to as many patients as possible as quickly as possible [see Simon (1977) for a review].

Zelen (1979) introduced a daring new idea into the design of clinical trials in seeking to overcome the reluctance of many patients and the physicians responsible for their care to consent to participate in a randomized experiment. This reluctance is especially prevalent when the patients suffer from life-threatening conditions such as cancer or heart disease. Zelen's idea is to randomize eligible patients into one of the two groups characterized as follows.

All patients in the first group will receive the currently accepted standard treatment. Because this is the treatment they would receive in any event, their informed consent to participate in an experiment is not required.

All patients in the second group will be asked whether they consent to be treated with the experimental treatment. If they consent, they will be so treated. If they refuse, they will be treated with the standard treatment. At the end of the trial, the results for *all* patients in the second group, regardless of treatment, will be compared with the results for all those in the first.

Operating to reduce the power associated with Zelen's proposed design is the fact that the response rate estimated for group two is an average of the rate for those patients receiving the experimental treatment and the rate for those patients receiving the standard treatment. If the experimental treatment is superior to the standard, this average will be closer to the response rate in group one than the response rate would be in a group of patients treated exclusively with the experimental treatment.

Operating in favor of the proposed design is the likelihood that much larger numbers of patients will be studied. The traditional design requires that an eligible patient give informed consent to participate in the trial before that patient is randomized. The proposed design permits all eligible patients to be randomized; a patient's refusal to give informed consent does not result in that patient's being dropped from the study. The increased power due to the larger numbers of patients might overcome the decreased power due to the attenuated difference between the two groups.

Whether Zelen's proposed design becomes accepted depends on what proportions of patients consent to receive the experimental treatment and on how strong the biasing effects are of the patient's and his physician's knowing which treatment he is receiving. This information will come only from experience with the proposed design.

Problem 7.1. Suppose that the two treatments contrasted in Table 7.1 are compared in a second hospital, with the following results.

	Number of Patients	Proportion Improved
Treatment 1	100	.35
Treatment 2	100	.75
Overall	200	.55

(a) Is the difference between the improvement rates significant in the second hospital?

(b) What is the simple difference [see (7.2)] between the two improvement rates in the second hospital? What is its standard error [see (7.3)]? Is the difference found in the second hospital significantly different from that found in the first? (*Hint*. Denote the difference and its standard error found in (7.4) and (7.5) by d_1 and s.e.(d_1), and denote the corresponding statistics just calculated by d_2 and s.e.(d_2). Refer the value of

$$z=\frac{|d_2-d_1|}{\sqrt{[\text{s.e.}(d_1)]^2+[\text{s.e.}(d_2)]^2}}$$

to Table A.2 of the normal distribution.)

(c) What is the relative difference [see (7.10)] between the two improvement rates in the second hospital? Is it significantly different from the relative difference found in the first hospital? (*Hint*. Define $L_1=\ln(1-$

$p_{e(1)}$) and $L_2 = \ln(1 - p_{e(2)})$. Refer the value of

$$z = \frac{|L_1 - L_2|}{\sqrt{[\text{s.e.}(L_1)]^2 + [\text{s.e.}(L_2)]^2}}$$

to Table A.2.)

Problem 7.2. Suppose two treatments were compared in a two-period crossover design, with results as follows.

Order AB			Order BA		
Response to A	Response to B Good	Poor	Response to A	Response to B Good	Poor
Good	20	15	Good	30	10
Poor	5	10	Poor	5	5

(a) Consider first the sample of patients given the treatments in the order AB. What is the value of n (i.e., how many of these patients gave responses that were informative about a difference between A and B)? What is the value of p_1 (i.e., what proportion of these n patients had a good response to the treatment given first)?

(b) Consider next the sample of patients given the treatments in the order BA. What is the value of m? What is the value of p_2?

(c) Test whether treatments A and B are equally effective by comparing p_1 with p_2.

REFERENCES

Anscombe, F. J. (1963). Sequential medical trials. *J. Am. Stat. Assoc.*, **58**, 365–384.

Anturane Reinfarction Trial Research Group (1978). Sulfinpyrazone in the prevention of cardiac death after myocardial infarction: The anturane reinfarction trial. *New Engl. J. Med.*, **298**, 290–295.

Byar, D. P., Simon, R. H., Friedewald, W. T., Schlesselman, J. J., DeMets, D. L., Ellenberg, J. H., Gail, M. H., and Ware, J. H. (1976). Randomized clinical trials: Perspectives on some recent ideas. *New Engl. J. Med.*, **295**, 74–80.

Canner, P. L. (1970). Selecting one of two treatments when the responses are dichotomous. *J Am. Stat. Assoc.*, **65**, 293–306.

Colton, T. (1963). A model for selecting one of two medical treatments. *J. Am. Stat. Assoc.*, **58**, 388–401.

Cornfield, J., Halperin, M., and Greenhouse, S. W. (1969). An adaptive procedure for sequential clinical trials. *J. Am. Stat. Assoc.*, **64**, 759–770.

Dunnett, C. W. and Gent, M. (1977). Significance testing to establish equivalence between treatments, with special reference to data in the form of 2 × 2 tables. *Biometrics*, **33**, 593–602.

Efron, B. (1971). Forcing a sequential experiment to be balanced. *Biometrika*, **58**, 403–417.

Fox, T. F. (1959). The ethics of clinical trials. Pp. 222–229 in D. R. Laurence (Ed.). *Quantitative methods in human pharmacology and therapeutics*. New York: Pergamon Press.

Freedman, L. S. and White, S. J. (1976). On the use of Pocock and Simon's method for balancing treatment numbers over prognostic factors in the controlled clinical trial. *Biometrics*, **32**, 691–694.

Frieman, J. A., Chalmers, T. C., Smith, H., and Kuebler, R. R. (1978). The importance of Beta, the type II error and sample size in the design and interpretation of the randomized control trial: Survey of 71 "negative" trials. *New Engl. J. Med.*, **299**, 690–694.

Gart, J. J. (1969). An exact test for comparing matched proportions in crossover designs. *Biometrika*, **56**, 75–80.

Gehan, E. A. and Freireich, E. J. (1974). Non-randomized controls in cancer clinical trials. *New Engl. J. Med.*, **290**, 198–203.

Grizzle, J. E. (1965). The two-period change-over design and its use in clinical trials. *Biometrics*, **21**, 467–480.

Hill, A. B. (1962). *Statistical methods in clinical and preventive medicine*. New York: Oxford University Press.

Mainland, D. (1960). The clinical trial—some difficulties and suggestions. *J. Chronic Dis.*, **11**, 484–496.

Meier, P. (1975). Statistics and medical experimentation. *Biometrics*, **31**, 511–529.

Meier, P. (1979). Terminating a trial—The ethical problem. *Clin. Pharmacol. Ther.*, **25**, 633–640.

Meier, P., Free, S. M., and Jackson, G. L. (1958). Reconsideration of methodology in studies of pain relief. *Biometrics*, **14**, 330–342.

Peto, R., Pike, M. C., Armitage, P., Breslow, N. E., Cox, D. R., Howard, S. V., Mantel, N., McPherson, K., Peto, J., and Smith, P. G. Design and analysis of randomized clinical trials requiring prolonged observation of each patient. I. Introduction and design. *Brit. J. Cancer*, **34**, 585–612.

Pocock, S. J. and Simon, R. (1975). Sequential treatment assignment with balancing for prognostic factors in the controlled clinical trial. *Biometrics*, **31**, 103–115.

Pocock, S. J. (1979). Allocation of patients to treatment in clinical trials. *Biometrics*, **35**, 183–197.

Report of the Committee for the Assessment of Biometric Aspects of Controlled Trials of Hypoglycemic Agents. (1975). *J. Am. Med. Assoc.*, **231**, 583–608.

Robbins, H. (1974). A sequential test for two binomial populations. *Proc. Natl. Acad. Sci.*, **71**, 4435–4436.

Roth, H. P. and Gordon, R. S. (Eds.) (1979). Proceedings of the National Conference on Clinical Trials Methodology. *Clin. Pharmacol. Ther.*, **25**, 629–766.

Sheps, M. C. (1958). Shall we count the living or the dead? *New Engl. J. Med.*, **259**, 1210–1214.

Sheps, M. C. (1959). An examination of some methods of comparing several rates or proportions. *Biometrics*, **15**, 87–97.

Simon, R. H. (1977). Adaptive treatment assignment methods and clinical trials. *Biometrics*, **33**, 743–749.

Simon R. (1979). Restricted randomization designs in clinical trials. *Biometrics*, **35**, 503–512.

Taves, D. R. (1974). Minimization: A new method of assigning patients to treatment and control groups. *Clin. Pharmacol. Ther.*, **15**, 443–453.

Walter, S. D. (1975). The distribution of Levin's measure of attributable risk. *Biometrika*, **62**, 371–374.

Weinstein, M. C. (1974). Allocation of subjects in medical experiments. *New Engl. J. Med.*, **291**, 1278–1285.

Zelen, M. (1969). Play the winner rule and the controlled clinical trial. *J. Am. Stat. Assoc.*, **64**, 131–146.

Zelen, M. (1979). A new design for randomized clinical trials. *New Engl. J. Med.*, **300**, 1242–1245.

CHAPTER 8

The Analysis of Data from Matched Samples

A device often employed in controlled trials (sampling method III) is to match subjects on the basis of characteristics that are associated with the response being studied, and to randomize the treatment assignments independently within each matched group. Matched pairs of subjects are used for comparing two treatments, matched triples for comparing three treatments, and in general matched *m-tuples* for comparing *m* treatments. The purpose of matching in controlled trials is to increase the precision of the comparisons among the treatments (Hill, 1962, p. 21).

Matching is also frequently employed in comparative prospective and retrospective studies (sampling method II), but more for increasing the validity of the inferences by controlling for *confounding* factors than for increasing precision (see Bross, 1969, and Miettinen, 1970a, for a debate on this point). Age and sex, for example, are possible confounding factors in the study of the association between cigarette smoking and lung cancer, because age and sex are associated both with smoking and with the risk of lung cancer. In a retrospective study, therefore, these factors might be controlled by matching each case of lung cancer with a control subject of the same sex and of a similar age. Because the cases and controls would then be similar on sex and age, any difference between the two samples would have to be attributable to other factors. Section 10.6 presents another device for the control of confounding factors.

Sampling method I does not lend itself to matching.

Section 8.1 is devoted to the analysis of data from matched pairs when only a dichotomous (yes-no) outcome is of interest, and Section 8.2 to the analysis of data from matched pairs when more than a dichotomous outcome is of interest. Section 8.3 considers the analysis of data resulting from the study of cases matched with multiple controls when the controls all form a single sample. Section 8.4 considers the analysis of data from a study comparing samples from more than two populations when the members of the several samples form matched sets. Some comments on the advantages and disadvantages of matching are made in Section 8.5.

8.1. MATCHED PAIRS: DICHOTOMOUS OUTCOME

Suppose that a retrospective study has been conducted in which each case has been matched with a single control and in which the relative frequency of an antecedent factor among the cases is to be compared with that among the controls. Because of the matching of cases with controls, the proper unit of analysis is the matched pair rather than the individual subject. Table 8.1 gives the appropriate means for presenting the resulting data.

Table 8.1. Data on two outcomes from matched pairs

	Controls		
Cases	Factor Present	Factor Absent	Total
Factor present	a	b	$a+b$
Factor absent	c	d	$c+d$
Total	$a+c$	$b+d$	n

Each frequency in Table 8.1 represents a number of *pairs*. Thus there were n pairs studied in all. Of these, a were such that both members (the case and his matched control) had the antecedent factor; b were such that the case had the factor but the control did not; c were such that the control had the factor but the case did not; and d were such that neither member had the factor.

The proportion of controls who had the factor is

$$p_1 = \frac{a+c}{n},$$

and the proportion of cases who had the factor is

$$p_2 = \frac{a+b}{n}.$$

The number of pairs in which both the case and his matched control had the factor, a, clearly does not affect the difference between the two proportions,

$$p_2 - p_1 = \frac{b-c}{n}. \tag{8.1}$$

As McNemar (1947) has shown, neither a nor d, the numbers of pairs both of whose members were similar with respect to the antecedent factor, contributes explicitly to the standard error of the difference when the two underlying proportions are equal. In fact,

$$\text{s.e.}(p_2 - p_1) = \frac{\sqrt{b+c}}{n}. \tag{8.2}$$

The square of the ratio of (8.1) to (8.2) may, with a correction for continuity, be used to test for the statistical significance of the difference between p_1 and p_2. The correction, due to Edwards (1948), yields the statistic

$$\chi^2 = \left\{ \frac{|p_2 - p_1| - 1/n}{\text{s.e.}(p_2 - p_1)} \right\}^2 = \frac{(|b-c|-1)^2}{b+c}. \tag{8.3}$$

The value of χ^2 may be referred to tables of chi square with one degree of freedom (see McNemar, 1947; Mosteller, 1952; and Stuart, 1957). If χ^2 is large, the inference can be made that the cases and controls differ in the proportion having the antecedent factor. It is noteworthy that only the pairs in which the members differ in the antecedent factor contribute to the test statistic. The power of this test has been studied by Miettinen (1968) and by Bennett and Underwood (1970).

The test based on (8.3), termed *McNemar's test*, is illustrated on the hypothetical data of Table 8.2.

Table 8.2. Hypothetical data to illustrate McNemar's test

	Controls		
Cases	Factor Present	Factor Absent	Total
Factor present	15	20	35
Factor absent	5	60	65
Total	20	80	100

The proportion of controls having the factor is $p_1 = 20/100 = .20$, and the proportion of cases having the factor is $p_2 = 35/100 = .35$. The standard error of the difference [see (8.2)] is $\text{s.e.}(p_2 - p_1) = \sqrt{20+5}\,/100 =$

$5/100 = .05$, and the test statistic [see (8.3)] has the value

$$\chi^2 = \left(\frac{|.35 - .20| - .01}{.05} \right)^2 = \left(\frac{.14}{.05} \right)^2 = 7.84,$$

equal to the value obtained by comparing the numbers of pairs whose members differed on the factor,

$$\chi^2 = \frac{(|20 - 5| - 1)^2}{20 + 5} = \frac{196}{25} = 7.84.$$

Since χ^2 exceeds 6.63, the value needed for significance at the .01 level, the conclusion can be drawn that the cases and controls differ in the presence of the antecedent factor.

As pointed out in Chapters 5 and 6, the *odds ratio* (the odds of the disease when the factor is present relative to the odds when the factor is absent) is an important measure of the degree of association between the antecedent factor and the disease. Mantel and Haenszel (1959) and Cornfield and Haenszel (1960) have investigated the proper method for estimating the odds ratio when matched pairs have been studied. When the data are arrayed as in Table 8.1, the estimate is simply

$$o = \frac{b}{c}, \tag{8.4}$$

and its standard error is estimated by

$$\text{s.e.}(o) = o\sqrt{\frac{1}{b} + \frac{1}{c}} \tag{8.5}$$

(see Ejigou and McHugh, 1977). For the frequencies of Table 8.2, the estimated odds ratio is $o = 20/5 = 4.0$ and its estimated standard error is

$$\text{s.e.}(o) = 4\sqrt{\frac{1}{20} + \frac{1}{5}} = 2.0.$$

An approximate confidence interval for ω, the underlying odds ratio, may be obtained as follows. Define

$$P = \frac{\omega}{\omega + 1}. \tag{8.6}$$

An estimate of P is

$$p = \frac{b}{b + c}, \tag{8.7}$$

based on a sample size of $b+c$. The methods described in Section 1.4 may be applied to find an approximate $100(1-\alpha)\%$ confidence interval for P, say

$$P_L \leqq P \leqq P_U, \tag{8.8}$$

and equation 8.6 inverted to find the desired interval for ω:

$$\frac{P_L}{1-P_L} \leqq \omega \leqq \frac{P_U}{1-P_U}. \tag{8.9}$$

For the data of Table 8.2, the sample size is $20+5=25$ and $p=20/25 = .80$. Formulas (1.26) and (1.27) yield

$$.587 \leqq P \leqq .924 \tag{8.10}$$

as an approximate 95% confidence interval for P, and

$$1.42 \leqq \omega \leqq 12.16 \tag{8.11}$$

as an approximate 95% confidence interval for the underlying odds ratio.

The simpler approach based on (1.29) yields

$$.623 \leqq P \leqq .977 \tag{8.12}$$

as an approximate 95% confidence interval for P, and

$$1.65 \leqq \omega \leqq 42.48 \tag{8.13}$$

as an approximate 95% confidence interval for ω. The lower limits of (8.11) and (8.13) agree well, but the upper limits are greatly different. Because, in this case, $p=.80$, which is outside the interval suggested in Section 1.4 for close correspondence between the two approaches ($.3 \leqq p \leqq .7$), the result given in (8.11) is preferred.

So far the analysis of the fourfold table resulting from matched pairs has been presented in the context of a comparative retrospective study. The analysis involving McNemar's test and the estimation of the odds ratio may also be applied to a comparative prospective study with matched pairs. In the analysis of a controlled trial with matched pairs, however, the finding of a significant difference by McNemar's test should be followed by point and interval estimation of either the simple or the relative difference between the two outcome proportions.

Table 8.3 presents the proper means for presenting the data from such a trial, in which we suppose that a new treatment was compared with a standard. As was the case for Table 8.1, each frequency represents a number of pairs.

Table 8.3. Data from a controlled trial with matched pairs

	Standard Treatment		
New Treatment	Recovered	Not Recovered	Total
Recovered	a	b	$a+b$
Not recovered	c	d	$c+d$
Total	$a+c$	$b+d$	n

The proportion of cases who recovered under the standard treatment is

$$p_1 = \frac{a+c}{n}$$

and the proportion who recovered under the new treatment is

$$p_2 = \frac{a+b}{n}.$$

The simple difference between p_2 and p_1 is

$$p_2 - p_1 = \frac{b-c}{n},$$

and an estimate of its standard error appropriate when the two underlying proportions are not hypothesized to be equal is

$$\text{s.e.}(p_2 - p_1) = \frac{\sqrt{n(b+c) - (b-c)^2}}{n\sqrt{n}} = \frac{\sqrt{(a+d)(b+c) + 4bc}}{n\sqrt{n}}. \quad (8.14)$$

An approximate $100(1-\alpha)\%$ confidence interval for the difference between the two underlying rates of recovery is

$$(p_2 - p_1) - c_{\alpha/2}\,\text{s.e.}(p_2 - p_1) - \frac{1}{n} \leqq P_2 - P_1$$

$$\leqq (p_2 - p_1) + c_{\alpha/2}\,\text{s.e.}(p_2 - p_1) + \frac{1}{n}. \quad (8.15)$$

Note that all four cell frequencies contribute explicitly to the estimated standard error in (8.14), unlike the standard error in (8.2) which is appropriate only for testing the hypothesis that the underlying proportions are equal.

Under the assumption that the new treatment can benefit only those patients who fail to improve under the standard treatment, the relative value of the new treatment may be estimated by the relative difference,

$$p_e = \frac{p_2 - p_1}{1 - p_1} = \frac{b-c}{b+d}. \tag{8.16}$$

The standard error of the relative difference may be estimated by

$$\text{s.e.}(p_e) = \frac{1}{(b+d)^2}\sqrt{(b+c+d)(bc+bd+cd) - bcd}\,. \tag{8.17}$$

Note that a, the number of pairs both of whose members recovered, contributes neither to the estimation of the relative difference nor to the estimation of its standard error. An approximate $100(1-\alpha)\%$ confidence interval for the underlying parameter is given by $p_e \pm c_{\alpha/2}\,\text{s.e.}(p_e)$.

Table 8.4 presents some hypothetical data. Of the patients who were

Table 8.4. Hypothetical data from a controlled trial with matched pairs

	Standard Treatment		
New Treatment	Recovered	Not Recovered	Total
Recovered	40	25	65
Not recovered	10	0	10
Total	50	25	75

given the standard treatment, the proportion who recovered was $p_1 = 50/75 = .67$. Of those who were given the new treatment, the proportion who recovered was $p_2 = 65/75 = .87$. The value of McNemar's chi square statistic [see (8.3)] for assessing the significance of the difference between these two proportions is

$$\chi^2 = \frac{(|25-10|-1)^2}{25+10} = 5.60.$$

The difference is therefore statistically significant at the .05 level.

The difference between the two proportions is

$$p_2 - p_1 = \frac{25-10}{75} = .20$$

and an estimate of its standard error is, by (8.14),

$$\mathrm{s.e.}(p_2 - p_1) = \frac{1}{75\sqrt{75}}\sqrt{(40+0)(25+10)+4\times 25\times 10} = .08.$$

An approximate 95% confidence interval for $P_2 - P_1$ is, by (8.15),

$$.20 - 1.96\times .08 - \frac{1}{75} \leqq P_2 - P_1 \leqq .20 + 1.96\times .08 + \frac{1}{75},$$

or

$$.03 \leqq P_2 - P_1 \leqq .37.$$

The value of the relative difference in (8.16) is

$$p_e = \frac{25-10}{25} = .60,$$

which means that, of every 100 patients who fail to recover under the standard treatment, 60 might be expected to recover under the new treatment. The estimated standard error of the relative difference in (8.17)

$$\mathrm{s.e.}(p_e) = \frac{[illegible]}{(25+0)^2} \times$$

$$\sqrt{(25+10+0)(25\times 10 + 25\times 0 + 10\times 0) - 25\times 10\times 0}$$

$$= .15.$$

An approximate 95% confidence interval for the parameter is $.60 \pm 1.96 \times .15$, or the interval from .30 to .90.

8.2. MATCHED PAIRS: MORE THAN DICHOTOMOUS OUTCOME

Often the response of a subject to treatment or the degree to which he or she possesses a factor may be graded more finely than on the simple presence-absence dichotomy considered in the preceding section. Response to treatment, for example, may be graded as improvement, essentially no change, or worsening. Extent of cigarette smoking, as another example, may be graded as none at all, between one and ten cigarettes per day, between 11 and 20 cigarettes per day, and 21 or more cigarettes per day. When the samples being compared are not matched, the methods of Chapter 9 may be applied. Here we consider the case of matched pairs, both members of which are classified into one of $k(>2)$ mutually exclusive categories.

Table 8.5 demonstrates the appropriate presentation of the data. Each entry in the table represents a number of pairs. For example, $n_{..}$ is the total number of matched pairs, $n_{1.}$ is the number of pairs in which the case was in category 1, $n_{.2}$ is the number in which the control was in category 2, and n_{12} is the number in which the case was in category 1 and the control in category 2. The differences between the cases and controls are represented by the k differences $d_1=(n_{1.}-n_{.1})$, $d_2=(n_{2.}-n_{.2}),\ldots,d_k=(n_{k.}-n_{.k})$. Clearly, these differences do not depend on the quantities $n_{11}, n_{22},\ldots,n_{kk}$, the numbers of pairs both of whose members had outcomes in the same category.

Table 8.5. Data from a study of matched pairs with k mutually exclusive outcome categories

Outcome Category for Cases	Outcome Category for Controls				
	1	2	...	k	Total
1	n_{11}	n_{12}	...	n_{1k}	$n_{1.}$
2	n_{21}	n_{22}	...	n_{2k}	$n_{2.}$
⋮					
k	n_{k1}	n_{k2}	...	n_{kk}	$n_{k.}$
Total	$n_{.1}$	$n_{.2}$	...	$n_{.k}$	$n_{..}$

Complicated test statistics for assessing the significance of the k differences $d_1, d_2,\ldots,d_k$ have been proposed by Bhapkar (1966), Grizzle, Starmer, and Koch (1969), and Ireland, Ku, and Kullback (1969). A simpler test statistic, but one that still requires the inversion of a matrix, has been proposed by Stuart (1955) and Maxwell (1970). A simple expression for the Stuart-Maxwell statistic when $k=3$ has been derived by Fleiss and Everitt (1971).

For $k=3$, define

$$\bar{n}_{ij}=\frac{n_{ij}+n_{ji}}{2}. \tag{8.18}$$

The statistic

$$\chi^2=\frac{\bar{n}_{23}d_1^2+\bar{n}_{13}d_2^2+\bar{n}_{12}d_3^2}{2(\bar{n}_{12}\bar{n}_{13}+\bar{n}_{12}\bar{n}_{23}+\bar{n}_{13}\bar{n}_{23})} \tag{8.19}$$

may be referred to tables of chi square with two degrees of freedom. If χ^2 is significantly large, the inference would be made that the distribution across the categories for the cases differs from the distribution for the controls.

Consider the hypothetical data in Table 8.6, in which it is assumed that two diagnosticians independently diagnosed each of a sample of 100 mental patients. The value of the Stuart-Maxwell χ^2 statistic in (8.19) is

Table 8.6. Hypothetical data to illustrate the Stuart–Maxwell test

	Diagnostician *A*			
Diagnostician *B*	Schizophrenia	Affective	Other	Total
Schizophrenia	35	5	0	40
Affective	15	20	5	40
Other	10	5	5	20
Total	60	30	10	100

$$\chi^2 = \frac{\frac{5+5}{2}(40-60)^2 + \frac{0+10}{2}(40-30)^2 + \frac{5+15}{2}(\text{[illegible]}}{2\left(\frac{5+15}{2}\times\frac{0+10}{2} + \frac{5+15}{2}\times\frac{5+5}{2} + \frac{0+10}{2}\times\frac{5+5}{2}\right)}$$

$$= \frac{3500}{2\times 125} = 14.00,$$

which, with 2 degrees of freedom, is significant beyond the .001 level. It may therefore be concluded that the diagnostic distribution of diagnostician *A* is different from that of diagnostician *B*.

When, as in this example, a significant difference is found between the two distributions, the next step in the analysis would be to find those single categories (in the case of more than three categories, possibly those combinations of categories) for which the differences are significant [see Fleiss and Everitt (1971) for a general discussion]. One need only collapse the original table into a 2×2 table and apply McNemar's statistic (8.3). The test for significance, however, must incorporate a control over the fact that the chances of erroneously declaring a difference to be significant increase when a number of tests are applied to the same data. An appropriate control in the case we are considering (see Miller, 1966, Section 6.2) is to refer McNemar's chi square statistic to the critical value of chi square with $k-1$ degrees of freedom.

We illustrate the search for those categories with a significant difference using the data of Table 8.6. To determine whether the proportions who were diagnosed as having schizophrenia by the two diagnosticians were different, we form the 2×2 table in Table 8.7. Sixty percent of the patients were diagnosed as having schizophrenia by A, while only 40% were so

Table 8.7. Two-by-two table for comparing rates of schizophrenia by diagnosticians A and B

	Diagnostician A		
Diagnostician B	Schizophrenia	Not Schizophrenia	Total
Schizophrenia	35	5	40
Not schizophrenia	25	35	60
Total	60	40	100

diagnosed by B. The value of McNemar's statistic is

$$\chi^2 = \frac{(|5 - 25| - 1)^2}{5 + 25} = \frac{19^2}{30} = 12.03.$$

The critical value of chi square with two degrees of freedom for a significance level of .05 (see Table A.1) is 5.99. Since the obtained value of McNemar's chi square exceeds 5.99, we may infer that A is more likely to diagnose schizophrenia than B.

Problem 8.1 calls for comparing the proportions of patients diagnosed as affectively ill and diagnosed as having a disorder other than schizophrenia or affective illness by A and B.

If the k outcome categories are ordered (as in the two examples cited at the beginning of this section), the analysis of the data should somehow take the ordering into account. Consider the hypothetical data in Table 8.8, in which it is assumed that the treatments were assigned to the members of each matched pair at random. The value of the Stuart–Maxwell

Table 8.8. Hypothetical data to illustrate the analysis of an ordered outcome variable

	Standard Treatment			
New Treatment	Improved	No Change	Worse	Total
Improved	40	20	10	70
No change	6	6	8	20
Worse	4	4	2	10
Total	50	30	20	100

chi square statistic is significant [see Problem 8.2(*a*)], so that a more detailed analysis of the data is in order.

The kind of category-by-category analysis illustrated above could be performed, but it would be inefficient in that it would ignore the ordering inherent in grading response to treatment. The following method of analysis is appropriate when interest is in whether one treatment tends to produce more responses at one end of the ordered scale and fewer at the other compared to the second treatment.

Consider the difference $d_1 - d_3$. If the new treatment is better than the standard in the sense that it has associated with it more improvement (so that d_1 is positive) and less worsening (so that d_3 is negative), $d_1 - d_3$ will be large in the positive direction. If the new treatment is poorer than the standard, $d_1 - d_3$ will be large in the negative direction. In either case, the hypothesis that the treatments do not differ at the two ends of the scale may be tested by referring the value of

$$\chi^2 = \frac{(d_1 - d_3)^2}{2(\bar{n}_{12} + 4\bar{n}_{13} + \bar{n}_{23})} \tag{8.20}$$

to the chi square distribution with 1 degree of freedom if this particular comparison was planned before the data were examined, and to the chi square distribution with two degrees of freedom if the comparison was suggested by the data. This test and the more general one when the number of outcome categories exceeds three were derived by Fleiss and Everitt (1971). Problem 8.2(*b*) calls for applying it to the data of Table 8.8.

8.3. THE CASE OF MULTIPLE MATCHED CONTROLS

Occasionally, two matched samples may be generated by matching each case (or each patient given a new treatment) with more than one control (or with more than one patient given a standard treatment). Matching with multiple controls is especially advantageous when the number of potential control subjects is large relative to the number of available cases and when little effort needs to be expended in obtaining the necessary information.

We assume that each subject is characterized by either the presence or the absence of some factor or outcome. A general method of analysis, valid even when the number of controls varies from one case to another, was originally derived by Mantel and Haenszel (1959). An alternative but more complex method of analysis in the general case is due to Cox (1966). Here we assume that each case is matched with the same number, say $m - 1$, of controls, and we consider only the Mantel–Haenszel method.

Suppose that there are a total of N matched m-tuples, each containing one case and $m-1$ controls. In the ith m-tuple ($i=1,\ldots,N$), let x_i denote the number of controls who had the factor (so that x_i may equal $0, 1, \ldots,$ or $m-1$), and let n_i denote the total number of subjects—including the case and controls—who had the factor. Thus if the case in the ith m-tuple had the factor, then $n_i = x_i + 1$; if he or she did not have the factor, then $n_i = x_i$.

Define

$$A = \sum_{i=1}^{N} x_i, \tag{8.21}$$

the total number of control subjects who had the factor, and define

$$B = \sum_{i=1}^{N} n_i, \tag{8.22}$$

the total number of either kind of subject who had the factor. Note that the total number of cases who had the factor is $B-A$. The rate at which the factor is present among the controls is

$$p_1 = \frac{A}{N(m-1)}, \tag{8.23}$$

and the rate at which it is present among the cases is

$$p_2 = \frac{B-A}{N}. \tag{8.24}$$

In order to test the significance of the difference between p_1 and p_2, the statistic

$$\chi^2 = \left(\frac{p_2 - p_1}{\text{s.e.}(p_2 - p_1)}\right)^2 = \frac{[(m-1)B - mA]^2}{mB - \sum_{i=1}^{N} n_i^2} \tag{8.25}$$

may be referred to tables of chi square with 1 degree of freedom (see Miettinen, 1969 and Pike and Morrow, 1970). Miettinen (1969) has studied the power of the test based on (8.25) and has given criteria (in terms of reducing cost) for deciding on an appropriate value for $m-1$, the number of controls per case.

The data in Table 8.9 are used to illustrate this analysis. Suppose that $N=10$ matched triples ($m=3$) were studied, implying $m-1=2$ controls

Table 8.9. Outcome data from matched triples

Triple	Case Has Factor[a]	Number of Controls with Factor ($=x_i$)	Total Having Factor ($=n_i$)	n_i^2
1	1	2	3	9
2	1	1	2	4
3	1	1	2	4
4	1	1	2	4
5	1	1	2	4
6	1	0	1	1
7	1	0	1	1
8	1	0	1	1
9	0	1	1	1
10	0	0	0	0
Total	$8(=B-A)$	$7(=A)$	$15(=B)$	29

[a] 1 = yes, 0 = no.

per case. The proportion of controls having the factor [see (8.23)] is

$$p_1 = \frac{7}{10 \times 2} = .35$$

and the proportion of cases having the factor [see (8.24)] is

$$p_2 = \frac{8}{10} = .80.$$

The value of the statistic in (8.25) for testing the significance of the difference between these two proportions is

$$\chi^2 = \frac{(2 \times 15 - 3 \times 7)^2}{3 \times 15 - 29} = \frac{9^2}{16} = 5.06.$$

Since this value exceeds 3.84, the value of chi square with one degree of freedom needed for significance at the .05 level, the inference may be drawn that the proportion of cases having the factor is larger than the proportion of controls having it.

The Mantel–Haenszel estimate of the assumed common odds ratio over the N m-tuples (1959, p. 736) is

$$o = \frac{(m-1)(B-A) - \sum_{i=1}^{N} x_i(n_i - x_i)}{A - \sum_{i=1}^{N} x_i(n_i - x_i)}. \tag{8.26}$$

The quantity $\sum x_i(n_i - x_i)$ is obtained by restricting attention to those m-tuples in which the case had the factor, and simply adding the numbers of controls in them who had the factor.

For the data of Table 8.9, only the first eight triples were such that the case had the factor. The total number of controls in those eight triples who had the factor is 6 $[= \sum x_i(n_i - x_i)]$, so that the estimated odds ratio in (8.26) is

$$o = \frac{2 \times 8 - 6}{7 - 6} = 10.0.$$

Miettinen (1970b) presents an alternative method for estimating the odds ratio in the case of matched m-tuples [but one more complicated than (8.26)] and gives approximate expressions for the standard error of the estimate.

The power of the chi square test based on (8.25) and the precision of the estimated odds ratio in (8.26) both increase as $m - 1$, the number of controls per case, increases. The improvement in power and precision is usually trivial, however, as soon as we get beyond three or four controls per case (Miettinen, 1969; Ury, 1975). The search for five or more controls per case therefore usually represents wasted effort.

8.4. THE COMPARISON OF *m* MATCHED SAMPLES

In the preceding section we considered the case where the $m - 1$ controls for each case formed a homogeneous group. In this section we consider the comparison of m distinct matched samples, but again restrict attention to the case where only two outcomes are of interest (see Koch and Reinfurt, 1971, for the general case).

The case being considered would arise in a comparative prospective study in which, for example, a number of quadruples ($m = 4$) of subjects would be matched on sex and age, under the restriction that one member did not smoke cigarettes, another smoked between one and ten cigarettes per day, a third smoked between 11 and 20 cigarettes per day, and a fourth smoked 21 or more cigarettes per day. The proportions of subjects from the four resulting matched samples who develop a disease would then be compared using the methods of this section. An example of a retrospective study with three matched samples (lung cancer patients, other patients, and community controls) is one by Doll and Hill (1952).

The methods of this section are also applicable to the results of a controlled trial in which $m > 2$ treatments are compared by grouping together a number of sets of m similar patients each and randomly assigning the treatments to the patients within each matched m-tuple. The methods are applicable, too, when each of a sample of subjects is studied under m different conditions. An example is the comparison of the proportions positive associated with m diagnostic tests, when each test is applied to each patient in the sample.

Table 8.10 illustrates the presentation of the data resulting from the study of m matched samples, with N observations in each sample. In Table

Table 8.10. Presentation of data from m matched samples

m-tuple	Sample				Total
	1	2	. . .	m	
1	X_{11}	X_{12}	. . .	X_{1m}	S_1
2	X_{21}	X_{22}	. . .	X_{2m}	S_2
⋮					
N	X_{N1}	X_{N2}	. . .	X_{Nm}	S_N
Total	T_1	T_2	. . .	T_m	T
Proportion	p_1	p_2	. . .	p_m	$\bar{p}$

8.10, each X is either 0 (if the response is negative) or 1 (if the response is positive). Thus, for example, S_1 represents the total number of positives from the first m-tuple, T_1 represents the total number of positives from the first sample, and T represents the overall total number of positives.

Define

$$p_j = \frac{T_j}{N}, \tag{8.27}$$

the proportion of subjects from the jth sample who were positive;

$$P_n = \frac{S_n}{m}, \tag{8.28}$$

the proportion of positives in the nth m-tuple; and

$$\bar{p} = \frac{1}{m}\sum_{j=1}^{m} p_j = \frac{1}{N}\sum_{n=1}^{N} P_n = \frac{T}{Nm}, \tag{8.29}$$

the overall proportion positive. Interest is in whether the proportions $p_1, \ldots, p_m$ differ significantly. The following statistic, due to Cochran (1950), may be used to test for the significance of the differences among the m proportions:

$$Q = \frac{N^2(m-1)}{m} \times \frac{\sum_{j=1}^{m} (p_j - \bar{p})^2}{N\bar{p}(1-\bar{p}) - \sum_{n=1}^{N} (P_n - \bar{p})^2}$$

$$= (m-1) \times \frac{m \sum_{j=1}^{m} T_j^2 - T^2}{mT - \sum_{n=1}^{N} S_n^2}. \tag{8.30}$$

The value of (8.30) may be referred to tables of chi square with $m-1$ degrees of freedom.

Consider the data of Table 8.11, originally reported by Fleiss (1965a). The proportions p_j of patients judged to have religious preoccupations vary

Table 8.11. Judgments by eight raters as to presence or absence[a] of religious preoccupations in eight patients

	Rater								
Patient	1	2	3	4	5	6	7	8	Total ($=S_n$)
1	0	0	0	0	0	0	0	0	0
2	0	0	0	0	1	0	0	0	1
3	0	0	0	0	0	0	0	0	0
4	0	0	0	0	0	0	0	0	0
5	0	0	1	0	0	1	0	1	3
6	0	0	1	1	1	1	0	1	5
7	0	0	0	0	0	0	0	0	0
8	1	0	1	1	1	1	0	1	6
Total ($=T_j$)	1	0	3	2	3	3	0	3	15 ($=T$)
Proportion ($=p_j$)	.125	0	.375	.250	.375	.375	0	.375	.234 ($=\bar{p}$)

[a] 1 or 0, respectively.

from a low of 0 to a high of .375. The value of Q (8.30) for testing whether this variation can be attributed to chance or whether it represents real differences among the raters is

$$Q = 7 \times \frac{8(1^2 + 0^2 + \cdots + 0^2 + 3^2) - 15^2}{8 \times 15 - (0^2 + 1^2 + \cdots + 0^2 + 6^2)} = 14.71.$$

The validity of calculating Q for many ratings on the same subjects has been established by Fleiss (1965b).

Referring to Table A.1 with $m - 1 = 7$ degrees of freedom, we find that Q must exceed 14.07 in order for the variation to be declared significant at the .05 level. Since our obtained value of 14.71 exceeds the critical value, we infer that the raters differ in their judgments of religious preoccupation.

Having found significant variation, our next step would be to try to identify those samples or groups of samples (in our example, those raters or groups of raters) that differed. A device that is frequently useful is to *partition* Q into separate components, each of which measures a specified source of variability. The general method for partitioning a chi square statistic is described by Everitt (1977, Chapter 3). Here we illustrate the method for the statistic in (8.30).

Suppose that the m samples represent two groups, with m_1 samples in the first group and m_2 in the second. In the first example given at the beginning of the section, one group consists of the single sample of nonsmokers (so that $m_1 = 1$) and the other of the three samples of cigarette smokers (so that $m_2 = 3$). Define

$$U_1 = \sum_{j=1}^{m_1} T_j \tag{8.31}$$

as the total number of positives in the first group of samples, and

$$U_2 = \sum_{j=m_1+1}^{m} T_j \tag{8.32}$$

as the total number of positives in the second group. The statistic for testing whether the proportion positive in the first group differs significantly from that in the second is

$$Q_{\text{diff}} = \frac{(m-1)}{m_1 m_2} \times \frac{(m_2 U_1 - m_1 U_2)^2}{mT - \sum_{n=1}^{N} S_n^2}. \tag{8.33}$$

The statistic Q_{diff} has 1 degree of freedom.

Consider again the data of Table 8.11. The first five raters were from New York and the last three were from Kentucky. They therefore form two natural groups, one containing $m_1 = 5$ raters and the other $m_2 = 3$. It is reasonable to inquire whether the two groups of raters differ in their judgments. The total number of positive ratings by the raters in the first group is

$$U_1 = 1 + 0 + 3 + 2 + 3 = 9,$$

and the total number of positive ratings by those in the second group is

$$U_2 = 3 + 0 + 3 = 6.$$

The value of Q_{diff} (8.33) is

$$Q_{\text{diff}} = \frac{7}{5 \times 3} \times \frac{(3 \times 9 - 5 \times 6)^2}{120 - 71} = 0.09,$$

indicating a negligible difference between the New York raters as a group and the Kentucky raters as a group.

The next step in the analysis would be to compare the m_1 samples within group 1 by means of the statistic

$$Q_{\text{group 1}} = \frac{m(m-1)}{m_1} \times \frac{m_1 \sum_{j=1}^{m_1} T_j^2 - U_1^2}{mT - \sum_{n=1}^{N} S_n^2}, \tag{8.34}$$

and to compare the m_2 samples within group 2 by means of the statistic

$$Q_{\text{group 2}} = \frac{m(m-1)}{m_2} \times \frac{m_2 \sum_{j=m_1+1}^{m} T_j^2 - U_2^2}{mT - \sum_{n=1}^{N} S_n^2}. \tag{8.35}$$

The statistic $Q_{\text{group 1}}$ has $m_1 - 1$ degrees of freedom and the statistic $Q_{\text{group 2}}$ has $m_2 - 1$ degrees of freedom. It may be checked that

$$Q = Q_{\text{diff}} + Q_{\text{group 1}} + Q_{\text{group 2}}.$$

In addition, note that the three degrees of freedom for the Q statistics of

(8.33)–(8.35), namely 1, $m_1 - 1$, and $m_2 - 1$, sum to $m_1 + m_2 - 1 = m - 1$, the degrees of freedom in the overall Q (8.30).

For the data of Table 8.11, the differences among the five New York raters are assessed by

$$Q_{\text{group 1}} = \frac{8 \times 7}{5} \times \frac{5(1^2 + 0^2 + 3^2 + 2^2 + 3^2) - 9^2}{120 - 71} = 7.77,$$

which with $5 - 1 = 4$ degrees of freedom fails to reach significance at the .05 level. The differences among the three Kentucky raters are assessed by

$$Q_{\text{group 2}} = \frac{8 \times 7}{3} \times \frac{3(3^2 + 0^2 + 3^2) - 6^2}{120 - 71} = 6.86,$$

which with $3 - 1 = 2$ degrees of freedom is significant at the .05 level. Note that

$$Q_{\text{diff}} + Q_{\text{group 1}} + Q_{\text{group 2}} = 0.09 + 7.77 + 6.86 = 14.72,$$

which equals, except for rounding errors, the overall value of Q, 14.71.

Cochran (1950, p. 265) suggests a slightly different approach to the partitioning of Q. Its effect is to reduce slightly the magnitudes of $Q_{\text{group 1}}$ and of $Q_{\text{group 2}}$ (see also Tate and Brown, 1970). The conclusions for the data of Table 8.11 are the same for both methods of partitioning: there are differences among the judgments of the eight raters, arising essentially from the variability among the three Kentucky raters.

On occasion, the m samples represent m separate levels of a quantitatively ordered variable (e.g., average number of cigarettes smoked per day, as in the example described at the beginning of this section). Let x_j denote the value of this variable for the jth sample ($j = 1, \ldots, m$), and define

$$b = \frac{\sum_{j=1}^{m} T_j(x_j - \bar{x})}{N \sum_{j=1}^{m} (x_j - \bar{x})^2}, \tag{8.36}$$

where

$$\bar{x} = \frac{1}{m} \sum_{j=1}^{m} x_j, \tag{8.37}$$

the mean value of x for these samples. The statistic b is the slope of the straight line fitted to the data. It describes the average change in the rate of occurrence of the event under study per unit change in x.

The statistical significance of b may be assessed by referring the value of

$$\chi^2_{\text{slope}} = \frac{m(m-1)N^2 \sum_{j=1}^{m} (x_j - \bar{x})^2}{mT - \sum_{n=1}^{N} S_n^2} b^2 \tag{8.38}$$

to the chi square distribution with 1 degree of freedom. If b is significant and positive (or negative), the inference would be that the proportions tend to increase (or decrease) with increasing values of x.

Suppose, contrary to fact and purely for illustrative purposes, that the numerals used in Table 8.11 to denote the several raters represent their numbers of years of experience. It might then be of interest to determine whether the likelihood of judging the presence of religious preoccupation varies systematically with length of experience. Note that $\bar{x}$ in Table 8.11 is equal to 4.5, and that $\Sigma(x_j - \bar{x})^2$ is equal to 42.

The slope of the straight line associating p with x is, by (8.36),

$$b = \frac{7.5}{8 \times 42} = .02,$$

indicating an average increase of .02 in the likelihood of judging the presence of religious preoccupation for each additional year of experience. The associated value of chi square from (8.38) is

$$\chi^2_{\text{slope}} = \frac{8 \times 7 \times 64 \times 42}{8 \times 15 - 71} (.02)^2 = 1.23,$$

which is not significant by any reasonable standard. No tendency for the proportions to vary systematically with length of experience can therefore be asserted.

The test statistics presented in this section, like those presented in the three preceding sections, are unaffected by the deletion of those m-tuples in which either all m responses were positive or all m were negative. Berger and Gold (1973) and Bhapkar and Somes (1977) have shown that the distribution of Q in large samples under the hypothesis of equal underlying probabilities is approximately chi square with $m-1$ degrees of freedom only if all pairwise probabilities $P(X_{ni} = 1 \text{ and } X_{nj} = 1)$, $i \neq j$, are equal. Seeger and Gabrielsson (1968) and Tate and Brown (1970) have studied the accuracy of the chi square approximation to the distribution of Q when this assumption holds and when the sample sizes are small. It seems that the approximation is adequate provided the product of the number of samples ($=m$) and the number of m-tuples remaining after the deletion of those in which all responses were the same is at least 24. For the data of Table 8.11, four patients (numbers 1, 3, 4, and 7) were such that the ratings were identical. The product of $m=8$ and the remaining number of patients ($=4$) is 32, indicating that the approximation was adequate.

A quite different approach from Cochran's (1950) to the comparison of m matched samples is due to Bennett (1967, 1968). The reader is referred to his two papers for the test statistics he derived (more complicated than the Q statistic except when $m=3$—see Mantel and Fleiss, 1975) and for their powers.

8.5. ADVANTAGES AND DISADVANTAGES OF MATCHING

In comparative prospective and retrospective studies, the matching of subjects is usually employed to assure that the samples being contrasted are similar with respect to characteristics associated with the factors being studied (see, e.g., Billewicz, 1965, and Miettinen, 1970a). The possible gain in efficiency due to the study of matched samples (i.e., increase in the power of the test of significance and increase in the precision of the estimated degree of association) therefore assumes lesser importance, but some results are available.

Cochran (1950) and Worcester (1964) have shown that matching is not guaranteed to increase efficiency. It can be expected to do so only when the characteristics being matched on are strongly associated with the factors under study. When the characteristics used for matching are only slightly, or are not at all associated with the factors under study, Youkeles (1963) has shown that efficiency may even be lost. When the number of matched sets exceeds 30, however, matching on irrelevant characteristics seems not to affect efficiency.

In the context of controlled comparative trials with random assignment of treatments to subjects, on the other hand, the purpose of matching is mainly the increase of efficiency. Chase (1968) has shown that matching is at least as efficient as no matching except for small sample sizes. Billewicz (1964, 1965), however, has indicated that the increase in efficiency is frequently only limited and may not be worth the effort involved in securing adequate matches.

He has shown how the length of time required to complete a study increases either as the number of matching characteristics increases or as the relative frequencies of some categories of the matching characteristics decrease. With too many matching characteristics, or with only a few but some containing too many categories, the investigator may find a large proportion of subjects left unmatched at the end of the study.

In view of the need to assure comparability in comparative prospective and retrospective studies, matching or some other method of control for biasing factors in these contexts is often necessary. In addition to matching, control may be effected by means of stratification or regression techniques (Cochran, 1968; Rubin, 1973; McKinlay, 1975a). When stratification is the method adopted, the techniques presented in Chapter 10 are applicable.

McKinlay (1977) has critically analyzed matching for the control of unwanted sources of variation and cites as its major advantages its easy understandability and the relative simplicity of the analysis of the resulting data. Its major disadvantages include the possibly excessive costs associated both with finding matches and with discarding subjects who could not be matched and the likely failure to detect *interaction*, the phenomenon in which the magnitude of difference or association varies across different subgroups. McKinlay (1975b, 1977) also points out that whereas matching may well yield greater precision than designs that do not control for sources of bias, it does not necessarily yield greater precision than other designs in which control is attempted.

It may be the case that an investigator nevertheless decides to employ matching instead of another method for control over biasing factors in a comparative prospective or retrospective study. In studies of hospitalized patients, for example, matching on the date of hospitalization is probably the best method for control of the effects of an epidemic. Matching should, however, be on a small number of characteristics (rarely more than four and preferably no more than two), with each defined by a small number of categories (with respect to age, e.g., matching by 10-year intervals should frequently suffice). If the investigator insists on controlling for a large number of biasing factors simultaneously, multivariate methods such as those proposed by Althouser and Rubin (1970) and Miettinen (1976) may have to be used.

Problem 8.1. Consider the hypothetical data of Table 8.6. Do the two diagnosticians differ significantly in the proportions of patients they diagnose affectively ill? other? (*Note*. Be sure to refer the value of McNemar's statistic to the critical value of chi square with 2 degrees of freedom.)

Problem 8.2. Consider the hypothetical data of Table 8.8.

(*a*) What is the value of the Stuart–Maxwell statistic given in (8.19)? Does the outcome distribution of patients given the new treatment differ significantly from that of patients given the standard treatment?

(*b*) What are the values of d_1, d_3, and $d_1 - d_3$? What does the direction of $d_1 - d_3$ suggest about the direction of the difference between the two treatments? What is the value of the test statistic in (8.20)? Is the new treatment significantly better than the standard?

Problem 8.3. When not corrected for continuity, McNemar's chi square statistic is given by

$$\chi_u^2 = \frac{(b-c)^2}{b+c}.$$

Prove that, when $m=2$, the expression for χ^2 given by (8.25) is equal to that of χ_u^2. (*Hint*. Refer to Table 8.1 for notation. Prove that, when $m=2$, A (8.21) equals $a+c$, B (8.22) equals $2a+b+c$, and Σn_i^2 equals $4a+b+c$.)

Problem 8.4. Prove that, when $m=2$, the value of Q given by (8.30) is equal to that of χ_u^2. (*Hint*. Prove that, when $m=2$, $T_1 = a+c$, $T_2 = a+b$, $T = 2a+b+c$, and $\Sigma S_n^2 = 4a+b+c$.)

REFERENCES

Althouser, R. P. and Rubin, D. B. (1970). The computerized construction of a matched sample. *Am. J. Sociol.*, **76**, 325–346.

Bennett, B. M. (1967). Tests of hypotheses concerning matched samples. *J. R. Stat. Soc., Ser. B*, **29**, 468–474.

Bennett, B. M. (1968). Note on X^2 tests for matched samples. *J. R. Stat. Soc., Ser. B*, **30**, 368–370.

Bennett, B. M. and Underwood, R. E. (1970). On McNemar's test for the 2 × 2 table and its power function. *Biometrics*, **26**, 339–343.

Berger, A. and Gold, R. Z. (1973). Note on Cochran's Q-test for the comparison of correlated proportions. *J. Am. Stat. Assoc.*, **68**, 989–993.

Bhapkar, V. P. (1966). A note on the equivalence of two test criteria for hypotheses in categorical data. *J. Am. Stat. Assoc.*, **61**, 228–235.

Bhapkar, V. P. and Somes, G. W. (1977). Distribution of Q when testing equality of matched proportions. *J. Am. Stat. Assoc.*, **72**, 658–661.

Billewicz, W. Z. (1964). Matched samples in medical investigations. *Brit. J. Prev. Soc. Med.*, **18**, 167–173.

Billewicz, W. Z. (1965). The efficiency of matched samples: An empirical investigation. *Biometrics*, **21**, 623–644.

Bross, I. D. J. (1969). How case-for-case matching can improve design efficiency. *Am. J. Epidemiol.*, **89**, 359–363.

Chase, G. R. (1968). On the efficiency of matched pairs in Bernoulli trials. *Biometrika*, **55**, 365–369.

Cochran, W. G. (1950). The comparison of percentages in matched samples. *Biometrika*, **37**, 256–266.

Cochran, W. G. (1968). The effectiveness of subclassification in removing bias in observational studies. *Biometrics*, **24**, 295–313.

Cornfield, J. and Haenszel, W. (1960). Some aspects of retrospective studies. *J. Chronic Dis.*, **11**, 523–534.

Cox, D. R. (1966). A simple example of a comparison involving quantal data. *Biometrika*, **53**, 215–220.

Doll, R. and Hill, A. B. (1952). A study of the etiology of carcinoma of the lung. *Brit. Med. J.*, **2**, 1271–1286.

Edwards, A. L. (1948). Note on the "correction for continuity" in testing the significance of the difference between correlated proportions. *Psychometrika*, **13**, 185–187.

Ejigou, A. and McHugh, R. (1977). Estimation of relative risk from matched pairs in epidemiologic research. *Biometrics*, **33**, 552–556.

Everitt, B. S. (1977). *The analysis of contingency tables*. London: Chapman and Hall.

Fleiss, J. L. (1965a). Estimating the accuracy of dichotomous judgments. *Psychometrika*, **30**, 469–479.

Fleiss, J. L. (1965b). A note on Cochran's Q test. *Biometrics*, **21**, 1008–1010.

Fleiss, J. L. and Everitt, B. S. (1971). Comparing the marginal totals of square contingency tables. *Brit. J. Math. Stat. Psychol.*, **24**, 117–123.

Grizzle, J. E., Starmer, C. F., and Koch, G. G. (1969). Analysis of categorical data by linear models. *Biometrics*, **25**, 489–504.

Hill, A. B. (1962). *Statistical methods in clinical and preventive medicine*. New York: Oxford University Press.

Ireland, C. T., Ku, H. H., and Kullback, S. (1969). Symmetry and marginal homogeneity of an $r \times r$ contingency table. *J. Am. Stat. Assoc.*, **64**, 1323–1341.

Koch, G. G. and Reinfurt, D. W. (1971). The analysis of categorical data from mixed models. *Biometrics*, **27**, 157–173.

McKinlay, S. M. (1975a). The design and analysis of the observational study: A review. *J. Am. Stat. Assoc.*, **70**, 503–520.

McKinlay, S. M. (1975b). A note on the chi-square test for pair-matched samples. *Biometrics*, **31**, 731–735.

McKinlay, S. M. (1977). Pair-matching: A reappraisal of a popular technique. *Biometrics*, **33**, 725–735.

McNemar, Q. (1947). Note on the sampling error of the difference between correlated proportions or percentages. *Psychometrika*, **12**, 153–157.

Mantel, N. and Fleiss, J. L. (1975). The equivalence of the generalized McNemar tests for marginal homogeneity in 2^3 and 3^2 tables. *Biometrics*, **31**, 727–729.

Mantel, N. and Haenszel, W. (1959). Statistical aspects of the analysis of data from retrospective studies of disease. *J. Natl. Cancer Inst.*, **22**, 719–748.

Maxwell, A. E. (1970). Comparing the classification of subjects by two independent judges. *Brit. J. Psychiatry*, **116**, 651–655.

Miettinen, O. S. (1968). The matched pairs design in the case of all-or-none responses. *Biometrics*, **24**, 339–352.

Miettinen, O. S. (1969). Individual matching with multiple controls in the case of all-or-none responses. *Biometrics*, **25**, 339–355.

Miettinen, O. S. (1970a). Matching and design efficiency in retrospective studies. *Am. J. Epidemiol.*, **91**, 111–118.

Miettinen, O. S. (1970b). Estimation of relative risk from individually matched series. *Biometrics*, **26**, 75–86.

Miettinen, O. S. (1976). Stratification by a multivariate confounder score. *Am. J. Epidemiol.*, **104**, 609–620.

Miller, R. G. (1966). *Simultaneous statistical inference*. New York: McGraw-Hill.

Mosteller, F. (1952). Some statistical problems in measuring the subjective response to drugs. *Biometrics*, **8**, 220–226.

Pike, M. C. and Morrow, R. H. (1970). Statistical analysis of patient-control studies in epidemiology: Factor under investigation an all-or-none variable. *Brit. J. Prev. Soc. Med.*, **24**, 42–44.

Rubin, D. B. (1973). The use of matched sampling and regression adjustment to remove bias in observational studies. *Biometrics*, **29**, 185–203.

Seeger, P. and Gabrielsson, A. (1968). Applicability of the Cochran Q test and the F test for statistical analysis of dichotomous data for dependent samples. *Psychol. Bull.*, **69**, 269–277.

Stuart, A. (1955). A test for homogeneity of the marginal distribution in a two-way classification. *Biometrika*, **42**, 412–416.

Stuart, A. (1957). The comparison of frequencies in matched samples. *Brit. J. Stat. Psychol.*, **10**, 29–32.

Tate, M. W. and Brown, S. M. (1970). Note on the Cochran Q test. *J. Am. Stat. Assoc.*, **65**, 155–160.

Ury, H. K. (1975). Efficiency of case-control studies with multiple controls per case: Continuous or dichotomous data. *Biometrics*, **31**, 643–649.

Worcester, J. (1964). Matched samples in epidemiologic studies. *Biometrics*, **20**, 840–848.

Youkeles, L. H. (1963). Loss of power through ineffective pairing of observations in small two-treatment all-or-none experiments. *Biometrics*, **19**, 175–180.

CHAPTER 9

The Comparison of Proportions from Several Independent Samples

With only a few exceptions, we have restricted our attention to the comparison of two proportions. In this chapter we consider the comparison of a number of proportions. In Section 9.1 we study the analysis of an $m \times 2$ contingency table, where $m > 2$ and where there is no necessary ordering to the m groups. Sections 9.2 and 9.3 are devoted to the case where an intrinsic ordering to the m groups exists. We consider in Section 9.2 the hypothesis that the proportions vary monotonically (i.e., steadily increase or steadily decrease) with m quantitatively ordered groups, and in Section 9.3 that they vary monotonically with m qualitatively ordered groups. In Section 9.4 we consider the comparison of two or more groups on a scale with qualitatively ordered categories.

The procedures of this chapter are appropriate to each of the three methods of sampling presented previously (see Section 2.1). In method III sampling, the m samples represent groups treated by m different treatments, with subjects assigned to groups at random. In method II sampling, the investigator selects either prespecified numbers of subjects from each of the m groups or prespecified numbers with and without the outcome characteristic. In method I sampling, these numbers become known only after the study is completed. As was the case for the comparison of $m = 2$ samples (see Sections 6.1 and 6.2), method II sampling with equal sample sizes is superior in terms of power and precision to method I sampling when $m > 2$.

9.1. THE COMPARISON OF m PROPORTIONS

Suppose that m independent samples of subjects are studied, with each subject characterized by the presence or absence of some characteristic.

Table 9.1. Proportions from m independent samples

Sample	Total in Sample	Number with Characteristic	Number without Characteristic	Proportion with Characteristic
1	$n_{1.}$	n_{11}	n_{12}	p_1
2	$n_{2.}$	n_{21}	n_{22}	p_2
⋮				
m	$n_{m.}$	n_{m1}	n_{m2}	p_m
Overall	$n_{..}$	$n_{.1}$	$n_{.2}$	$\bar{p}$

The resulting data might be presented as in Table 9.1. In Table 9.1,

$$p_i = \frac{n_{i1}}{n_{i.}} \tag{9.1}$$

and

$$\bar{p} = \frac{n_{.1}}{n_{..}} = \frac{\Sigma n_{i.} p_i}{\Sigma n_{i.}}. \tag{9.2}$$

For testing the significance of the differences among the m proportions, the value of

$$\chi^2 = \sum_{i=1}^{m} \sum_{j=1}^{2} \frac{(n_{ij} - n_{i.}n_{.j}/n_{..})^2}{n_{i.}n_{.j}/n_{..}} \tag{9.3}$$

may be referred to tables of chi square (see Table A.1) with $m-1$ degrees of freedom. An equivalent and more suggestive formula for the test statistic is

$$\chi^2 = \frac{1}{\bar{p}\bar{q}} \sum_{i=1}^{m} n_{i.}(p_i - \bar{p})^2, \tag{9.4}$$

where $\bar{q} = 1 - \bar{p}$.

Consider, as an example, the data in Table 9.2 from four studies cited by Dorn (1954). In each study, the number of smokers among lung cancer

Table 9.2. Smoking status among lung cancer patients in four studies

Study	Number of Patients	Number of Smokers	Proportion of Smokers
1	86 $(=n_{1.})$	83	.965 $(=p_1)$
2	93 $(=n_{2.})$	90	.968 $(=p_2)$
3	136 $(=n_{3.})$	129	.949 $(=p_3)$
4	82 $(=n_{4.})$	70	.854 $(=p_4)$
Overall	397 $(=n_{..})$	372	.937 $(=\bar{p})$

patients was recorded. For these data, the value of χ^2 (9.4) is

$$\chi^2 = \frac{1}{.937 \times .063}\Big[86 \times (.965 - .937)^2 + 93 \times (.968 - .937)^2$$

$$+ 136 \times (.949 - .937)^2 + 82 \times (.854 - .937)^2\Big]$$

$$= 12.56 \tag{9.5}$$

which, with three degrees of freedom, is significant at the .01 level.

Having found the proportions to differ significantly, one would next proceed to identify the samples or groups of samples that contributed to the significant difference. Methods for isolating sources of significant differences in the context of a general contingency table are given by Irwin (1949), Lancaster (1950), Kimball (1954), Kastenbaum (1960), Castellan (1965), and Knoke (1976). Here we illustrate the simplest method for the $m \times 2$ table.

Suppose that the m samples are partitioned into two groups, the first containing m_1 samples and the second m_2, where $m_1 + m_2 = m$. Define

$$n_{(1)} = \sum_{i=1}^{m_1} n_{i.} \tag{9.6}$$

to be the total number of subjects in the first group of samples and

$$n_{(2)} = \sum_{i=m_1+1}^{m} n_{i.} \tag{9.7}$$

to be the total number of subjects in the second group.

Let the proportion in the first group be denoted $\bar{p}_1$, where

$$\bar{p}_1 = \frac{\sum_{i=1}^{m_1} n_{i.} p_i}{n_{(1)}}, \tag{9.8}$$

and that in the second group be denoted $\bar{p}_2$, where

$$\bar{p}_2 = \frac{\sum_{i=m_1+1}^{m} n_{i.} p_i}{n_{(2)}}. \tag{9.9}$$

Then

$$\chi^2_{\text{diff}} = \frac{1}{\bar{p}\bar{q}} \times \frac{n_{(1)}n_{(2)}}{n_{..}} (\bar{p}_1 - \bar{p}_2)^2, \tag{9.10}$$

with 1 degree of freedom, may be used to test for the significance of the difference between $\bar{p}_1$ and $\bar{p}_2$. Note that χ^2_{diff} is identical to the chi square, without the continuity correction, that one would calculate on the fourfold table obtained by combining all the data from the first m_1 samples into one single set and all the data from the remaining m_2 samples into a second.

The statistic

$$\chi^2_{\text{group 1}} = \frac{1}{\bar{p}\bar{q}} \sum_{i=1}^{m_1} n_{i.}(p_i - \bar{p}_1)^2, \tag{9.11}$$

with $m_1 - 1$ degrees of freedom, may be used to test the significance of the differences among the m_1 proportions in the first group, and the statistic

$$\chi^2_{\text{group 2}} = \frac{1}{\bar{p}\bar{q}} \sum_{i=m_1+1}^{m} n_{i.}(p_i - \bar{p}_2)^2, \tag{9.12}$$

with $m_2 - 1$ degrees of freedom, may be used to test the significance of the differences among the m_2 proportions in the second group. It may be checked that the three statistics given by (9.10)–(9.12) sum to the overall value of χ^2 in (9.4).

If $\bar{p}_1$ and $\bar{p}_2$ differ appreciably, then the product $\bar{p}_1\bar{q}_1 = \bar{p}_1(1 - \bar{p}_1)$ should replace $\bar{p}\bar{q}$ in (9.11), and $\bar{p}_2\bar{q}_2 = \bar{p}_2(1 - \bar{p}_2)$ should replace $\bar{p}\bar{q}$ in (9.12). These adjustments have little effect on the magnitudes of χ^2, but the sum of the adjusted chi squares plus the chi square in (9.10) will no longer generally recapture the overall value of chi square in (9.4).

A more serious modification is called for, however, if the partitioning of the samples into groups is suggested by the data instead of being planned beforehand. Of the four samples in Table 9.2, for example, the first three appear, on the basis of the similarity of their proportions, to form one homogeneous group, whereas the fourth sample seems to stand by itself as a second group. To control for the erroneous inferences possible by making comparisons suggested by the data, each of the χ^2 values in (9.10) to (9.12) should be referred to the critical value of chi square with $m - 1$ degrees of freedom and *not* to the critical values of chi square with 1, $m_1 - 1$, and $m_2 - 1$ degrees of freedom (Miller, 1966, Section 6.2).

For the data of Table 9.2, for example, the first set of $m_1 = 3$ studies consists of

$$n_{(1)} = 86 + 93 + 136 = 315$$

lung cancer patients, of whom the proportion smoking is

$$\bar{p}_1 = \frac{83 + 90 + 129}{315} = .959.$$

The second set of $m_2 = 1$ study alone consists of $n_{(2)} = 82$ patients, of whom the proportion smoking is $\bar{p}_2 = .854$.

The significance of the difference between $\bar{p}_1$ and $\bar{p}_2$ is assessed by the magnitude of χ^2_{diff} [see (9.10)]:

$$\chi^2_{\text{diff}} = \frac{1}{.937 \times .063} \times \frac{315 \times 82}{397}(.959 - .854)^2$$
$$= 12.15. \tag{9.13}$$

The significance of the differences among p_1, p_2 and p_3—all from group 1—is assessed by the magnitude of $\chi^2_{\text{group 1}}$ [see (9.11)]:

$$\chi^2_{\text{group 1}} = \frac{1}{.937 \times .063}\left[86 \times (.965 - .959)^2 + 93 \times (.968 - .959)^2 + 136 \times (.949 - .959)^2\right]$$
$$= 0.41. \tag{9.14}$$

Because group 2 consists of but a single study sample, the statistic $\chi^2_{\text{group 2}}$ [see (9.12)] is inapplicable here.

Note first of all that

$$\chi^2_{\text{diff}} + \chi^2_{\text{group 1}} = 12.15 + 0.41 = 12.56,$$

which is equal to the value of the overall chi square statistic given in (9.5). Note next that, with $\bar{p}_1\bar{q}_1 = .959 \times .041$ replacing $.937 \times .063$ in (9.14), the value of $\chi^2_{\text{group 1}}$ increases only slightly, to 0.62. Recall, finally, that the partitioning was suggested by the data and not planned *a priori*. The values of both χ^2_{diff} and $\chi^2_{\text{group 1}}$ must therefore be referred to the critical value of chi square with $m - 1 = 4 - 1 = 3$ degrees of freedom. Since the critical value for a significance level of .05 is 7.81, the conclusion would be that the proportion of smokers among the patients in study 4 differed from the proportions in studies 1 to 3 (because $\chi^2_{\text{diff}} = 12.15 > 7.81$), but that there were no differences among the proportions in studies 1 to 3 (because $\chi^2_{\text{group 1}} = 0.41 < 7.81$).

9.2. GRADIENT IN PROPORTIONS: SAMPLES QUANTITATIVELY ORDERED

The analysis of the preceding section is of quite general validity, but lacks sensitivity when the m samples possess an intrinsic ordering. We assume in this section that the ordering is quantitative; specifically, that a measurement x_i is naturally associated with the ith sample. Data from the National Center for Health Statistics (1970, Tables 1 and 6) are used for illustration (Table 9.3).

Table 9.3. Prevalence of reported insomnia among adult women by age

Age Interval	Number in Interval ($=n_{i.}$)	Proportion Reporting Insomnia ($=p_i$)	Midpoint Age ($=x_i$)
18–24	534	.280	21.5
25–34	746	.335	30.0
35–44	784	.337	40.0
45–54	705	.428	50.0
55–64	443	.538	60.0
65–74	299	.590	70.0
Overall	3511 ($=n_{..}$)	.393 ($=\bar{p}$)	42.15 ($=\bar{x}$)

Different methods of analysis are called for depending on how the proportions are hypothesized to vary with x (Yates, 1948). Here we consider only the simplest kind of variation, a linear one. Let P_i denote the proportion in the population from which the ith sample was drawn. We hypothesize that

$$P_i = \alpha + \beta x_i, \tag{9.15}$$

where β, the slope of the line, indicates the amount of change in the proportion per unit change in x and α, the intercept, indicates the proportion expected when $x = 0$.

The two parameters of (9.15) may be estimated as follows. Define

$$\bar{x} = \sum_{i=1}^{m} \frac{n_{i.} x_i}{n_{..}}, \tag{9.16}$$

the mean value of x in the given series of data. The slope is estimated as

$$b = \frac{\sum_{i=1}^{m} n_{i.}(p_i - \bar{p})(x_i - \bar{x})}{\sum_{i=1}^{m} n_{i.}(x_i - \bar{x})^2} \tag{9.17}$$

and the intercept as

$$a = \bar{p} - b\bar{x}. \tag{9.18}$$

The calculation of b is simplified somewhat by noting that its numerator is

$$\text{numerator}(b) = \sum_{i=1}^{m} n_{i.} p_i x_i - n_{..}\bar{p}\bar{x} \tag{9.19}$$

and that its denominator is

$$\text{denominator}(b) = \sum_{i=1}^{m} n_{i.} x_i^2 - n_{..}\bar{x}^2. \tag{9.20}$$

A simple expression for the fitted line is

$$\hat{p}_i = \bar{p} + b(x_i - \bar{x}). \tag{9.21}$$

For the data of Table 9.3, $\bar{p} = .393$, $\bar{x} = 42.15$, and

$$b = .0064. \tag{9.22}$$

The fitted straight line becomes

$$\hat{p}_i = .393 + .0064(x_i - 42.15), \tag{9.23}$$

implying an increase of .64% in the proportion of adult women reporting insomnia per yearly increase in age.

It is useful to calculate the estimated proportion corresponding to each x_i in order to compare it with the actual proportion, p_i. If p_i and $\hat{p}_i$ are close in magnitude for all or most categories, then one can conclude that (9.15) provides a good fit to the data, that is, P_i tends to vary linearly with x_i. If p_i and $\hat{p}_i$ tend to differ, then the conclusion is that the association between P_i and x_i is more complicated than a linear one. Having the differences $p_i - \hat{p}_i$ available serves also to identify those categories for which the departures from linearity are greatest.

Table 9.4 contrasts the actual proportions of Table 9.3 with those yielded by (9.23). The fit appears to be a good one.

Table 9.4. Observed and linearly predicted age-specific rates of insomnia

x_i	$n_{i.}$	p_i	$\hat{p}_i$
21.5	534	.280	.261
30.0	746	.335	.315
40.0	784	.337	.379
50.0	705	.428	.443
60.0	443	.538	.507
70.0	299	.590	.571

A chi square statistic due to Cochran (1954) and Armitage (1955) is available for testing whether the association between P_i and x_i is a linear one. This chi square statistic is

$$\chi^2_{\text{linearity}} = \sum_{i=1}^{m} n_{i.}(p_i - \hat{p}_i)^2/\bar{p}\bar{q}. \tag{9.24}$$

$\chi^2_{\text{linearity}}$ has $m-2$ degrees of freedom, and the hypothesis of linearity would be rejected if $\chi^2_{\text{linearity}}$ were found to be large. The power of this test was studied by Chapman and Nam (1968).

The calculation of $\chi^2_{\text{linearity}}$ is simplified if one first calculates the statistic

$$\chi^2_{\text{slope}} = b^2 \sum_{i=1}^{m} n_{i.}(x_i - \bar{x})^2/\bar{p}\bar{q} \tag{9.25}$$

because it may be shown that

$$\chi^2_{\text{linearity}} = \chi^2 - \chi^2_{\text{slope}}, \tag{9.26}$$

where χ^2 is given by (9.4). The statistic χ^2_{slope} has 1 degree of freedom and may be used to test the significance of the slope, b. If χ^2_{slope} is large, the inference is that the slope is significantly different from zero, indicating that there is a tendency for increasing values of x_i to be associated with increasing values of P_i if b is positive or with decreasing values of P_i if b is negative.

For the data of Table 9.3, the value of the overall chi square statistic in (9.4) for testing the hypothesis that the proportion reporting insomnia is constant for all age groups is

$$\chi^2 = 140.72. \tag{9.27}$$

The magnitude of this chi square, which has five degrees of freedom, indicates highly significant differences among the age-specific proportions, but fails to describe the steady increase with age of the proportion reporting insomnia.

The chi square statistic for linearity [see (9.24)] is, for the data from Table 9.4,

$$\chi^2_{\text{linearity}} = \frac{534 \times (.280 - .261)^2 + \cdots + 299 \times (.590 - .571)^2}{.393 \times .607}$$

$$= 10.76 \tag{9.28}$$

which, with 4 degrees of freedom, is significant at the .05 level. The association with age of the proportion of women reporting insomnia is thus not precisely a linear one, but the departures from linearity (i.e., the differences between the observed and linearly predicted proportions) are sufficiently small to make the hypothesis of linearity reasonable.

The chi square statistic of (9.25) assumes the value

$$\chi^2_{\text{slope}} = \frac{.0064^2 \times 757{,}964.7975}{.393 \times .607} = 130.15, \tag{9.29}$$

which, with 1 degree of freedom, indicates that the slope of the fitted line, $b = .0064$, is significantly different from zero. The difference between the overall chi square of 140.72 and the chi square for testing the significance of the slope, 130.15, should, by (9.26), be equal to the chi square for linearity, 10.76. Except for errors due to rounding, this is seen to be the case.

The inferences to be drawn from this more detailed chi square analysis are that there is a significant tendency for the proportion of women reporting insomnia to increase steadily with age and that this tendency is, effectively, a linear one. Had the chi square for linearity been significant at, say, the .01 or .005 level instead of merely at the .05 level, the latter inference would not have been warranted.

Slightly different versions of the estimators and of the test statistics given above have been derived by Mantel (1963), Chapman and Nam (1968), and Wood (1978), who also consider the comparison and combination of fitted regression lines across several independent samples. The procedure presented here is valid when, as in the example, the p_i's are not close to 0 or to 1.

9.3. GRADIENT IN PROPORTIONS: SAMPLES QUALITATIVELY ORDERED

We assumed in Section 9.2 that the *m* samples could be ordered on a quantitative scale. We assume in this section that the ordering is merely qualitative. Suppose, for example, that one has data as in Table 9.5. The value of χ^2 in (9.4) for these data is

$$\chi^2 = 28.74 \tag{9.30}$$

with 3 degrees of freedom, clearly significant beyond the .001 level (see Table A.1).

Table 9.5. Hypothetical one-month release rates as a function of initial severity

Initial Severity	Total	Number Released within One Month	Proportion Released within One Month
Mild	30 $(=n_{1.})$	25	.83 $(=p_1)$
Moderate	25 $(=n_{2.})$	22	.88 $(=p_2)$
Serious	20 $(=n_{3.})$	12	.60 $(=p_3)$
Extreme	25 $(=n_{4.})$	6	.24 $(=p_4)$
Overall	100 $(=n_{..})$	65	.65 $(=\bar{p})$

The inference that the four release rates differ significantly is a valid one, but is clearly insufficient in that it fails to describe the almost steady decline in release rates as initial severity worsens. Because it would have been reasonable to hypothesize beforehand a gradient of release rate with severity, an alternative method of analysis is called for. The method of the preceding section is not appropriate because no numerical values can naturally be assigned to the four levels of severity.

Chassan (1960, 1962) proposed a simple test of the hypothesis that *m* proportions were arrayed in a prespecified order, but his test was shown by Bartholomew (1963) to lack adequate power. Specifically, Chassan's test may be applied only when the sample proportions are arrayed *without exception* in the same order as hypothesized. It would therefore be inapplicable whenever there were slight departures (as, e.g., for p_1 and p_2 in Table 9.5) from the hypothesized order. A more powerful procedure due to Bartholomew (1959a, 1959b) will be described.

Suppose that the hypothesis predicts that the ordering $p_1 > p_2 > \cdots > p_m$ should obtain, but that departures from this ordering are observed. For the proportions in Table 9.5, for example, the ordering $p_1 > p_2 > p_3 > p_4$ was predicted, but instead we obtained $p_1 < p_2$ and then, as predicted, $p_2 > p_3 > p_4$.

When departures are found, weighted averages of those adjacent proportions that are out of order are taken until, when the averages replace the original proportions, the hypothesized ordering is observed. The revised proportions are denoted p'. For the proportions in Table 9.5, the weighted average of p_1 and p_2 must be taken. It is

$$\bar{p}_{1,2} = \frac{30 \times .83 + 25 \times .88}{30 + 25} = .85. \tag{9.31}$$

When p_1 and p_2 are replaced by $\bar{p}_{1,2}$, Table 9.6 results.

Table 9.6. Proportions from Table 9.5 revised to be in hypothesized order

Initial Severity	Total	Revised Proportion
Mild	30 $(=n_{1.})$	.85 $(=p_1')$
Moderate	25 $(=n_{2.})$	.85 $(=p_2')$
Serious	20 $(=n_{3.})$	.60 $(=p_3=p_3')$
Extreme	25 $(=n_{4.})$	.24 $(=p_4=p_4')$
Overall	100 $(=n_{..})$	.65 $(=\bar{p})$

The revised proportions are no longer out of order. If they were, the process would have to be continued. When the process has been completed, the statistic

$$\bar{\chi}^2 = \frac{1}{\bar{p}\bar{q}} \sum_{i=1}^{m} n_{i.}(p_i' - \bar{p})^2 \tag{9.32}$$

is calculated. For the revised proportions of Table 9.6,

$$\bar{\chi}^2 = \frac{1}{.65 \times .35} \left[30 \times (.85 - .65)^2 + 25 \times (.85 - .65)^2 + 20 \times (.60 - .65)^2 + 25 \times (.24 - .65)^2 \right] = 28.27. \tag{9.33}$$

The value of $\bar{\chi}^2$ may no longer be referred to tables of chi square, however. Instead, Tables A.6 to A.8 are to be used. When $m = 3$ proportions are compared, calculate

$$c = \sqrt{\frac{n_{1.}n_{3.}}{(n_{1.} + n_{2.})(n_{2.} + n_{3.})}} \tag{9.34}$$

and enter Table A.6 under the desired significance level, interpolating if necessary. When $m = 4$, calculate

$$c_1 = \sqrt{\frac{n_{1.}n_{3.}}{(n_{1.}+n_{2.})(n_{2.}+n_{3.})}} \tag{9.35}$$

and

$$c_2 = \sqrt{\frac{n_{2.}n_{4.}}{(n_{2.}+n_{3.})(n_{3.}+n_{4.})}}, \tag{9.36}$$

and enter Table A.7 under the desired significance level, interpolating in both c_1 and c_2 if necessary. If all sample sizes are equal, and if $m \leqq 12$, Table A.8 may be used.

For the data of Table 9.6, for which $m = 4$,

$$c_1 = \sqrt{\frac{30 \times 20}{(30+25)(25+20)}} = .49$$

and

$$c_2 = \sqrt{\frac{25 \times 25}{(25+20)(20+25)}} = .56.$$

Visual interpolation in Table A.7 (c_1 is approximately equal to .5 and c_2 is nearly midway between .5 and .6) shows that $\bar{\chi}^2$ would have to exceed 9.0 in order for significance to be declared at the .005 level. The obtained value of $\bar{\chi}^2 = 28.27$ from (9.33) is far beyond this critical value.

What is noteworthy, however, is the comparison of the value just found from Table A.7 with the corresponding value from Table A.1 for the standard chi square test with $m - 1 = 3$ degrees of freedom. If no ordering is hypothesized, χ^2 would have to exceed 12.8 (instead of 9.0) for significance to be declared at the .005 level. Thus, *if the hypothesized ordering actually obtains in the population*, Bartholomew's test is more powerful than the standard chi square test. If the hypothesized ordering is not true, however, the averaging process necessary before the calculation of $\bar{\chi}^2$ in (9.32) could well reduce its magnitude to insignificance. Further analyses and generalizations of Bartholomew's test have been made by Barlow, Bartholomew, Bremner, and Brunk (1972).

9.4. RIDIT ANALYSIS

Suppose that one has data available from two or more samples, with the subjects from each sample distributed across a number of ordered categories. Let k denote the number of categories. For example, let us consider automobile accidents, with the phenomenon studied being the degree of injury sustained by the driver. The degree of injury might be graded from none through severe to fatal. Such a grading is clearly subjective and probably not too reliable. It nevertheless seems preferable to the adoption of the simple dichotomy, little or no injury versus severe or fatal injury, because it both possesses some degree of reliability and succeeds in describing the phenomenon more completely than the cruder yes-no system.

There exists the problem, however, of summarizing the data and making comparisons among different samples in an intelligible way. When two samples are being compared, the data may be arrayed as in Table 9.7. The proportions $(p_{11},\ldots,p_{k1})$ represent the frequency distribution in sample 1,

Table 9.7. Relative frequency distributions from two samples

Outcome Category	Sample 1 (sample size $= n_1$)	Sample 2 (sample size $= n_2$)	Combined Sample (sample size $= n_.$)
1	p_{11}	p_{12}	$\bar{p}_1$
2	p_{21}	p_{22}	$\bar{p}_2$
⋮			
k	p_{k1}	p_{k2}	$\bar{p}_k$
Total	1	1	1

and the proportions $(p_{12},\ldots,p_{k2})$ represent the frequency distribution in sample 2. The frequency distribution in the combined sample is $(\bar{p}_1,\ldots,\bar{p}_k)$, where

$$\bar{p}_i = \frac{n_1 p_{i1} + n_2 p_{i2}}{n_.} \tag{9.37}$$

$(i = 1,\ldots,k)$ and $n_. = n_1 + n_2$, the total sample size. The value of chi square with $k-1$ degrees of freedom may be found using the formula

$$\chi^2 = \frac{n_1 n_2}{n_.} \sum_{i=1}^{k} \frac{(p_{i1} - p_{i2})^2}{\bar{p}_i} \tag{9.38}$$

(see problem 9.4), but crucial information on the natural ordering of the k categories would be lost.

A frequently employed device is to number the categories from 0 for the least serious to some highest number for the most serious, and then calculate means and standard deviations and apply t tests or analyses of variance. This device of concocting a seemingly numerical measurement system has many drawbacks. For one thing, one is giving the impression of greater accuracy than really exists. For another, the results one gets depend on the particular system of numbers employed. The choice of a system is by no means a simple one.

Consider again the study of automobile accidents, and suppose that we have seven categories of injury, the first two being None and Mild, and the last two, Critical and Fatal. The straightforward system of numbering assigns the seven integers from 0 to 6 successively to the seven categories. This system is hard to justify, for it implies that the difference between no injury and a mild one is equivalent to the difference between a critical injury and a fatal one. The latter difference is obviously more important, but this greater importance can be picked up only by assigning a value in excess of 6 to the final category. Just what this value should be can, however, only be decided arbitrarily. If an underlying logistic model (see Section 5.4) may be assumed, a procedure due to Snell (1964) is appropriate.

Let us abandon the attempt to quantify the categories and instead agree to work only with the natural ordering that exists. A technique that takes advantage of this natural ordering is *ridit analysis*. Virtually the only assumption made in ridit analysis is that the discrete categories represent intervals of an underlying but unobservable continuous distribution. No assumption is made about normality or any other form for the distribution.

Ridit analysis is due to Bross (1958) and has been applied to the study of automobile accidents (Bross, 1960), of cancer (Wynder, Bross, and Hirayama, 1960), and of schizophrenia (Spitzer, et al., 1965). A mathematical study of ridit analysis was made by Kantor, Winkelstein, and Ibrahim (1968). A critique of ridit analysis has been offered by Mantel (1979).

Ridit analysis begins with the selection of a population to serve as a standard or reference group. The term ridit is derived from the initials of "relative to an identified distribution." For the reference group, we estimate the proportion of all individuals with a value on the underlying continuum falling at or below the midpoint of each interval, that is, each interval's *ridit*. This initial arithmetic is illustrated in Table 9.8, using data from Bross (1958, p. 20).

1. In general, column 1 contains the distribution over the various categories for the reference group. In Table 9.8, the distribution is over seven categories of injury for the 179 members of a selected sample.

Table 9.8. An illustration of the calculation of ridits for degrees of injury

Severity	(1)	(2)	(3)	(4)	(5) = ridit
None	17	8.5	0	8.5	.047
Minor	54	27.0	17	44.0	.246
Moderate	60	30.0	71	101.0	.564
Severe	19	9.5	131	140.5	.785
Serious	9	4.5	150	154.5	.863
Critical	6	3.0	159	162.0	.905
Fatal	14	7.0	165	172.0	.961

2. The entries in column 2 are simply half the corresponding entries in column 1.

3. The entries in column 3 are the accumulated entries in column 1, but displaced one category downwards.

4. The entries in column 4 are the sums of the corresponding entries in columns 2 and 3.

5. The entries in column 5, finally, are those in column 4 divided by the total sample size, in this case 179.

The final values are the *ridits* associated with the various categories. The ridit for a category, then, is nothing but the proportion of all subjects from the reference group falling in the lower ranking categories plus half the proportion falling in the given category. If, in the model of an underlying continuum, we assume that the distribution is uniform in each interval, then a category's ridit is the proportion of all subjects from the reference group with an underlying value at or below the midpoint of the corresponding interval.

Given the distribution of any other group over the same categories, the mean ridit for that group may be calculated. The resulting mean value is interpretable as a probability. The mean ridit for a group is the probability that a randomly selected individual from it has a value indicating greater severity or seriousness than a randomly selected individual from the standard group.

In our example, if this probability is .50, we infer that the comparison group tends to sustain neither more nor less serious injuries than the reference group. For the reference group itself, by the way, the mean ridit is necessarily .50. This is consistent with the fact that, if two subjects are randomly selected from the same population, then the second subject will have a more extreme value half the time and will have a less extreme value also half the time.

If the mean ridit for a comparison group is greater than .50, then more than half of the time a randomly selected subject from it will have a more extreme value than a randomly selected subject from the reference group. In our example, we would infer that the comparison group tends to sustain more serious injuries than the reference group. If, finally, a comparison group's mean ridit is less than .50, we would infer that its subjects tend to have less extreme values than the subjects of the reference group.

As an example, consider the hypothetical data of Table 9.9, giving the distribution of seriousness of injury to the driver when he was involved in an accident and had been slightly intoxicated.

Table 9.9. Seriousness of injury sustained by slightly intoxicated drivers of automobiles involved in accidents

Severity	Number	Ridit	Product
None	5	.047	0.235
Minor	10	.246	2.460
Moderate	16	.564	9.024
Severe	5	.785	3.925
Serious	3	.863	2.589
Critical	6	.905	5.430
Fatal	5	.961	4.805
Total	50		28.468

The mean ridit for a group is simply the sum of the products of observed frequencies times corresponding ridits, divided by the total frequency. For slightly intoxicated drivers the mean is

$$\bar{r} = \frac{28.468}{50} = .57. \tag{9.39}$$

Thus the odds are 4 to 3 (= .57/.43) that a slightly intoxicated driver will sustain a more serious injury than a driver from the reference group if both are involved in accidents.

Selvin (1977) has shown how ridit analysis is closely connected with so-called rank order analysis used in nonparametric statistics and thus how standard errors of mean ridits can be found. Let N_j denote the number of individuals from the reference group in category j, $N = \Sigma N_j$ the total number of individuals in the reference group, n_j the number of individuals from the comparison group in category j, and $n = \Sigma n_j$ the total number of individuals in the comparison group. If the reference group is not too much larger than the comparison group, the standard error of the mean for the comparison group is given by

$$\text{s.e.}(\bar{r}) = \frac{1}{2\sqrt{3n}}\sqrt{1 + \frac{n+1}{N} + \frac{1}{N(N+n-1)} - \frac{\Sigma(N_j + n_j)^3}{N(N+n)(N+n-1)}}\ . \tag{9.40}$$

The two frequency distributions for the current problem are presented in Table 9.10. The standard error of the mean ridit for slightly intoxicated

Table 9.10. Frequency distributions from reference and comparison groups

Severity	Reference Group (N_j)	Comparison Group (n_j)	Total ($N_j + n_j$)
None	17	5	22
Minor	54	10	64
Moderate	60	16	76
Severe	19	5	24
Serious	9	3	12
Critical	6	6	12
Fatal	14	5	19
Total	179 ($= N$)	50 ($= n$)	229 ($= N + n$)

drivers involved in accidents is, by (9.40),

$$\text{s.e.}(\bar{r}) = \frac{1}{2\sqrt{150}}\sqrt{1 + \frac{51}{179} + \frac{1}{179 \times 228} - \frac{735{,}907}{179 \times 229 \times 228}} = .045. \tag{9.41}$$

The significance of the difference between an obtained mean ridit and the standard value of .5 may be tested by referring the value of

$$z = \frac{\bar{r} - .5}{\text{s.e.}(\bar{r})} \tag{9.42}$$

to Table A.2 of the normal distribution. For our example,

$$z = \frac{.57 - .50}{.045} = 1.56. \tag{9.43}$$

Because z failed to reach significance, we would have to conclude that the seriousness of injuries to slightly intoxicated drivers might equal that of the injuries to members of the reference group.

When N, the size of the reference group, is very large relative to that of any possible comparison group, the standard error of the mean ridit simplifies to

$$\text{s.e.}(\bar{r}) = \frac{1}{2\sqrt{3n}}. \tag{9.44}$$

For the current data, this simple approximation yields an estimated standard error of .041, slightly smaller than the correct value of .045 from (9.41).

As another example of the use of ridit analysis, suppose we have data on a sample of 50 extremely intoxicated drivers who were involved in accidents, and suppose that their mean ridit is .73. An important comparison is between slightly and extremely intoxicated drivers. Instead of identifying a new reference group, all we need to do is to subtract one of the two mean ridits from the other and add .50. Thus we obtain $(.73 - .57) + .50 = .66$ as the chances that a driver who is extremely intoxicated will sustain a more severe injury than one who is slightly so, when they are involved in accidents.

If one mean ridit is based on N_1 subjects and the other on N_2, the standard error is approximately

$$\text{s.e.}(\bar{r}_2 - \bar{r}_1) = \frac{\sqrt{N_1 + N_2}}{2\sqrt{3N_1N_2}}. \tag{9.45}$$

This formula provides a good approximation to the standard error of the mean ridit for a single comparison group when N and n are of comparable magnitudes (see expression 9.40), provided N and n replace N_1 and N_2. It *overestimates* the standard error, but usually by only a very slight amount [see Problem 9.5, parts (*c*) and (*d*)].

With $N_1 = N_2 = 50$, the approximate standard error is

$$\text{s.e.}(\bar{r}_2 - \bar{r}_1) = \frac{\sqrt{100}}{2\sqrt{3 \times 50 \times 50}} = .06. \tag{9.46}$$

The significance of the difference between $\bar{r}_1$ and $\bar{r}_2$ may be tested by referring the value of

$$z = \frac{\bar{r}_2 - \bar{r}_1}{\text{s.e.}(\bar{r}_2 - \bar{r}_1)} \tag{9.47}$$

to Table A.2. For our example,

$$z = \frac{.73 - .57}{.06} = 2.67, \tag{9.48}$$

which indicates a difference significant at the .01 level. We can therefore infer that extremely intoxicated drivers involved in accidents tend to sustain more serious injuries than slightly intoxicated drivers involved in accidents.

The reader is warned of the possibility of an anomalous result using the approach just described for contrasting two comparison groups, that is, the estimated probability may be less than zero or greater than unity. Consider the hypothetical data in Table 9.11, where one frequency distribution is the mirror image of the other.

Table 9.11. Hypothetical data on seriousness of injury in two comparison groups

Severity	Group A	Group B
None	46	1
Minor	34	2
Moderate	9	3
Severe	5	5
Serious	3	9
Critical	2	34
Fatal	1	46
Total	100	100

It is easily checked using the ridits of Table 9.8 that the two mean ridits are $\bar{r}_A = .25$ and $\bar{r}_B = .89$. The above approach yields an impossible value of $(.89 - .25) + .50 = 1.14$ for the probability that a randomly selected member of group B will sustain a more severe injury than a randomly selected member of group A!

When, as in this hypothetical case, frequency distributions in the two contrasted comparison groups are widely different, it is appropriate to ignore the original reference group entirely and to calculate the desired probability as a mean ridit, with either of the two groups serving as the ad hoc reference group. Problem 9.5 is devoted to the appropriate analysis of the data from Table 9.11.

Problem 9.1. Prove the equality of expressions 9.3 and 9.4 for χ^2.

Problem 9.2. The estimate of the slope b is given by (9.17). Prove that its numerator is given by (9.19) and that its denominator is given by (9.20).

Problem 9.3. Three samples of New York mental hospital patients were studied as part of a collaborative project (Cooper, et al., 1972). The numbers of hospital diagnoses of affective disorders were as follows:

Sample	Age Range	Number of Patients	Number Diagnosed Affective	Proportion
1	20–34	105	2	
2	20–59	192	13	
3	35–59	145	24	
Overall		442	39	

(*a*) Calculate the proportions diagnosed affective and test whether they differ significantly.

(*b*) Test whether the proportion diagnosed affective in the first two samples combined differs significantly from the proportion in the third sample. Test whether the proportions in the first two samples differ.

(*c*) The patients in sample 1 tend to be younger than those in sample 2, who in turn tend to be younger than those in sample 3. Because the chances of an affective disorder increase with age, it might be hypothesized that p_1 should be less than p_2, and that p_2 in turn should be less than p_3. Are the proportions in this order? What is the value of $\bar{\chi}^2$ in (9.32)? What is the value of c in (9.34)? Refer to Table A.6 to test the hypothesized ordering.

Problem 9.4. Let the frequencies underlying the data in Table 9.7 be $(n_{11},\dots,n_{k1})$ and $(n_{12},\dots,n_{k2})$, so that $p_{i1}=n_{i1}/n_{\cdot 1}$ and $p_{i2}=n_{i2}/n_{\cdot 2}$, $i=1,\dots,k$. Define $n_{i.}=n_{i1}+n_{i2}$, so that $\bar{p}_i=n_{i.}/n_{..}$. The classic formula for chi square is

$$\chi^2=\sum_{i=1}^{k}\frac{\left(n_{i1}-\dfrac{n_{i.}n_{\cdot 1}}{n_{..}}\right)^2}{\dfrac{n_{i.}n_{\cdot 1}}{n_{..}}}+\sum_{i=1}^{k}\frac{\left(n_{i2}-\dfrac{n_{i.}n_{\cdot 2}}{n_{..}}\right)^2}{\dfrac{n_{i.}n_{\cdot 2}}{n_{..}}}.$$

Prove that this formula is equal to the one in (9.38).

Problem 9.5. Consider the data of Table 9.11.

(*a*) Taking group A as the reference group, find the mean ridit for group B.

(*b*) Taking group B as the reference group, find the mean ridit for group A. What is the relation between the answers in (*a*) and (*b*)?

(*c*) Use formula (9.40) to find the standard error of the mean ridit contrasting groups A and B.

(*d*) Use formula (9.45) to estimate the same standard error. How do the values in (*c*) and (*d*) compare?

REFERENCES

Armitage, P. (1955). Tests for linear trends in proportions and frequencies. *Biometrics*, **11**, 375–385.

Barlow, R. E., Bartholomew, D. J., Bremner, J. M., and Brunk, H. D. (1972). *Statistical inference under order restrictions*. New York: Wiley.

Bartholomew, D. J. (1959a). A test of homogeneity for ordered alternatives. *Biometrika*, **46**, 36–48.

Bartholomew, D. J. (1959b). A test of homogeneity for ordered alternatives II. *Biometrika*, **46**, 328–335.

Bartholomew, D. J. (1963). On Chassan's test for order. *Biometrics*, **19**, 188–191.

Bross, I. D. J. (1958). How to use ridit analysis. *Biometrics*, **14**, 18–38.

Bross, I. D. J. (1960). How to cut the highway toll in half in the next ten years. *Public Health Rep.*, **75**, 573–581.

Castellan, N. J. (1965). On the partitioning of contingency tables. *Psychol. Bull.*, **64**, 330–338.

Chapman, D. G. and Nam, J. (1968). Asymptotic power of chi square tests for linear trends in proportions. *Biometrics*, **24**, 315–327.

Chassan, J. B. (1960). On a test for order. *Biometrics*, **16**, 119–121.

Chassan, J. B. (1962). An extension of a test for order. *Biometrics*, **18**, 245–247.

Cochran, W. G. (1954). Some methods of strengthening the common χ^2 tests. *Biometrics*, **10**, 417–451.

Cooper, J. E., Kendell, R. E., Gurland, B. J., Sharpe, L., Copeland, J. R. M., and Simon, R. (1972). *Psychiatric diagnosis in New York and London*. London: Oxford University Press.

Dorn, H. F. (1954). The relationship of cancer of the lung and the use of tobacco. *Am. Stat.*, **8**, 7–13.

Irwin, J. O. (1949). A note on the subdivision of chi-square into components. *Biometrika*, **36**, 130–134.

Kantor, S., Winkelstein, W., and Ibrahim, M. A. (1968). A note on the interpretation of the ridit as a quantile rank. *Am. J. Epidemiol.*, **87**, 609–615.

Kastenbaum, M. A. (1960). A note on the additive partitioning of chi-square in contingency tables. *Biometrics*, **16**, 416–422.

Kimball, A. W. (1954). Short-cut formulas for the exact partition of χ^2 in contingency tables. *Biometrics*, **10**, 452–458.

Knoke, J. D. (1976). Multiple comparisons with dichotomous data. *J. Am. Stat. Assoc.*, **71**, 849–853.

Lancaster, H. O. (1950). The exact partitioning of chi-square and its application to the problem of pooling small expectations. *Biometrika*, **37**, 267–270.

Mantel, N. (1963). Chi-square tests with one degree of freedom: Extensions of the Mantel-Haenszel procedure. *J. Am. Stat. Assoc.*, **58**, 690–700.

Mantel, N. (1979). Ridit analysis and related ranking procedures—Use at your own risk. *Am. J. Epidemiol.*, **109**, 25–29.

Miller, R. G. (1966). *Simultaneous statistical inference*. New York: McGraw-Hill.

National Center for Health Statistics (1970). Selected symptoms of psychological distress in the United States. *Data from National Health Survey*, Series 11, No. 37.

Selvin, S. (1977). A further note on the interpretation of ridit analysis. *Am. J. Epidemiol.*, **105**, 16–20.

Snell, E. J. (1964). A scaling procedure for ordered categorical data. *Biometrics*, **20**, 592–607.

Spitzer, R. L., Fleiss, J. L., Kernohan, W., Lee, J., and Baldwin, I. T. (1965). The Mental Status Schedule: Comparing Kentucky and New York schizophrenics. *Arch. Gen. Psychiatry*, **12**, 448–455.

Wood, C. L. (1978). Comparison of linear trends in binomial proportions. *Biometrics*, **34**, 496–504.

Wynder, E. L., Bross, I. D. J., and Hirayama, T. (1960). A study of the epidemiology of cancer of the breast. *Cancer*, **13**, 559–601.

Yates, F. (1948). The analysis of contingency tables with groupings based on quantitative characteristics. *Biometrika*, **35**, 176–181.

CHAPTER 10

Combining Evidence from Fourfold Tables

There are a number of ways in which data relevant to the association between a risk factor A and a disease B might end up arrayed in several fourfold tables. If the possibility of an association between a factor A and a disease B is strong, interesting, or important enough, it is virtually guaranteed that a number of investigators will study the association. Similarly, if an association has been found to exist in one kind of population, it is to be expected that the possibility of association in other kinds of populations will be studied. As a final example, a single given study might call for stratifying the samples being compared on variables known to be associated with the outcome variable under investigation; each stratum would then provide its own fourfold table.

Suppose that the association between A and B has been studied in each of g groups, with each group generating its own fourfold table. The following questions can be asked:

1. Is there evidence that the degree of association, whatever its magnitude, is consistent from one group to another?
2. Assuming that the degree of association is found to be consistent, is the common degree of association statistically significant?
3. Assuming that the common degree of association is significant, what is the best estimate of the common value for the measure of association? What is its standard error? How does one construct a confidence interval for the underlying measure?

Section 10.1 provides a simple statistical framework within which these questions can be answered. Section 10.2 describes a method for use with the logarithm of the odds ratio, Section 10.3 describes a method based on results of Cornfield and Gart, and Section 10.4 describes the Mantel-Haenszel method. Section 10.5 compares these methods for different kinds of study designs, and Section 10.6 indicates how they can be used as alternatives to matching in the control of confounding factors. Section 10.7 describes some popular but generally invalid methods for comparing and combining data from several fourfold tables.

The methods reviewed in this chapter are special cases of those available for the analysis of complex cross-classification tables. The texts by Cox (1970), Bishop, Fienberg, and Holland (1975), Everitt (1977), and Fienberg (1977) are excellent references to the general methods of log linear and logistic regression analysis.

10.1. THE CONSTRUCTION AND INTERPRETATION OF SOME CHI SQUARE TESTS

To answer the questions posed above, some knowledge of the theory of chi square tests is necessary. For a single one of the g groups, say the ith, let y_i denote the value of the chosen measure of association. The measure might be the difference between two proportions, the logarithm of the odds ratio, and so on.

Whatever y_i is, let s.e.(y_i) denote its standard error and define

$$w_i = \frac{1}{[\text{s.e.}(y_i)]^2}, \tag{10.1}$$

so that w_i is the reciprocal of the squared standard error of y_i. The quantity w_i is the weight to be attached to y_i. If the standard error of y_i is large, implying that y_i is not too precise, then w_i is small. This is reasonable, since imprecise estimates should not be given great weight. If, on the other hand, the standard error of y_i is small, implying that y_i is rather precise, then w_i is large. This, too, is reasonable, since precise estimates should be given great weight.

Let us suppose that y_i is such that a value of zero indicates no association. Then, when the hypothesis of no association in the ith group is true, the quantity

$$\chi_i = \frac{y_i}{\text{s.e.}(y_i)} = y_i\sqrt{w_i} \tag{10.2}$$

has, approximately, the standard normal distribution, and the quantity

$$\chi_i^2 = w_i y_i^2 \tag{10.3}$$

has, approximately, the chi square distribution with 1 degree of freedom. If the hypothesis of no association in the ith group is false, χ_i^2 may be expected to be large, so that the hypothesis is likely to be rejected if a chi square test is applied.

We are not so much interested in the ith group, or in any other single group, however, as in all the groups together. The analysis of all groups conveniently begins with the calculation of

$$\chi^2_{\text{total}} = \sum_{i=1}^{g} \chi_i^2 = \sum_{i=1}^{g} w_i y_i^2. \tag{10.4}$$

If there is no association in any of the g groups, then χ^2_{total} has a chi square distribution with g degrees of freedom. This follows because the sum of g independent chi squares, each with one degree of freedom, has a chi square distribution with g degrees of freedom and because the g groups are assumed to be independent.

If we calculate χ^2_{total} and find it to be significantly large, we may validly conclude that there is association somewhere within the g groups. We would not, however, know whether the association was consistent across all groups or whether it varied from one group to another. χ^2_{total} is not, therefore, informative by itself. Rather, its calculation serves the purpose of simplifying other calculations, as will now be indicated.

χ^2_{total} is subdivided, or partitioned, into two components,

$$\chi^2_{\text{total}} = \chi^2_{\text{homog}} + \chi^2_{\text{assoc}}. \tag{10.5}$$

The quantity χ^2_{homog} assesses the degree of homogeneity, or equality, among the g measures of association, and the quantity χ^2_{assoc} assesses the significance of the average degree of association. The subdivision indicated by (10.5) is most easily effected by calculating χ^2_{assoc} first and determining χ^2_{homog} by simple subtraction.

The term χ^2_{assoc} is calculated as follows. The overall measure of association across all groups is taken as the weighted average of the g individual measures, with the weights being those defined in (10.1):

$$\bar{y} = \frac{\sum_{i=1}^{g} w_i y_i}{\sum_{i=1}^{g} w_i}. \tag{10.6}$$

Under the hypothesis that the overall association is zero, $\bar{y}$ has an average value of zero and a standard error of

$$\text{s.e.}(\bar{y}) = \frac{1}{\sqrt{\sum_{i=1}^{g} w_i}}. \tag{10.7}$$

Hence

$$\chi_{\text{assoc}} = \frac{\bar{y}}{\text{s.e.}(\bar{y})} = \frac{\sum_{i=1}^{g} w_i y_i}{\sqrt{\sum_{i=1}^{g} w_i}} \tag{10.8}$$

is distributed approximately as a standard normal variate under the hypothesis, and

$$\chi^2_{\text{assoc}} = \bar{y}^2 \sum_{i=1}^{g} w_i = \frac{\left(\sum_{i=1}^{g} w_i y_i\right)^2}{\sum_{i=1}^{g} w_i} \tag{10.9}$$

is distributed approximately as chi square with 1 degree of freedom.

The term χ^2_{homog} is then easily obtained by subtraction:

$$\chi^2_{\text{homog}} = \chi^2_{\text{total}} - \chi^2_{\text{assoc}} = \sum_{i=1}^{g} w_i y_i^2 - \bar{y}^2 \sum_{i=1}^{g} w_i. \tag{10.10}$$

An equivalent expression for χ^2_{homog} is

$$\chi^2_{\text{homog}} = \sum_{i=1}^{g} w_i (y_i - \bar{y})^2. \tag{10.11}$$

This expression for χ^2_{homog} is useful for two purposes. One is to provide a numerical check on the arithmetic. The other is to point out that χ^2_{homog} actually measures the degree of variability among the separate values of y_i. χ^2_{homog} is approximately distributed as chi square with $g-1$ degrees of freedom under the hypothesis of consistent (homogeneous) association.

Means are therefore provided for answering the three questions posed at the beginning of this chapter.

(1) Consistency of association can be tested by referring χ^2_{homog} to tables of chi square with $g-1$ degrees of freedom. If χ^2_{homog} is significant,

the next step in the analysis would be to partition χ^2_{homog} into appropriate components in order to identify those groups in which the association is different from that in the remaining groups (see Problem 10.1).

(2) If χ^2_{homog} is not significant, the significance of the overall degree of association can be tested by referring χ^2_{assoc} to tables of chi square with 1 degree of freedom.

(3) The best estimate of the overall degree of association is $\bar{y}$ [see (10.6)]. Its standard error is given by (10.7). An approximate $100(1-\alpha)\%$ confidence interval for the underlying overall degree of association is

$$\bar{y} \pm c_{\alpha/2}\text{s.e.}(\bar{y}), \tag{10.12}$$

where $c_{\alpha/2}$ is the value cutting off the proportion $\alpha/2$ in the upper tail of the standard normal curve.

In general, one hopes to find that the value of χ^2_{homog} is small, so that homogeneous association may be inferred, and that the value of χ^2_{assoc} is large, so that real overall association may be inferred.

The issue of whether a test of the hypothesis of homogeneous association should ever be performed has been debated. Bishop, Fienberg, and Holland (1975, p. 147), for example, state that homogeneity of association must always be verified before inferences are made about a purportedly common degree of association. Mantel, Brown, and Byar (1977), on the other hand, suggest caution in interpreting the results of such tests; they point out that the presence or absence of homogeneous association is strongly dependent on which measure of association is used.

In practice, it would seem prudent before proceeding too far with the analysis both to inspect the data to confirm that the g measures are at least pointing to association in the same direction even if not of exactly the same magnitude, and to test χ^2_{homog} at a conservative level of significance (say .01) to confirm that the several measures of association are not widely divergent and that subsequent inferences about a supposedly common measure actually tend to apply to the individual groups.

What remains, then, is to apply these results to particular choices of the measure of association. The following notation will be used consistently in this chapter. In the ith group, n_{i1} is the number of observations in the first sample and p_{i1} is the proportion of the first sample having the studied characteristic. The quantity n_{i2} is the number of observations in the second sample, and p_{i2} is the proportion of the second sample having the studied characteristic. The total number of observations in the ith group is denoted by $n_{i.} = n_{i1} + n_{i2}$, and the overall proportion in the ith group having the characteristic is denoted by

$$\bar{p}_i = \frac{n_{i1}p_{i1} + n_{i2}p_{i2}}{n_{i.}}. \tag{10.13}$$

The complementary proportion is $\bar{q}_i = 1 - \bar{p}_i$.

10.2. COMBINING THE LOGARITHMS OF ODDS RATIOS

The odds ratio itself,

$$o_i = \frac{p_{i1}(1-p_{i2})}{p_{i2}(1-p_{i1})}, \tag{10.14}$$

does not have the property that a value of zero indicates no association. The logarithm of the odds ratio does have this property. Thus consider taking as the measure of association

$$y_i = L_i = \ln(o_i). \tag{10.15}$$

The squared standard error of L_i is approximately

$$[\text{s.e.}(L_i)]^2 = \frac{1}{w_i} = \frac{1}{n_{i1}p_{i1}(1-p_{i1})} + \frac{1}{n_{i2}p_{i2}(1-p_{i2})}, \tag{10.16}$$

which is equal to the sum of the reciprocals of the frequencies within the cells. The weight w_i is then the reciprocal of this sum of reciprocals.

The chi square analyses of Section 10.1 have been applied to the logarithm of the odds ratio by Gart (1962) and Sheehe (1966). They will now be illustrated with the data of Table 10.1, which presents the proportions of patients diagnosed as schizophrenic by resident hospital psychiatrists in New York and London (see Cooper, et al., 1972).

Table 10.1. Diagnoses of schizophrenia by resident hospital psychiatrists in three studies in New York and London

	New York		London	
Study	n_{i1}	p_{i1}	n_{i2}	p_{i2}
$i=1$ (ages 20–34)	105	.771	105	.324
$i=2$ (ages 20–59)	192	.615	174	.397
$i=3$ (ages 35–59)	145	.566	145	.359

Table 10.2 outlines the arithmetic required to perform the analysis of the log odds ratios. To reduce bias (Naylor, 1967; Gart, 1970, 1971), the constant .5 was added to each frequency as in (5.20) and (5.33) in calculating $L_i' = \ln(o_i')$ and $w_i' = 1/[\text{s.e.}(L_i')]^2$. The values of the individual chi squares $[= w_i'(L_i')^2]$ are all approximately equal to the values given by the standard 1 degree of freedom chi squares incorporating the continuity correction and are all highly significant.

Table 10.2. Analysis of logarithms of odds ratios applied to data of Table 10.1

Study	o_i'	L_i'	w_i'	$w_i'L_i'$	$w_i'(L_i')^2$
1	6.894	1.931	10.410	20.102	38.816
2	2.415	0.881	21.868	19.266	16.973
3	2.314	0.839	17.357	14.563	12.218
Total			49.635	53.931	68.007

The value of the total chi square is

$$\chi^2_{\text{total}} = \sum_{i=1}^{3} w_i'(L_i')^2 = 68.01, \tag{10.17}$$

and the value of the chi square statistic for testing the homogeneity of the odds ratios is

$$\begin{aligned}\chi^2_{\text{homog}} &= \sum_{i=1}^{3} w_i'(L_i')^2 - \frac{\left(\sum_{i=1}^{3} w_i'L_i'\right)^2}{\sum_{i=1}^{3} w_i'} \\ &= 68.01 - \frac{(53.931)^2}{49.635} \\ &= 9.41 \end{aligned} \tag{10.18}$$

with 2 degrees of freedom, indicating the existence of significant differences at the .01 level among the three odds ratios. Problem 10.1 is devoted to a detailed analysis of the heterogeneity of the odds ratios in Table 10.2.

While not equal, the three odds ratios are at least all in the same direction. Further analysis of all the data, in terms of a hypothetical common underlying odds ratio, might be justified if the conclusions are understood to apply to hospital diagnoses made on psychiatric patients in New York and London in general, not to diagnoses made on patients of any specific age group. Problem 10.2 carries the following kind of analysis forward on the data from studies $i=2$ and $i=3$ only, where the patients tended to be older and the odds ratios were similar.

The estimate of the logarithm of the supposedly common odds ratio is

$$\bar{L}' = \frac{\sum_{i=1}^{3} w_i' L_i'}{\sum_{i=1}^{3} w_i'} = \frac{53.931}{49.635} = 1.087, \tag{10.19}$$

with an estimated standard error of

$$\text{s.e.}(\bar{L}') = \frac{1}{\sqrt{\sum_{i=1}^{3} w_i'}} = \frac{1}{\sqrt{49.635}} = 0.142. \tag{10.20}$$

The value of chi square for testing the significance of the mean log odds ratio is then

$$\chi^2_{\text{assoc}} = \left[\frac{\bar{L}'}{\text{s.e.}(\bar{L}')}\right]^2 = \left(\frac{1.087}{.142}\right)^2 = 58.60 \tag{10.21}$$

with 1 degree of freedom, which is obviously highly significant. The inference may therefore be drawn that the odds that a mental patient hospitalized in New York will be diagnosed schizophrenic by a hospital psychiatrist are significantly greater than the corresponding odds for a mental patient hospitalized in London.

An approximate 95% confidence interval for λ, the logarithm of the supposed common odds ratio, is

$$\bar{L}' - 1.96\,\text{s.e.}(\bar{L}') \leqq \lambda \leqq \bar{L}' + 1.96\,\text{s.e.}(\bar{L}'),$$

$$1.087 - 1.96 \times .142 \leqq \lambda \leqq 1.087 + 1.96 \times .142,$$

or

$$0.809 \leqq \lambda \leqq 1.365. \tag{10.22}$$

It is usually desirable to report the final results in terms of the odds ratio itself rather than in terms of its logarithm. The mean odds ratio is estimated by

$$\bar{o}' = e^{\bar{L}'} = \text{antilog}(\bar{L}'), \tag{10.23}$$

and an approximate 95% confidence interval for ω, the supposed common odds ratio, is given by

$$\text{antilog}\left[\bar{L}' - 1.96\,\text{s.e.}(\bar{L}')\right] \leqq \omega \leqq \text{antilog}\left[\bar{L}' + 1.96\,\text{s.e.}(\bar{L}')\right]. \tag{10.24}$$

For the data at hand,

$$\bar{o}' = \text{antilog}(1.087) = 2.97, \tag{10.25}$$

and an approximate 95% confidence interval for ω is, from (10.22),

$$\text{antilog}(.809) \leqq \omega \leqq \text{antilog}(1.365),$$

or

$$2.25 < \omega < 3.92. \tag{10.26}$$

10.3. METHOD DUE TO CORNFIELD AND GART

An outline of the theory to be applied here was presented in Section 5.5. The theory of Section 10.1 does not apply in this case. Let the data from group i be reexpressed as in Table 10.3. As proven by Cornfield (1956), if all four marginal frequencies are held fixed and if ω is the underlying odds

Table 10.3. Notation for data from group i

	Outcome Variable		
Sample	Present	Absent	Total
1	X_i	$n_{i1} - X_i$	n_{i1}
2	$m_i - X_i$	$n_{i2} - m_i + X_i$	n_{i2}
Total	m_i	$n_{i.} - m_i$	$n_{i.}$

ratio, then X_i is approximately normally distributed with mean x_i and standard error

$$\text{s.e.}(X_i) = \frac{1}{\sqrt{W_i(x_i)}}. \qquad (10.27)$$

In (10.27),

$$W_i(x_i) = \frac{1}{x_i} + \frac{1}{n_{i1} - x_i} + \frac{1}{m_i - x_i} + \frac{1}{n_{i2} - m_i + x_i} \qquad (10.28)$$

and x_i is the unique root in the admissible interval

$$\text{larger of } (0, m_i - n_{i2}) \leqq x_i \leqq \text{ smaller of } (n_{i1}, m_i) \qquad (10.29)$$

of the quadratic equation

$$\frac{x_i(n_{i2} - m_i + x_i)}{(n_{i1} - x_i)(m_i - x_i)} = \omega. \qquad (10.30)$$

Explicitly, the quadratic equation is

$$x_i^2(\omega - 1) - x_i[\omega(n_{i1} + m_i) + n_{i2} - m_i] + \omega n_{i1} m_i = 0. \qquad (10.31)$$

Gart (1970) has taken these results and extended them to the case of several fourfold tables. An examination of the data for heterogeneous odds ratios begins by estimating the hypothesized common odds ratio, say $\hat{\omega}$. The appropriate estimate, which cannot be obtained by means of an explicit equation, is such that, when $\hat{x}_i$ is the admissible root of

$$\frac{\hat{x}_i(n_{i2} - m_i + \hat{x}_i)}{(n_{i1} - \hat{x}_i)(m_i - \hat{x}_i)} = \hat{\omega} \qquad (10.32)$$

$(i = 1, \ldots, g)$, then

$$\sum_{i=1}^{g} X_i = \sum_{i=1}^{g} \hat{x}_i. \qquad (10.33)$$

The estimate $\hat{\omega}$ may be found either by trial and error or by any one of the standard iteration methods available for the solution of complicated equations; either $\bar{o}'$ from (10.23) or the Mantel-Haenszel estimate (see Section 10.4) can serve as the initial approximation to $\hat{\omega}$.

For the data of Table 10.1, the desired estimate is

$$\hat{\omega} = 3.04, \tag{10.34}$$

which is only slightly larger than the estimate based on the log odds ratio given in (10.25). Table 10.4 presents the associated values of $\hat{x}_i$ and $W_i(\hat{x}_i)$. Note that $\sum X_i = \sum \hat{x}_i = 281$.

Table 10.4. Values associated with $\hat{\omega} = 3.04$ for data of Table 10.1

Study	n_{i1}	n_{i2}	m_i	X_i	$\hat{x}_i$	$W_i(\hat{x}_i)$
1	105	105	115	81	71.601	.0832
2	192	174	187	118	122.856	.0473
3	145	145	134	82	86.543	.0600
Total				281	281.000	

The hypothesis that the underlying odds ratios are equal may be tested by referring the quantity

$$\chi^2_{\text{homog}} = \sum_{i=1}^{g} W_i(\hat{x}_i)(X_i - \hat{x}_i)^2 \tag{10.35}$$

to the chi square distribution with $g - 1$ degrees of freedom. For the data of Table 10.4,

$$\chi^2_{\text{homog}} = 9.70 \tag{10.36}$$

with 2 degrees of freedom, indicating statistically significant differences ($p < .01$) among the odds ratios for the three studies of Table 10.1. The chi square value in (10.36) happens to be slightly larger than the corresponding value in (10.18) based on the log odds ratios.

A test for the significance of the overall odds ratio is performed as follows. If ω, the underlying supposed common odds ratio, is equal to unity, (10.31) becomes linear, and its unique solution is

$$x_i = \frac{n_{i1}m_i}{n_{i.}}. \tag{10.37}$$

The corresponding value of $W_i(x_i)$ is

$$W_i(x_i) = \frac{n_{i.}^3}{n_{i1}n_{i2}m_i(n_{i.} - m_i)}. \tag{10.38}$$

Under the hypothesis that $\omega = 1$, the quantity

$$\chi^2_{\text{assoc}} = \frac{\left(\left| \sum_{i=1}^{g} X_i - \sum_{i=1}^{g} x_i \right| - .5 \right)^2}{\sum_{i=1}^{g} \frac{1}{W_i(x_i)}} \tag{10.39}$$

has, approximately, a chi square distribution with 1 degree of freedom. As pointed out in Section 10.4, this statistic is closely related to the Mantel-Haenszel chi square statistic.

The values of x_i and $W_i(x_i)$ associated with the hypothesis that $\omega = 1$ are presented in Table 10.5. The value of the statistic in (10.39) for testing

Table 10.5. Values associated with hypothesis that $\omega = 1$ for data of Table 10.1

Study	n_{i1}	n_{i2}	m_i	X_i	x_i	$W_i(x_i)$	$1/W_i(x_i)$
1	105	105	115	81	57.500	.0769	13.006
2	192	174	187	118	98.098	.0438	22.809
3	145	145	134	82	67.000	.0555	18.021
Total				281	222.598		53.836

whether the overall degree of association is significant is

$$\chi^2_{\text{assoc}} = \frac{(|281 - 222.598| - .5)^2}{53.836} = 62.28. \tag{10.40}$$

The inference may therefore be drawn that the supposed common value of the underlying odds ratio is different from unity. The value of chi square in (10.40) happens to be somewhat larger than the value of the corresponding chi square statistic in (10.21) based on the log odds ratio.

An approximate $100(1-\alpha)\%$ confidence interval for the supposed common underlying odds ratio is determined as follows. The lower confidence limit, say ω_L, is such that, if x_{iL} is the admissible root of

$$\frac{x_{iL}(n_{i2} - m_i + x_{iL})}{(n_{i1} - x_{iL})(m_i - x_{iL})} = \omega_L \tag{10.41}$$

and if

$$W_i(x_{iL}) = \frac{1}{x_{iL}} + \frac{1}{n_{i1} - x_{iL}} + \frac{1}{m_i - x_{iL}} + \frac{1}{n_{i2} - m_i + x_{iL}}, \tag{10.42}$$

then

$$\frac{\left[\left(\sum_{i=1}^{g} X_i - \sum_{i=1}^{g} x_{iL}\right) - .5\right]^2}{\sum_{i=1}^{g} \frac{1}{W_i(x_{iL})}} = c_{\alpha/2}^2. \tag{10.43}$$

The upper limit, say ω_U, is found similarly, except that the continuity correction in (10.43) is taken as $+.5$ instead of $-.5$.

As with the estimation of the common odds ratio described earlier in this section, the upper and lower limits must be found either by trial and error or by means of a formal iterative procedure. The limits based on the log odds ratio, given in (10.24), may be used as first approximations.

Table 10.6 presents values associated with the lower 95% confidence limit, $\omega_L = 2.28$. Note that

$$\frac{[(281 - 266.421) - .5]^2}{51.606} = 3.84, \tag{10.44}$$

the value required for a confidence coefficient of 95%.

Table 10.6. Values associated with lower 95% confidence limit, $\omega_L = 2.28$, for data of Table 10.1

Study	n_{i1}	n_{i2}	m_i	X_i	x_{iL}	$W_i(x_{iL})$	$1/W_i(x_{iL})$
1	105	105	115	81	68.083	.0803	12.453
2	192	174	187	118	116.672	.0457	21.882
3	145	145	134	82	81.666	.0579	17.271
Total				281	266.421		51.606

Table 10.7 presents values associated with the upper 95% confidence limit, $\omega_U = 4.06$. Note that

$$\frac{[(281 - 295.033) + .5]^2}{47.686} = 3.84, \tag{10.45}$$

as required.

Table 10.7. Values associated with upper 95% confidence limit, $\omega_U = 4.06$, for data of Table 10.1

Study	n_{i1}	n_{i2}	m_i	X_i	x_{iU}	$W_i(x_{iU})$	$1/W_i(x_{iU})$
1	105	105	115	81	74.985	.0870	11.494
2	192	174	187	118	128.811	.0494	20.243
3	145	145	134	82	91.237	.0627	15.949
Total				281	295.033		47.686

Using Cornfield's results, then, the approximate 95% confidence interval for the supposed common odds ratio is

$$2.28 \leqq \omega \leqq 4.06. \tag{10.46}$$

This interval is shifted slightly to the right of, and is slightly wider than, the interval based on the log odds ratio given in (10.26).

10.4. THE MANTEL-HAENSZEL METHOD

A procedure due to Mantel and Haenszel (1959), and extended by Mantel (1963), permits one to estimate the assumed common odds ratio and to test whether the overall degree of association is significant. Curiously, it is not the odds ratio but another measure of association that directly underlies the test for association; Radhakrishna (1965) has shown that such an approach is valid.

The Mantel-Haenszel summary estimate of the odds ratio is, say,

$$\bar{o}_{\text{MH}} = \frac{\sum_{i=1}^{g} \frac{n_{i1}n_{i2}}{n_{i.}} p_{i1}(1-p_{i2})}{\sum_{i=1}^{g} \frac{n_{i1}n_{i2}}{n_{i.}} p_{i2}(1-p_{i1})}; \tag{10.47}$$

$\bar{o}_{\text{MH}}$ is a weighted average of the separate odds ratios from the g groups (see Problem 10.3). For the data of Table 10.1,

$$\bar{o}_{\text{MH}} = \frac{87.516}{29.143} = 3.00, \tag{10.48}$$

which happens to be slightly greater than the estimate given in (10.25) based on the log odds ratio, and slightly smaller than the estimate given in (10.34) based on the Cornfield-Gart approach.

The Mantel-Haenszel chi square test for the significance of the overall degree of association is based on a weighted average of the g differences between proportions, say

$$\bar{d} = \sum_{i=1}^{g} \frac{n_{i1}n_{i2}}{n_{i.}}(p_{i1} - p_{i2}) \Big/ \sum_{i=1}^{g} \frac{n_{i1}n_{i2}}{n_{i.}}. \tag{10.49}$$

The Mantel-Haenszel chi square statistic is given by

$$\chi^2_{\text{MH}} = \frac{\left(\left|\sum_{i=1}^{g} \frac{n_{i1}n_{i2}}{n_{i.}}(p_{i1} - p_{i2})\right| - .5\right)^2}{\sum_{i=1}^{g} \frac{n_{i1}n_{i2}}{n_{i.} - 1}\bar{p}_i\bar{q}_i}, \tag{10.50}$$

with 1 degree of freedom. For the data of Table 10.1, its value is

$$\chi^2_{\text{MH}} = \frac{(|58.374| - .5)^2}{54.024} = 62.00, \tag{10.51}$$

which is slightly smaller than the value given in (10.40) for the chi square statistic for association based on the Cornfield-Gart approach. In fact, if $n_{i.} - 1$ is replaced by $n_{i.}$ in the denominator of (10.50), the two chi square statistics would be identical.

Closely related to the Mantel-Haenszel chi square statistic is one due to Cochran (1954), say

$$\chi^2_C = \frac{\left(\sum_{i=1}^{g} \frac{n_{i1}n_{i2}}{n_{i.}}(p_{i1} - p_{i2})\right)^2}{\sum_{i=1}^{g} \frac{n_{i1}n_{i2}}{n_{i.}}\bar{p}_i\bar{q}_i}. \tag{10.52}$$

The statistics in (10.50) and (10.52) differ not only in the former's inclusion of the continuity correction but, more seriously, in the former's taking $n_{i.} - 1$ rather than $n_{i.}$ in the denominator. When the sample sizes in the g groups are all large, the difference is trivial. When, however, the sample sizes in the g groups are small, the difference becomes substantial.

Consider, for example, the extreme case in which, as in the study of matched pairs, each group consists of two individuals, one from each sample. It is easy to check that McNemar's chi square statistic, given in (8.3), is identical to the Mantel-Haenszel chi square statistic given in (10.50). Cochran's statistic in (10.52) with the continuity correction, on the other hand, would yield a value twice as large as McNemar's statistic.

Mantel (1966) has indicated how the Mantel-Haenszel chi square test may be used in comparing independent life tables. Mantel (1977) has also described how the Mantel-Haenszel chi square statistic can be modified to provide approximate confidence limits for the supposed common odds ratio. The details of these procedures are too complicated for description here. The reader is referred to Mantel (1966, 1977) and to Mantel and Hankey (1975) for details.

The classic criterion for whether the sample sizes in a fourfold table are large enough to warrant referring the value of the standard chi square statistic to tables of the chi square distribution with 1 degree of freedom is that each expected cell frequency be at least equal to five (see Section 2.2). A similar criterion has been proposed for the Mantel-Haenszel chi square statistic by Mantel and Fleiss (1980). It is that each of the four sums of expected cell frequencies,

$$\sum_{i=1}^{g} n_{i1}\bar{p}_i, \qquad \sum_{i=1}^{g} n_{i2}\bar{p}_i, \qquad \sum_{i=1}^{g} n_{i1}\bar{q}_i, \qquad \sum_{i=1}^{g} n_{i2}\bar{q}_i,$$

differ by five or more from its maximum and from its minimum.

It is therefore not necessary that each table have large marginal frequencies in order for the statistic in (10.50) to be safely referred to the chi square distribution with 1 degree of freedom; in fact, as in the case of matched pairs, the total frequency in each table can be as small as two. All that is required is that there be sufficiently many tables so that each sum of cell expectations is large.

10.5. A COMPARISON OF THE THREE PROCEDURES

Gart (1962, 1970), Odoroff (1970), and McKinlay (1975b, 1978) have compared the procedures described in the three preceding sections as well as procedures due to Birch (1964) and Goodman (1969). Two cases must be distinguished.

In one case, the number of groups or strata is small or moderate, and the sample sizes within each are large. This would be the case if the samples being compared were stratified into a limited number of strata, with additional subjects being assigned to existing strata, or if a limited number of replicate studies were being analyzed. For this case, procedures based on the logarithms of the odds ratios either perform better than or only slightly poorer than their competitors. Given their fair to good precision and accuracy, together with their relative simplicity, the methods of Section 10.2 are recommended for addressing, in a unified manner, all the major inference problems for the odds ratios in each of a small number of strata.

In the second case, each group or stratum is of small or moderate size, but the number of groups or strata is large. This would be the case if the samples being compared were stratified (usually after the data were collected) on several dimensions, or if matching were employed and the number of matched individuals possibly varied (e.g., some matched sets consisting of a pair of individuals, others of a single member from one sample and two members from the other, etc.), or, in general, if the recruitment of additional subjects meant the creation of additional groups or strata.

In this case, the Mantel-Haenszel estimate of the overall odds ratio [see (10.47)], the Mantel-Haenszel 1 degree of freedom chi square statistic for testing its significance [see (10.50)], and the Cornfield-Gart confidence interval for the overall odds ratio [described in (10.41)–(10.45)] are the methods of choice. Testing for the equality of the several odds ratios is less important in this case. In contrast to the case of few groups each of large size, procedures based on the log odds ratio perform terribly in the case of many groups each of small size.

10.6. ALTERNATIVES TO MATCHING

McKinlay (1975a) has conducted a historical review of methods that have been used to control for biasing factors in nonrandomized studies such as comparative prospective and comparative retrospective studies, and has also reviewed statistical studies of these methods. Her bibliography consists of 165 items, and Fienberg adds others in his Comment on her paper. There are three relatively simple methods available for the control of biasing factors: matching (for which the analytic procedures described in Chapter 8 are appropriate), stratification (for which the procedures described in this chapter are appropriate), and covariance or regression control (not discussed here; see Rubin, 1973).

Suppose, for example, that a retrospective study is contemplated of the association between cigarette smoking and lung cancer, with control for the possible confounding effects of age and sex. One method of control is to pair each lung cancer case with one or more controls of the same sex and of a similar age and to apply the methods of Section 8.1 or 8.3.

Another method of control is to draw a cross-sectional sample of cases and a cross-sectional sample of controls, to stratify the two samples by sex and by age intervals, and then, separately for each resulting stratum, to set up a fourfold table contrasting the rates of smoking for the cases and controls. If there are, say, five age intervals, the total number of fourfold tables is $g = 10$: five for the males plus five for the females. The resulting set of tables may be viewed as coming from g distinct groups, and the methods of Sections 10.2 to 10.4 may be applied.

If only a small number (two or three) of biasing factors out of several are actually controlled, the possible effects of those factors not controlled (and perhaps not even measured) may be assessed by criteria suggested by Bross (1966) and Schlesselman (1978). If simultaneous control over several biasing factors (more than three) is desired, the composite "multivariate confounder score" of Miettinen (1976) may be used as the basis for stratification; Miettinen suggests five as a reasonable number of strata. A multivariate procedure such as discriminant function analysis must be applied first, however, in order to determine how the composite score is to be calculated.

Matching has the advantage of assuring that the two samples are comparable on the factors used for matching, but has as a major disadvantage the practical difficulty of finding a matched control for each case if the number of cases is large. Other disadvantages are cited in Section 8.5.

Stratifying the samples after they have been drawn has the advantage of not requiring a specification beforehand of the composition of the two samples, as well as the advantage of permitting an examination of the consistency of association across the various strata. A disadvantage is that, if the sample sizes are not large, the number of individuals in a stratum from one sample may be small compared to the number of individuals in it from the other sample. The power and precision of the comparisons may therefore suffer.

Research on the effectiveness of matching versus the effectiveness of stratification in controlling for confounding factors has been performed by Cochran (1968) and Rubin (1973) for quantitative measurements and by McKinlay (1975c) for dichotomous measurements. Based on their results, we may view matching as the method of choice only for moderate sample sizes and cross-sectional sampling followed by stratification as the method of choice for large sample sizes.

10.7. METHODS TO BE AVOIDED

A Test Described by Fleiss

In the first edition of this book, a test for homogeneity originally proposed by Yates (1955) was described. It calls for summing the values of the standard single degree of freedom chi squares for the individual groups (without the continuity correction) and for subtracting from this sum the value of Cochran's (1954) single degree of freedom summary chi square given in (10.52). The procedure is based on the following application of the theory of Section 10.1.

Let the measure of association in the ith group be the so-called *standardized difference*,

$$y_i = d_i = \frac{p_{i1} - p_{i2}}{\bar{p}_i \bar{q}_i}. \tag{10.53}$$

Its squared standard error is

$$[\text{s.e.}(d_i)]^2 = \frac{1}{\bar{p}_i \bar{q}_i}\left(\frac{n_{i.}}{n_{i1} n_{i2}}\right) \tag{10.54}$$

so that

$$w_i = \frac{\bar{p}_i \bar{q}_i n_{i1} n_{i2}}{n_{i.}} \tag{10.55}$$

and

$$\chi_i^2 = w_i d_i^2 = \frac{(p_{i1} - p_{i2})^2}{\bar{p}_i \bar{q}_i (1/n_{i1} + 1/n_{i2})}, \tag{10.56}$$

which is precisely the usual chi square value, aside from the continuity correction.

The mean standardized difference is

$$\bar{d} = \frac{\sum_{i=1}^{g} \frac{(p_{i1} - p_{i2}) n_{i1} n_{i2}}{n_{i.}}}{\sum_{i=1}^{g} \frac{\bar{p}_i \bar{q}_i n_{i1} n_{i2}}{n_{i.}}} \tag{10.57}$$

with a squared standard error of

$$\left[\text{s.e.}(\bar{d})\right]^2 = \frac{1}{\sum_{i=1}^{g} \frac{\bar{p}_i \bar{q}_i n_{i1} n_{i2}}{n_{i.}}}, \tag{10.58}$$

so the chi square statistic for testing overall association is

$$\chi^2_{\text{assoc}} = \frac{\bar{d}^2}{\left[\text{s.e.}(\bar{d})\right]^2} = \frac{\left(\sum_{i=1}^{g} \frac{(p_{i1} - p_{i2}) n_{i1} n_{i2}}{n_{i.}}\right)^2}{\sum_{i=1}^{g} \frac{\bar{p}_i \bar{q}_i n_{i1} n_{i2}}{n_{i.}}}, \tag{10.59}$$

which is identical to Cochran's χ_C^2 in (10.52).

The error in the first edition was in suggesting that

$$\chi^2_{\text{homog}} = \sum_{i=1}^{g} w_i d_i^2 - \chi_C^2, \tag{10.60}$$

with $g - 1$ degrees of freedom, always formed the basis of a valid test of homogeneity. Mantel, Brown, and Byar (1977) have shown that the standardized difference in (10.53), and therefore the test statistic in (10.60), is sensitive to the ratio of sample sizes as well as to the underlying degree of association, so that the test based on (10.60) may sometimes be invalid.

Consider the data in Table 10.8, taken from Mantel, Brown, and Byar (1977). It is easily checked that the odds ratio in both groups is equal to

Table 10.8. Data to illustrate previously suggested test for homogeneity

	Sample 1		Sample 2		Overall	
Group	n_{i1}	p_{i1}	n_{i2}	p_{i2}	$n_{i.}$	$\bar{p}_i$
$i=1$	230	.87	50	.20	280	.75
$i=2$	40	.25	810	.0123	850	.0235

26.7. Nevertheless, as shown in Table 10.9, the two standardized differences are markedly different, and the test based on (10.60) suggests,

Table 10.9. Summarization of data in Table 10.8

Group	d_i	w_i	$w_i d_i$	$w_i d_i^2$
$i=1$	3.57	7.70	27.489	98.136
$i=2$	10.36	0.87	9.013	93.375
Total		8.57	36.502	191.511

incorrectly, that association differs between the two groups:

$$\chi^2_{\text{homog}} = 191.511 - \frac{36.502^2}{8.57} = 36.04 \tag{10.61}$$

with $g-1=1$ degree of freedom, which is highly significant.

Mantel, Brown, and Byar (1977) illustrate other possible anomalies associated with the statistic given in (10.60) (e.g., the odds ratios may vary markedly across the g groups, but χ^2_{homog} might nevertheless equal zero). The problem with χ^2_{homog} is the dependence of the standardized differences, which this statistic compares, on the sample sizes n_{i1} and n_{i2}. The sample sizes affect the values of $\bar{p}_i$ and $\bar{q}_i$; therefore, as seen in (10.53), they affect the value of d_i. Because of the above difficulties, the test for homogeneity based on (10.60) should be avoided.

The Summation of Chi Procedure

One of the more frequently employed methods for combining data from different fourfold tables is of the form outlined in Section 10.1, although not obviously so. The method, usually referred to as the summation of chi procedure, has long been known to have serious defects but keeps reappearing nevertheless (see, e.g., Finney, 1965). The method in effect takes as the measure of association

$$y_i = z_i = \frac{p_{i1} - p_{i2}}{\sqrt{\bar{p}_i \bar{q}_i (1/n_{i1} + 1/n_{i2})}}. \tag{10.62}$$

Because z_i has been standardized to have a standard error of unity, therefore

$$w_i = \frac{1}{[\text{s.e.}(z_i)]^2} = 1. \tag{10.63}$$

The word "chi" in the name of the procedure derives from z_i's being the square root of a chi square variate (without the correction for continuity), and hence being a *chi* variate.

When y_i is defined by (10.62),

$$\bar{y} = \frac{\sum_{i=1}^{g} z_i}{g} = \bar{z} \tag{10.64}$$

and

$$\sum_{i=1}^{g} w_i = g, \tag{10.65}$$

by (10.63). Therefore, by (10.9),

$$\chi^2_{\text{assoc}} = \frac{\left(\sum_{i=1}^{g} z_i\right)^2}{g} = g\bar{z}^2. \tag{10.66}$$

There is a serious flaw inherent in this chi square statistic (see, e.g., Pasternack and Mantel, 1966). Consider the numerical example of Table 10.10.

Table 10.10. Data to illustrate summation of chi procedure

	Sample 1		Sample 2		Overall		
Group	n_{i1}	p_{i1}	n_{i2}	p_{i2}	$n_{i.}$	$\bar{p}_i$	z_i
$i=1$	100	.60	100	.40	200	.50	2.83
$i=2$	1000	.60	1000	.40	2000	.50	8.94

For group 1,

$$\chi^2_1 = z^2_1 = 8.00; \tag{10.67}$$

for group 2,

$$\chi^2_2 = z^2_2 = 80.00. \tag{10.68}$$

The average value of z is

$$\bar{z} = \tfrac{1}{2}(2.83 + 8.94) = 5.88, \tag{10.69}$$

so that, by (10.66),

$$\chi^2_{\text{assoc}} = 2 \times (5.88)^2 = 69.15 \tag{10.70}$$

with 1 degree of freedom. What is disquieting about this value for the overall test of association is that it is less than the value for one of the individual chi squares for association, $\chi_2^2 = 80$. The addition of the evidence from group 1, in which the association was really the same as in group 2, would be expected to increase the statistical significance of the association. The summation of chi procedure failed to do so.

Any procedure for which an accumulation of evidence for association may lead to a reduction in chi square is to be avoided. So be it with the summation of chi procedure.

Summation Observed Versus Summation Expected

A relative lack of sensitivity to added evidence for association in a given direction characterizes the following method, too. Although it can be cast into the terms of the theory of Section 10.1, it would not be an aid to understanding.

The method calls first for generating a total fourfold table by summing the frequencies across the g individual tables. Let the observed frequencies for $g = 2$ groups be as in Table 10.11. Chi square is calculated without the continuity correction. The association between A and B, measured by an odds ratio of 26.7, is the same in both groups.

Table 10.11. Data to illustrate procedure based on summation observed versus summation expected

	Group 1				Group 2		
	B	$\bar{B}$	Total		B	$\bar{B}$	Total
A	200	30	230	A	40	120	160
$\bar{A}$	10	40	50	$\bar{A}$	10	800	810
Total	210	70	280	Total	50	920	970
		$\chi_1^2 = 98.20$				$\chi_2^2 = 154.35$	

The table of total frequencies is as shown in Table 10.12.

Table 10.12. Sum of frequencies for groups 1 and 2

	B	$\bar{B}$	Total
A	240	150	390
$\bar{A}$	20	840	860
Total	260	990	1250

The next step is to determine, for each group, the set of frequencies expected under the hypothesis of no association. The expected frequency for a cell is calculated as the product of the total frequencies in a cell's row and column divided by the overall frequency in the table. Thus the expected frequency in the (A, B) cell for group 2 is $160 \times 50/970 = 8.25$. All expected frequencies are shown in Table 10.13.

Table 10.13. Expected frequencies for groups 1 and 2

	Group 1				Group 2		
	B	$\bar{B}$	Total		B	$\bar{B}$	Total
A	172.5	57.5	230	A	8.25	151.75	160
$\bar{A}$	37.5	12.5	50	$\bar{A}$	41.75	768.25	810
Total	210	70	280	Total	50	920	970

Next, generate an overall table of expected frequencies by summing, as in Table 10.14, across the individual tables just determined.

Table 10.14. Sum of expected frequencies for groups 1 and 2

	B	$\bar{B}$	Total
A	180.75	209.25	390
$\bar{A}$	79.25	780.75	860
Total	260	990	1250

Finally, calculate the summary chi square for association by taking, for each cell, the difference between the total observed and total expected frequencies, squaring, dividing by the total expected frequency, and summing across all four cells. Thus from Tables 10.12 and 10.14,

$$\chi^2_{\text{assoc}} = \frac{(240 - 180.75)^2}{180.75} + \frac{(150 - 209.25)^2}{209.25} + \frac{(20 - 79.25)^2}{79.25} + \frac{(840 - 780.75)^2}{780.75} = 84.99. \quad (10.71)$$

This value for chi square is less than either of the two original chi square values shown in Table 10.11. The procedure based on comparing the sums of observed frequencies with the sums of expected frequencies therefore suffers from the same deficiency as the summation of chi procedure and is to be avoided for the same reason.

Chi Square on the Table of Totals

A defect opposite in nature to that of the two preceding methods characterizes the following procedure for testing overall association. The method cannot be described in terms of the theory of Section 10.1. It calls merely for generating the table of total observed frequencies as described in the preceding section and then calculating a straightforward chi square on it.

The method works quite well on the data of the two previous sections, and in general for groups in which corresponding proportions are nearly equal. Such a state of affairs is exceptional, however. Consider the data of Table 10.15. No association between A and B exists in either group, although the basic rates are different in the two groups.

Table 10.15. Association between A and B in two groups

	Group 1				Group 2		
	B	$\bar{B}$	Total		B	$\bar{B}$	Total
A	10	40	50	A	60	40	100
$\bar{A}$	20	80	100	$\bar{A}$	30	20	50
Total	30	120	150	Total	90	60	150
		$\chi_1^2 = 0$				$\chi_2^2 = 0$	

The table of total frequencies is as shown in Table 10.16. Its associated chi square is 5.01, indicating an association significant at the .05 level. The

Table 10.16. Sum of frequencies for groups 1 and 2

	B	$\bar{B}$	Total
A	70	80	150
$\bar{A}$	50	100	150
Total	120	180	300

combination of data from tables with unequal proportions and with unequal ratios of sample sizes, n_{i1}/n_{i2}, has created the impression of association where none basically existed.

The procedures just described should be avoided for the reasons indicated (see also Gart, 1962, and Sheehe, 1966). This necessarily means that the calculations become more complicated, as was seen in Sections 10.2 to 10.4, but this is the price one must pay for a valid analysis.

Problem 10.1. It was found in Section 10.2 that the odds ratios in the three studies summarized in Table 10.1 were significantly different.

(*a*) The odds ratios o_2' and o_3' appear to be similar (see Table 10.2). Test whether they differ significantly, basing the test on the value of

$$\chi^2_{2\,\mathrm{vs}\,3} = \frac{w_2' w_3'}{w_2' + w_3'}(L_2' - L_3')^2.$$

(*b*) The odds ratios o_2' and o_3' differ by less than either does from o_1'. Test whether the mean of o_2' and o_3' differs significantly from o_1'. (*Hint.* The mean of L_2' and L_3' is $\bar{L}_{2,3} = (w_2' L_2' + w_3' L_3')/(w_2' + w_3')$. Refer the value of

$$\chi^2_{1\,\mathrm{vs}\,(2,3)} = \frac{w_1'(w_2' + w_3')}{w_1' + w_2' + w_3'}\left(L_1' - \bar{L}_{2,3}\right)^2$$

to critical values of chi square with 2 degrees of freedom rather than 1 because the comparison was suggested by the data.)

(*c*) How does the sum of the chi square statistics determined in (*a*) and (*b*) compare with the value of χ^2_{homog} found in (10.18)?

Problem 10.2. Apply the methods of Section 10.2 to the data of groups 2 and 3 only in Tables 10.1 and 10.2. Specifically,

(*a*) What is the mean log odds ratio? What is its standard error? Is the mean log odds ratio significantly different from zero?

(*b*) Find an approximate 95% confidence interval for the underlying log odds ratio.

(*c*) What is the mean odds ratio? What is the approximate 95% confidence interval for the underlying odds ratio corresponding to the interval found in (*b*)?

Problem 10.3. While it perhaps is not obvious, $\bar{o}_{\mathrm{MH}}$ (see 10.47) is actually a weighted average of the g individual odds ratios,

$$o_i = \frac{p_{i1}(1 - p_{i2})}{p_{i2}(1 - p_{i1})}, \qquad i = 1, \ldots, g.$$

Show that this is so by finding a set of weights, $w_1, \ldots, w_g$, so that, with

$\bar{o}_{\text{MH}}$ given by (10.47),

$$\bar{o}_{\text{MH}} = \frac{\sum_{i=1}^{g} o_i w_i}{\sum_{i=1}^{g} w_i}.$$

Problem 10.4. Prove that, when $n_{i1} = n_{i2} = 1$, as in the study of matched pairs, the Mantel-Haenszel chi square statistic given in (10.50) is identical to McNemar's chi square statistic given in (8.3).

REFERENCES

Birch, M. W. (1964). The detection of partial association. I: The 2 × 2 case. *J. R. Stat. Soc., Ser. B*, **26**, 313–324.

Bishop, Y. M. M., Fienberg, S. E., and Holland, P. W. (1975). *Discrete multivariate analysis: Theory and practice*. Cambridge, Mass.: M.I.T. Press.

Bross, I. D. J. (1966). Spurious effects from an extraneous variable. *J. Chronic Dis.*, **19**, 637–647.

Cochran, W. G. (1954). Some methods of strengthening the common χ^2 tests. *Biometrics*, **10**, 417–451.

Cochran, W. G. (1968). The effectiveness of adjustment by subclassification in removing bias in observational studies. *Biometrics*, **24**, 295–313.

Cooper, J. E., Kendell, R. E., Gurland, B. J., Sharpe, L., Copeland, J. R. M., and Simon, R. (1972). *Psychiatric diagnosis in New York and London*. London: Oxford University Press.

Cornfield, J. (1956). A statistical problem arising from retrospective studies. Pp. 135–148 in J. Neyman (Ed.). *Proceedings of the third Berkeley symposium on mathematical statistics and probability*, Vol. 4. Berkeley: University of California Press.

Cox, D. R. (1970). *The analysis of binary data*. London: Methuen.

Everitt, B. S. (1977). *The analysis of contingency tables*. New York: Halsted Press.

Fienberg, S. E. (1977). *The analysis of cross-classified categorical data*. Cambridge, Mass.: M.I.T. Press.

Finney, D. J. (1965). The design and logic of a monitor of drug use. *J. Chronic Dis.*, **18**, 77–98.

Gart, J. J. (1962). On the combination of relative risks. *Biometrics*, **18**, 601–610.

Gart, J. J. (1970). Point and interval estimation of the common odds ratio in the combination of 2 × 2 tables with fixed marginals. *Biometrika*, **57**, 471–475.

Gart, J. J. (1971). The comparison of proportions: A review of significance tests, confidence intervals and adjustments for stratification. *Rev. Int. Stat. Inst.*, **39**, 16–37.

Goodman, L. A. (1969). On partitioning χ^2 and detecting partial association in three-way contingency tables. *J. R. Stat. Soc., Ser. B*, **31**, 486–498.

McKinlay, S. M. (1975a). The design and analysis of the observational study—A review. (With a Comment by S. E. Fienberg). *J. Am. Stat. Assoc.*, **70**, 503–523.

McKinlay, S. M. (1975b). The effect of bias on estimators of relative risk for pair-matched and stratified samples. *J. Am. Stat. Assoc.*, **70**, 859–864.

McKinlay, S. M. (1975c). A note on the chi-square test for pair-matched samples. *Biometrics*, **31**, 731–735.

McKinlay, S. M. (1978). The effect of non-zero second order interaction on combined estimators of the odds-ratio. *Biometrika*, **65**, 191–202.

Mantel, N. (1963). Chi-square tests with one degree of freedom: Extensions of the Mantel-Haenszel procedure. *J. Am. Stat. Assoc.*, **58**, 690–700.

Mantel, N. (1966). Evaluation of survival data and two new rank order statistics arising in its consideration. *Cancer Chemother. Rep.*, **50**, 163–170.

Mantel, N. (1977). Tests and limits for the common odds ratio of several 2 × 2 contingency tables: Methods in analogy with the Mantel-Haenszel procedure. *J. Stat. Plann. Inf.*, **1**, 179–189.

Mantel, N., Brown, C., and Byar, D. P. (1977). Tests for homogeneity of effect in an epidemiologic investigation. *Am. J. Epidemiol.*, **106**, 125–129.

Mantel, N. and Fleiss, J. L. (1980). Minimum expected cell size requirements for the Mantel-Haenszel one-degree-of-freedom chi square test and a related rapid procedure. *Am. J. Epidemiol.*, **112**, 129–134.

Mantel, N. and Haenszel, W. (1959). Statistical aspects of the analysis of data from retrospective studies of disease. *J. Natl. Cancer Inst.*, **22**, 719–748.

Mantel, N. and Hankey, B. F. (1975). The odds ratios of a 2 × 2 contingency table. *Am. Stat.*, **29**, 143–145.

Miettinen, O. S. (1976). Stratification by a multivariate confounder score. *Am. J. Epidemiol.*, **104**, 609–620.

Naylor, A. F. (1967). Small sample considerations in combining 2 × 2 tables. *Biometrics*, **23**, 349–356.

Odoroff, C. L. (1970). A comparison of minimum logit chi-square estimation and maximum likelihood estimation in 2 × 2 × 2 and 3 × 2 × 2 contingency tables: Tests for interaction. *J. Am. Stat. Assoc.*, **65**, 1617–1631.

Pasternack, B. S. and Mantel, N. (1966). A deficiency in the summation of chi procedure. *Biometrics*, **22**, 407–409.

Radhakrishna, S. (1965). Combination of results from several 2 × 2 contingency tables. *Biometrics*, **21**, 86–98.

Rubin, D. B. (1973). The use of matched sampling and regression adjustment to remove bias in observational studies. *Biometrics*, **29**, 185–203.

Schlesselman, J. J. (1978). Assessing effects of confounding variables. *Am. J. Epidemiol.*, **108**, 3–8.

Sheehe, P. R. (1966). Combination of log relative risk in retrospective studies of disease. *Am. J. Public Health*, **56**, 1745–1750.

Yates, F. (1955). The use of transformations and maximum likelihood in the analysis of quantal experiments involving two treatments. *Biometrika*, **42**, 382–403.

CHAPTER 11

The Effects of Misclassification Errors

We have so far assumed that the assignment of a subject to the diseased or nondiseased category, and to the category for the presence or absence of the antecedent factor, is made without error. This assumption is assuredly not valid. Whenever such assignments are made by the response to a questionnaire, or by responses in an interview, or by an examination of a case record, or by a physical or chemical test, or by any means imaginable, the assignment can be in error. For unintentional reasons—chance misreadings, failure to hear a response, and so on—or because of unconscious biases, a subject having the disease may be recorded as not having it, and conversely. Exactly the same is true for the recording of the antecedent factor.

In sampling methods I and II, misclassification errors can operate on either of the two characteristics studied. In sampling method III, they can operate only on the response variable.

In this chapter we consider the effects of misclassification errors, and in the next two chapters give some methods for reducing and for estimating the extent of error. Section 11.1 presents in detail one example of the effects of misclassification. Section 11.2 describes algebraically the effects of misclassification errors in one variable on measures of association. Section 11.3 is devoted to the case where both variables are observed with error.

11.1. AN EXAMPLE OF THE EFFECTS OF MISCLASSIFICATION

Dern, Glynn, and Brewer (1963) studied the frequency of glucose-6-phosphate dehydrogenase (G-6-PD) deficiency in the erythrocytes of black male schizophrenic patients in the Chicago area. G-6-PD deficiency is inherited as a sex-linked error of metabolism, and is found in about 10% to 15% of the American black male population.

The deficiency is sometimes referred to as the fava bean disease because exposure to the fava bean by affected individuals tends to result in hemolysis, that is, in a breakdown of red blood cells. Antimalarial agents and other drugs also result in hemolysis in affected individuals.

The data provided by Dern, Glynn, and Brewer are summarized in Table 11.1. For these data, chi square = 7.95, which indicates an association significant at the .01 level.

Table 11.1. Association between G-6-PD deficiency and subtype of schizophrenia in Chicago

G-6-PD Status	Diagnosis: Catatonic	Diagnosis: Paranoid	Total
Deficient	15	6	21
Nondeficient	57	99	156
Total	72	105	177

The contrasted proportions are the proportion of the catatonics who are deficient, $p_C = 15/72 = .208$, and the proportion of the paranoids who are deficient, $p_P = 6/105 = .057$. The value of the odds ratio is

$$o = \frac{15 \times 99}{6 \times 57} = 4.34; \tag{11.1}$$

that is, the odds that a catatonic is G-6-PD deficient are over four times the odds that a paranoid is deficient.

Fieve, Brauninger, Fleiss, and Cohen (1965) repeated the study at five state hospitals in the New York City area. At four of the hospitals, the results were as given in Table 11.2.

Table 11.2. Association between G-6-PD deficiency and subtype of schizophrenia in four New York state hospitals

Hospital	Catatonic N	Catatonic % Deficient	Paranoid N	Paranoid % Deficient	o
Central Islip	32	15.6	80	12.5	1.30
Pilgrim	78	16.7	76	6.6	2.84
Brooklyn	13	30.8	18	11.1	3.56
Kings Park	55	10.9	96	6.3	1.84

The four individual odds ratios did not differ significantly; the mean odds ratio,

$$\bar{o}_{MH} = 2.09, \tag{11.2}$$

was significantly different from unity at the .05 level (see Chapter 10 for methods of comparing and combining different odds ratios). The findings from these four hospitals tend to confirm Dern, Glynn, and Brewer's (1963) original finding, but indicate a reduced degree of association.

At the fifth hospital, Rockland, the odds ratio was again significantly different from unity (Table 11.3). The problem at Rockland State Hospital,

Table 11.3. Association between G-6-PD deficiency and subtype of schizophrenia at Rockland State Hospital

Catatonic		Paranoid		
N	% Deficient	N	% Deficient	o
28	7.1	29	24.1	0.24

clearly, was that the odds ratio was significantly different from unity in the reverse direction from that found previously.

The investigators quickly checked back with the administration at Rockland and breathed a sigh of relief when they discovered that half of the schizophrenic patients had been withheld because they were subjects in other research investigations. The investigators therefore returned to Rockland and succeeded in studying all of the resident black male catatonics and paranoids. The results from the second survey were as shown in Table

Table 11.4. Association between G-6-PD deficiency and subtype of schizophrenia at Rockland State Hospital—second survey

Catatonic		Paranoid		
N	% Deficient	N	% Deficient	o
37	2.7	87	16.1	0.14

11.4. The odds ratio was again significantly different from unity at the .05 level, but even further below it than before.

Thus there is evidence for an association between an inherited enzyme deficiency and subtype of schizophrenia, but, unfortunately, for association in both possible directions. What are missing to complete a confusing picture are data from a large sample of patients showing no significant difference between paranoids and catatonics. Just such data were provided by a sample of 426 patients at a Veterans Administration hospital in Alabama (Bowman, et al., 1965). The data are presented in Table 11.5. For these data, $p_C = 10.4\%$, $p_P = 11.8\%$, $o = 0.87$, and chi square $= 0.07$.

Table 11.5. Association between G-6-PD deficiency and subtype of schizophrenia at a VA hospital in Alabama

	Diagnosis		
G-6-PD Status	Catatonic	Paranoid	Total
Deficient	17	31	48
Nondeficient	146	232	378
Total	163	263	426

There is, therefore, evidence in the literature for all three kinds of association: positive, negative, and zero. This conflicting evidence is summarized in Table 11.6.

Table 11.6. Evidence for various directions of association between G-6-PD deficiency and subtype of schizophrenia

Direction of Association	Source	o
Catatonics over paranoids	Chicago	4.34
	Four New York hospitals	2.09
Paranoids over catatonics	Fifth New York hospital	0.14
No difference	Alabama	0.87

In attempting to account for this confusion, the New York investigators looked first at the experimental techniques used in the three studies. The techniques, while different, were not sufficiently dissimilar to explain the discrepancies. At any rate, the technique used at Rockland was no different from that used at the other four hospitals in New York.

Differences in the drugs given the patients might conceivably have produced these discrepant findings. Within each study, therefore, odds ratios were calculated for each major category of drug administered at the time of the study. With only a few exceptions, on too few cases to have much impact, the odds ratios for the specific drug categories within a study were in the same direction as the overall value for the study.

Whatever the influences might be of differences in techniques of blood testing and of drugs, they pale when the unreliability of psychiatric diagnosis is considered. A large literature has accumulated indicating just how unreliable psychiatric diagnosis is (Zubin, 1967; Spitzer and Fleiss, 1974). With respect to schizophrenia, for example, it has been found that, typically, only about 70% of all patients given a diagnosis of schizophrenia by one diagnostician are given the same diagnosis by a second, and that perhaps 10% of patients given a diagnosis other than schizophrenia by one diagnostician are given a diagnosis of schizophrenia by a second. From the few published data for the subtypes of schizophrenia, it appears that reliability is less for paranoid and catatonic schizophrenia than for schizophrenia in general.

In each of the three studies cited of G-6-PD deficiency and schizophrenia, the then current hospital diagnoses were accepted uncritically, with no attempt made to verify their accuracy. It is thus likely that the single most important source of the discrepancies among the three studies, as well as between Rockland and the other four New York state hospitals, was the unreliability of psychiatric diagnosis.

Prima facie evidence for diagnostic differences among the five New York hospitals is afforded by the variability of the proportions of patients diagnosed catatonic among those diagnosed either catatonic or paranoid. There exist some differences among the five hospitals in the kinds of patients they receive, but not of sufficient magnitude to account for the differences in their proportions of catatonics. Problem 11.1 is devoted to these differences.

The example just given is from psychiatry, but the inference should not be drawn that psychiatry is unique in being plagued by inaccurate diagnoses. Unreliability exists in the diagnosis of childhood disorders (Derryberry, 1938; Bakwin, 1945); in the diagnosis of emphysema (Fletcher, 1952); in the interpretation of electrocardiograms (Davies, 1958); in the interpretation of X-rays (Yerushalmy, 1947; Cochrane and Garland, 1952); and in the certification of causes of death (Markush, Schaaf, and Seigel, 1967). Reviews of diagnostic unreliability in other branches of physical medicine are given by Garland (1960) and Koran (1975a, 1975b).

It may, in fact, be taken as axiomatic that the determination of the presence or absence of any disease or condition and the determination of the exact form of the condition when it is present are subject to error. Likewise, the determination of the presence or absence of an antecedent factor, with the possible exception of a subject's sex, is subject to error.

11.2. THE ALGEBRA OF MISCLASSIFICATION

Confusion exists about the effects of errors of misclassification. The facts of the matter are that such errors can turn a truly strong positive association into one that is less strongly positive or even apparently negative; one that is truly strongly negative into one that is less strongly negative or even apparently positive; and one that is nil into one that is apparently strong. These facts contradict the long-standing, but erroneous impression that errors of misclassification tend only to reduce the magnitude of association (Newell, 1962).

Assume for simplicity that the classification of a person as diseased or not is accurate, so that the only source of error is in the determination of whether or not the factor under study was present. To keep matters specific, let us consider the comparison of women aged 55–64 who were newly diagnosed as having lung cancer with similarly aged women newly diagnosed as having cancer of the breast, with respect to whether they ever smoked.

We assume that the diagnoses are accurate, but that the determination of smoking history is subject to error. Two sources of error exist, one residing with the informant and one with the person taking the history. With respect to the informant (e.g., the patient herself or a relative):

1. She may misunderstand the intent or phrasing of the question.

2. She may make an honest mistake in reporting what the patient's smoking status was.

3. She may deliberately misrepresent the patient's smoking status (more likely in claiming the patient never smoked when in fact she did than the reverse).

With respect to the person taking the history:

1. He may misunderstand the informant's answer.

2. He may make an honest coding error.

3. He may apply different standards in the recording of certain responses for one type of patient versus another. Suppose, for example, that the history-taker elects to bend over backwards in an attempt to control for a possible bias toward finding an association. He may then record the statement, "I smoked once in a while when I was a kid, but never since," as Never Smoked if made by a lung cancer patient but as Ever Smoked if made by a patient with breast cancer.

Horwitz and Lysgaard-Hansen (1975) have enumerated several other prevalent sources of error, and have given prescriptions for their control (see also Chapter 12). Here we study what the effects of these errors are. The analysis is that of Keys and Kihlberg (1963). Others who have analyzed the effects of misclassification on measures of association and on chi square tests are Rogot (1961), Mote and Anderson (1965), Assakul and Proctor (1967), Koch (1969), Goldberg (1975), and Copeland, et al. (1977).

Consider first the lung cancer patients, who, we are assuming, can be identified without error. Let P_L denote the true proportion of lung cancer patients who had ever smoked, so that $1 - P_L$ is the true proportion who had never smoked. Denote by E_L the complement of sensitivity and by F_L the complement of specificity for the lung cancer patients; that is, E_L is the probability that a lung cancer patient who actually had smoked is recorded as not having smoked, and F_L is the probability that a lung cancer patient who actually had never smoked is recorded as having smoked. Whereas one would wish to estimate P_L, the true proportion who had ever smoked, one can instead, from the recorded histories, only estimate

$$p_L = (1 - E_L)P_L + F_L(1 - P_L). \tag{11.3}$$

The estimated proportion of lung cancer patients who had ever smoked, p_L, is a fraction, $1 - E_L$, of those who truly ever smoked plus a fraction, F_L, of those who truly never smoked.

The observable proportion p_L may be less than, equal to, or greater than the true proportion P_L depending on the relative magnitudes of E_L and F_L. In fact,

$$p_L > P_L \quad \text{if} \quad \frac{F_L}{E_L + F_L} > P_L,$$

$$p_L = P_L \quad \text{if} \quad \frac{F_L}{E_L + F_L} = P_L,$$

and

$$p_L < P_L \quad \text{if} \quad \frac{F_L}{E_L + F_L} < P_L.$$

If E_L and F_L are of approximately the same magnitude, there will be overestimation if P_L is less than .5 and underestimation if P_L is greater than .5. Thus, even if the error rates are equal, the errors will not necessarily cancel out.

Let, now, P_B denote the true proportion of breast cancer patients who had ever smoked, E_B the complement of their sensitivity, and F_B the complement of their specificity. For the breast cancer patients, therefore, one can estimate only the proportion recorded as ever having smoked, say

$$p_B = (1 - E_B)P_B + F_B(1 - P_B). \qquad (11.4)$$

The algebra of the effects of errors on the odds ratio is complicated (Diamond and Lilienfeld, 1962a, 1962b; Goldberg, 1975; see Copeland, et al., 1977, for a graphic study). Suppose, therefore, that the association between smoking and type of cancer is measured simply by the difference between the proportions who had smoked. Instead of being able to estimate the true difference, say

$$D = P_L - P_B, \qquad (11.5)$$

we can only estimate $d = p_L - p_B$. This difference between the recorded proportions is easily seen to reduce algebraically to

$$d = D + (F_L - F_B) + P_B(E_B + F_B) - P_L(E_L + F_L), \qquad (11.6)$$

which indicates that d is typically a biased estimate of D.

The estimated difference d may be less than, equal to, or greater than the true difference D. It may even be of opposite sign, which means that an association that is actually in one direction may be estimated as being in the opposite direction.

This possibility of a reversal of the direction of association cannot arise in the special case in which the two sensitivities are equal,

$$1 - E_L = 1 - E_B = 1 - E, \qquad (11.7)$$

say, and in which the two specificities are equal,

$$1 - F_L = 1 - F_B = 1 - F, \qquad (11.8)$$

say. By substituting (11.7) and (11.8) into (11.6) and simplifying, we find that the difference between the recorded proportions is

$$d = D[1 - (E + F)]. \qquad (11.9)$$

The first point to notice about (11.9) is that d, the difference that can be calculated, cannot possibly equal D, the true difference, whenever either error rate is nonzero. The second point to notice is that, provided both E and F are less than $\frac{1}{2}$—that is, both error rates are less than 50%—the observed difference is in the same direction as the true difference, but is numerically smaller—that is, is closer to zero. This is the situation considered by Bross in a classic paper (1954) and the one that has led to the erroneous anticipation that misclassification errors *always* tend to deflate the difference between two rates. Equal sensitivities and equal specificities must, however, be considered unusual (see, e.g., Lilienfeld and Graham, 1958; Goldberg, 1975).

With respect to the odds ratio, the effect again can in general be anything. In the particular case just considered, where $E_L = E_B < \frac{1}{2}$, and where $F_L = F_B < \frac{1}{2}$, the same kind of underestimation as for the difference between rates occurs. Specifically, if ω is the true odds ratio and o the odds ratio estimated from misclassified data, then, if $\omega > 1$, we would expect to find $\omega > o > 1$. That is, the estimated odds ratio would also be greater than unity, but not by as much as the true value.

11.3. THE ALGEBRA OF MISCLASSIFICATION: BOTH VARIABLES IN ERROR

The previous section was concerned only with errors in one of the two variables under study. A general discussion of the more realistic situation in which both variables are subject to errors of misclassification is given by Keys and Kihlberg (1963). The following analysis and example, dealing with the determination of the odds ratio, are by Barron (1977).

Let A denote the presence and $\bar{A}$ the absence of one of the two variables under study, and B and $\bar{B}$ the presence and absence of the other variable. Let $P(AB)$, $P(A\bar{B})$, and so on denote the various probabilities of joint occurrence if both variables are measured without error, and let

$$\omega = \frac{P(AB)P(\bar{A}\bar{B})}{P(A\bar{B})P(\bar{A}B)} \tag{11.10}$$

denote the odds ratio accurately associating A and B.

Suppose, however, that both variables are subject to errors of misclassification, with probabilities of correct and incorrect classification given in Table 11.7. It is assumed that the two errors operate independently.

Table 11.7. Probabilities of correct and incorrect classification of A and B

	Classified Status			Classified Status	
True Status	A	$\bar{A}$	True Status	B	$\bar{B}$
A	a_1	$1-a_1$	B	b_1	$1-b_1$
$\bar{A}$	a_2	$1-a_2$	$\bar{B}$	b_2	$1-b_2$

Finally, let $p(AB)$, $p(A\bar{B})$, and so on denote the various probabilities of joint occurrence when the variables are ascertained with error. Explicitly,

$$p(AB) = a_1 b_1 P(AB) + a_1 b_2 P(A\bar{B}) + a_2 b_1 P(\bar{A}B) + a_2 b_2 P(\bar{A}\bar{B}), \tag{11.11}$$

$$p(A\bar{B}) = a_1(1-b_1)P(AB) + a_1(1-b_2)P(A\bar{B}) + a_2(1-b_1)P(\bar{A}B) + a_2(1-b_2)P(\bar{A}\bar{B}), \tag{11.12}$$

$$p(\bar{A}B) = (1-a_1)b_1 P(AB) + (1-a_1)b_2 P(A\bar{B}) + (1-a_2)b_1 P(\bar{A}B) + (1-a_2)b_2 P(\bar{A}\bar{B}), \tag{11.13}$$

and

$$p(\bar{A}\bar{B}) = (1-a_1)(1-b_1)P(AB) + (1-a_1)(1-b_2)P(A\bar{B}) + (1-a_2)(1-b_1)P(\bar{A}B) + (1-a_2)(1-b_2)P(\bar{A}\bar{B}). \tag{11.14}$$

The observable odds ratio is then

$$o = \frac{p(AB)p(\bar{A}\bar{B})}{p(A\bar{B})p(\bar{A}B)}, \tag{11.15}$$

which bears no necessary relation to ω in (11.10).

Suppose that, if hypertension (A) and endometrial cancer (B) could be observed without error in a hospitalized sample, their probabilities of joint occurrence were as in Table 11.8. Suppose further that the probability of

Table 11.8. Hypothetical joint probabilities of hypertension and endometrial cancer if both were observed without error

	Endometrial Cancer	
Hypertension	B	$\bar{B}$
A	.122	.060
$\bar{A}$	.211	.607

correctly ascertaining the presence of hypertension is $a_1 = .90$; of correctly ascertaining the absence of hypertension, $1 - a_2 = .98$; of correctly ascertaining the presence of endometrial cancer, $b_1 = .95$; and of correctly ascertaining the absence of endometrial cancer, $1 - b_2 = .98$. These rates of correct ascertainment were all empirically determined (see Barron, 1977 for references), and are all high.

The observable rates of joint occurrence are, by (11.11)–(11.14),

$$p(AB) = .110, \tag{11.16}$$

$$p(A\bar{B}) = .070, \tag{11.17}$$

$$p(\bar{A}B) = .220, \tag{11.18}$$

and

$$p(\bar{A}\bar{B}) = .600. \tag{11.19}$$

The observable value of the odds ratio is then

$$o = \frac{.110 \times .600}{.070 \times .220} = 4.29, \tag{11.20}$$

over 25% lower than the odds ratio from the accurate data of Table 11.8,

$$\omega = \frac{.122 \times .607}{.060 \times .211} = 5.85. \tag{11.21}$$

Problem 11.1. The text cited the differences among the five New York state hospitals in the proportions of patients diagnosed catatonic, out of all those diagnosed either catatonic or paranoid.

(a) The frequencies are given below. Calculate the indicated proportions.

Hospital	Catatonic	Paranoid	Total	Proportion Catatonic
Central Islip	32	80	112	$= p_1$
Pilgrim	78	76	154	$= p_2$
Brooklyn	13	18	31	$= p_3$
Kings Park	55	96	151	$= p_4$
Rockland	37	87	124	$= p_5$
Total	215	357	572	$= \bar{p}$

(*b*) The chi square statistic for comparing a series of proportions is given by (9.4). Calculate chi square for the proportions determined in (*a*).

(*c*) Refer the value just calculated to Table A.1 with 4 degrees of freedom. At what significance level would the hypothesis of no difference in proportions be rejected? What would you conclude about the standards for the differential diagnosis of catatonic and paranoid schizophrenia in the five hospitals?

Problem 11.2. Suppose that the rate of smoking among women aged 55–64 who were newly diagnosed as having lung cancer is $P_L = .50$, and suppose that the error rates are $E_L = .25$ and $F_L = .05$.

(*a*) What is the value of p_L [see (11.3)], the estimated proportion of such women who ever smoked?

Suppose that the rate of smoking among women aged 55–64 who were newly diagnosed as having breast cancer is $P_B = .40$, and suppose that the error rates are $E_B = F_B = .10$.

(*b*) What is the value of p_B [see (11.4)], the estimated proportion of such women who ever smoked?

(*c*) What is the value of $P_L - P_B$? What is the value of $p_L - p_B$? How do these compare?

(*d*) What is the odds ratio as a function of P_L and P_B? What is the odds ratio as a function of p_L and p_B? How do these compare?

REFERENCES

Assakul, K. and Proctor, C. H. (1967). Testing independence in two-way contingency tables with data subject to misclassification. *Psychometrika*, **32**, 67–76.

Bakwin, H. (1945). Pseudodoxia pediatrica. *New Engl. J. Med.*, **232**, 691–697.

Barron, B. A. (1977). The effects of misclassification on the estimation of relative risk. *Biometrics*, **33**, 414–418.

Bowman, J. E., Brewer, G. J., Frischer, H., Carter, J. L., Eisenstein, R. B., and Bayrakci, C. (1965). A re-evaluation of the relationship between glucose-6-phosphate dehydrogenase deficiency and the behavioral manifestations of schizophrenia. *J. Lab. Clin. Med.*, **65**, 222–227.

Bross, I. (1954). Misclassification in 2 × 2 tables. *Biometrics*, **10**, 478–486.

Cochrane, A. L. and Garland, L. H. (1952). Observer error in interpretation of chest films: International investigation. *Lancet*, **2**, 505–509.

Copeland, K. T., Checkoway, H., McMichael, A. J., and Holbrook, R. H. (1977). Bias due to misclassification in the estimation of relative risk. *Am. J. Epidemiol.*, **105**, 488–495.

Davies, L. G. (1958). Observer variation in reports on electrocardiograms. *Brit. Heart J.*, **20**, 153–161.

Dern, R. J., Glynn, M. F., and Brewer, G. J. (1963). Studies on the correlation of the genetically determined trait G-6-PD deficiency with behavioral manifestations in schizophrenia. *J. Lab. Clin. Med.*, **62**, 319–329.

Derryberry, M. (1938). Reliability of medical judgments on malnutrition. *Public Health Rep.*, **53**, 263–268.

Diamond, E. L. and Lilienfeld, A. M. (1962a). Effects of errors in classification and diagnosis in various types of epidemiological studies. *Am. J. Public Health*, **52**, 1137–1144.

Diamond, E. L. and Lilienfeld, A. M. (1962b). Misclassification errors in 2×2 tables with one margin fixed: Some further comments. *Am. J. Public Health*, **52**, 2106–2110.

Fieve, R. R., Brauninger, G., Fleiss, J. L., and Cohen, G. (1965). Glucose-6-phosphate dehydrogenase deficiency and schizophrenic behavior. *J. Psychiatr. Res.*, **3**, 255–262.

Fletcher, C. M. (1952). Clinical diagnosis of pulmonary emphysema—an experimental study. *Proc. R. Soc. Med.*, **45**, 577–584.

Garland, L. H. (1960). The problem of observer error. *Bull. N.Y. Acad. Med.*, **36**, 570–584.

Goldberg, J. D. (1975). The effects of misclassification on the bias in the difference between two proportions and the relative odds in the fourfold table. *J. Am. Stat. Assoc.*, **70**, 561–567.

Horwitz, O. and Lysgaard-Hansen, B. (1975). Medical observations and bias. *Am. J. Epidemiol.*, **101**, 391–399.

Keys, A. and Kihlberg, J. K. (1963). The effect of misclassification on estimated relative prevalence of a characteristic. *Am. J. Public Health*, **53**, 1656–1665.

Koch, G. G. (1969). The effect of non-sampling errors on measures of association in 2×2 contingency tables. *J. Am. Stat. Assoc.*, **64**, 852–863.

Koran, L. M. (1975a). The reliability of clinical methods, data and judgments, part 1. *New Engl. J. Med.*, **293**, 642–646.

Koran, L. M. (1975b). The reliability of clinical methods, data and judgments, part 2. *New Engl. J. Med.*, **293**, 695–701.

Lilienfeld, A. M. and Graham, S. (1958). Validity of determining circumcision status by questionnaire as related to epidemiological studies of cancer of the cervix. *J. Natl. Cancer Inst.*, **21**, 713–720.

Markush, R. E., Schaaf, W. E., and Seigel, D. G. (1967). The influence of the death certifier on the results of epidemiologic studies. *J. Natl. Med. Assoc.*, **59**, 105–113.

Mote, V. L. and Anderson, R. L. (1965). An investigation of the effect of misclassification on the properties of chi square tests in the analysis of categorical data. *Biometrika*, **52**, 95–109.

Newell, D. J. (1962). Errors in the interpretation of errors in epidemiology. *Am. J. Public Health*, **52**, 1925–1928.

Rogot, E. (1961). A note on measurement errors and detecting real differences. *J. Am. Stat. Assoc.*, **56**, 314–319.

Spitzer, R. L. and Fleiss, J. L. (1974). A re-analysis of the reliability of psychiatric diagnosis. *Br. J. Psychiatry*, **125**, 341–347.

Yerushalmy, J. (1947). Statistical problems in assessing methods of medical diagnosis, with special reference to X-ray techniques. *Public Health Rep.*, **62**, 1432–1449.

Zubin, J. (1967). Classification of the behavior disorders. Pp. 373–406 in P. R. Farnsworth, O. McNemar, and Q. McNemar (Eds.). *Annual review of psychology*. Palo Alto, Calif.: Annual Reviews.

CHAPTER 12

The Control of Misclassification Error

In the preceding chapter we considered some possible effects of misclassification errors. In Section 12.1 of this chapter we consider statistical means of controlling for error. Algebraic results for probabilistic control are presented in Section 12.2, and some techniques for the experimental control of error are discussed in Section 12.3.

12.1. STATISTICAL CONTROL FOR ERROR

Occasionally, an investigator has available two or more means for determining the status of a patient—one quite expensive but reliable (i.e., subject to little error), the others relatively inexpensive but unreliable. To plan a survey or comparative study on even a moderate scale, the investigator must, to keep the cost of the study as low as possible, employ one of the unreliable devices (Rubin, Rosenbaum, and Cobb, 1956).

If the investigator uses *only* an unreliable device, he or she runs the risk of obtaining the kinds of biased estimates described in the preceding chapter. By selecting a subsample of the total sample to be assessed by both the unreliable and more reliable devices, however, the investigator will be able to estimate, for only a relatively small added cost, the rates of misclassification and thus to correct for bias.

Consider, as an example, the determination of the current smoking habits of each of a sample of subjects. The investigator can choose to rely solely on the subject's report, but would be sacrificing reliability for simplicity. Reliance instead on a chemical test for the concentrations of thiocyanates in the subject's urine, saliva, and plasma (Densen, Davidow, Bass, and Jones, 1967) would mean sacrificing inexpensiveness for precision.

Suppose that a sample of N newly hospitalized women diagnosed as having lung cancer is to be evaluated for smoking habits, and suppose that the investigator chooses to rely on each woman's verbal report on her current smoking practice. For simplicity, each woman is characterized as either a heavy smoker (say, smoking ten or more cigarettes per day, on the average) or not. Let p_L denote the resulting proportion who report being heavy smokers.

The selected means of determining smoking status has the virtue of being inexpensive, but the drawback of being prone to possibly excessive error. Suppose, therefore, that the investigator decides to estimate the degree of error present in the patients' reports by taking a subsample of n out of the total of N lung cancer patients and testing, in addition, their plasma concentrations of thiocyanates. A positive result on the test is taken as indicative of the patient's being a heavy smoker, and a negative result as indicative of her not being a heavy smoker.

The results of this blood test can hardly be viewed as establishing the patient's true status, not only because the dichotomy between heavy and nonheavy smoking is imprecise but also because the results of the test are subject to random fluctuations themselves. Nevertheless, because of its greater reproducibility, the blood test may be viewed as a standard against which to compare verbal reports.

Let Table 12.1 represent the cross-classification of reported and tested smoking status for the subsample of n lung cancer patients. The notation

Table 12.1. Smoking status as determined by report and by the standard blood test

	Standard		
Report	Heavy	Not Heavy	Total
Heavy	n_{00}	n_{01}	n_0
Not heavy	n_{10}	n_{11}	n_1

is that of Tenenbein (1970, 1971). From these data we may estimate as n_{00}/n_0 the proportion, of those women whose reported status is that of heavy smoker, who would be assigned to the heavy smoking category by the standard, and as n_{10}/n_1 the proportion, of those women whose reported status is that of nonheavy smoker, who would be assigned to the heavy smoking category by the standard. Recalling that p_L is the overall proportion of women assigned to the heavy smoking category on the basis of verbal report, it is easily checked that an estimate of the overall proportion who would have been so assigned by the standard is

$$P_L = \frac{n_{00}}{n_0} p_L + \frac{n_{10}}{n_1}(1 - p_L). \tag{12.1}$$

Whereas the estimated standard error of p_L is simply $\sqrt{p_L(1 - p_L)/N}$, that of P_L is more complicated. It is, in fact, given by

$$\text{s.e.}(P_L) = \sqrt{\frac{P_L(1 - P_L)}{N}\left(1 + (1 - K)\frac{N - n}{n}\right)}, \tag{12.2}$$

where

$$K = \frac{1 - p_L}{p_L} \cdot \frac{(P_L - n_{10}/n_1)^2}{P_L(1 - P_L)}. \tag{12.3}$$

The estimate (12.1) and standard error (12.2) are derived by Tenenbein (1970, 1971), who also gives criteria for choosing a reasonable value of n.

Deming (1977) presents an interesting application of what is essentially Tenenbein's double sampling scheme to a problem in survey sampling and gives some further criteria for choosing a value of n. Chiacchierini and Arnold (1977) extend Tenenbein's scheme to the case where both variables are subject to error, and Hochberg (1977) extends it to the case of multidimensional cross-classification tables. Other approaches to the estimation of correction factors are described by Harper (1964), Bryson (1965), and Press (1968).

The following numerical example will illustrate the algebra presented above. Suppose that a total of 200 female lung cancer patients are interviewed and that 88 of them so respond that they are judged to be heavy smokers. The observed, but biased rate of heavy smoking among lung cancer patients is therefore

$$p_L = .44. \tag{12.4}$$

Suppose, however, that 50 of the 200 patients are tested for levels of serum thiocyanates as well as interviewed, and suppose that the resulting cross-classification is as given in Table 12.2. Then, $n_{00}/n_0 = 18/20 = .90$ and $n_{10}/n_1 = 6/30 = .20$ are the two correction factors, and substitution into (12.1) yields

$$P_L = .90 \times .44 + .20 \times .56 = .51 \tag{12.5}$$

as an improved estimate of the rate of heavy smoking in this group. Note that the rate given in (12.4) is an underestimate by more than 10%.

Table 12.2. Smoking status of 50 lung cancer patients as determined by report and by chemical test

Report	Chemical Test: Heavy	Chemical Test: Not Heavy	Total
Heavy	18	2	20
Not heavy	6	24	30

To determine the standard error of P_L, the quantity K given by (12.3) must first be calculated. It is

$$K = \frac{.56}{.44} \cdot \frac{(.51 - .20)^2}{.51 \times .49} = .4894. \tag{12.6}$$

Substitution of this value in (12.2) yields

$$\text{s.e.}(P_L) = \sqrt{\frac{.51 \times .49}{200}\left(1 + .5106 \times \frac{150}{50}\right)}$$

$$= \sqrt{.0032} = .06 \tag{12.7}$$

as the estimated standard error of P_L.

If the study is a comparative one, such as comparing the rates of heavy smoking among lung cancer and breast cancer patients, then a subsample of the breast cancer patients would also have to be administered the blood test. Problem 12.1 gives some comparative data for analysis.

12.2. PROBABILISTIC CONTROL FOR ERROR

On occasion, one can rely on external sources for information on the magnitude of error. In the example presented in Section 11.3, the rates of correct and incorrect ascertainment of hypertension and endometrial cancer were from sources other than those giving rise to the data on association between these two diseases. The notation here is the same as that used in Section 11.3, and the results are those of Barron (1977).

In practice, one obtains the fallible estimates $p(AB)$, $p(A\bar{B})$, and so on, and may have available from one source or another the values of a_1, a_2, b_1, and b_2 (see Table 11.7). Equations (11.11)–(11.14) may be inverted to yield values for the underlying correct probabilities as follows, where

$$p(A) = p(AB) + p(A\bar{B}) \tag{12.8}$$

and

$$p(B)=p(AB)+p(\bar{A}B). \tag{12.9}$$

The correct probabilities are

$$P(AB)=\frac{p(AB)+a_2b_2-a_2p(B)-b_2p(A)}{(a_1-a_2)(b_1-b_2)}, \tag{12.10}$$

$$P(A\bar{B})=\frac{-p(AB)-a_2b_1+a_2p(B)+b_1p(A)}{(a_1-a_2)(b_1-b_2)}, \tag{12.11}$$

$$P(\bar{A}B)=\frac{-p(AB)-a_1b_2+a_1p(B)+b_2p(A)}{(a_1-a_2)(b_1-b_2)}, \tag{12.12}$$

and

$$P(\bar{A}\bar{B})=\frac{p(AB)+a_1b_1-a_1p(B)-b_1p(A)}{(a_1-a_2)(b_1-b_2)}. \tag{12.13}$$

With $p(AB)=.110$, $p(A)=.110+.070=.180$, and $p(B)=.110+.220=.330$ [see (11.16)–(11.18)] and with $a_1=.9$, $a_2=.02$, $b_1=.95$, and $b_2=.02$, the correct probabilities are found to be

$$P(AB)=.122, \tag{12.14}$$

$$P(A\bar{B})=.060, \tag{12.15}$$

$$P(\bar{A}B)=.211, \tag{12.16}$$

and

$$P(\bar{A}\bar{B})=.607, \tag{12.17}$$

that is, identical to the values originally presented in Table 11.8.

When independent estimates of the probabilities of correct ascertainment are available, (12.10)–(12.13) may be applied to obtain the correct probabilities, and desired measures of association may be derived from these rather than from the probabilities based on the erroneous observations. Explicit formulas for standard errors have yet to be derived, however.

12.3. THE EXPERIMENTAL CONTROL OF ERROR

It is almost always possible to modify a contemplated research design so as to reduce the probable magnitude of error. Only a few of the large number of techniques and ideas that exist can be presented here. One procedure is modeled on the double-blind feature of a properly designed clinical trial. In a double-blind trial, the patient is ignorant as to which of the drugs being compared he or she is getting, and the person conducting the evaluation of the patient's response is likewise ignorant of the patient's treatment.

This idea of keeping both the patient and the evaluator in the dark has obvious relevance in studies in which the presence or absence of a disease and the presence or absence of an antecedent factor have to be determined at nearly the same time (e.g., the patients being studied may be admissions to an acute treatment ward in a hospital, and neither previous records nor the opportunity for follow-up may exist). One danger is that the diagnostician, knowing whether or not the factor is present, may be prejudiced in favor of one or another of the diagnoses under study. A form of control is to instruct the diagnostician not to ask about the antecedent factor unless it happens to be pathognomonic.

A second danger is that the person seeking to establish the presence or absence of the factor, knowing that the diagnosis has been made, may be prejudiced in favor of recording the factor as present, and vice versa. A form of control is to keep this second person ignorant of the diagnosis.

A third danger is that the patient may respond differently depending on whether he knows, or even believes, that he has the disease being studied. A form of control is to keep the patient ignorant of his diagnosis long enough for all the background information of interest to be collected. An ethical problem must be solved here: just how much can be withheld from a patient, and for how long, has to be determined on an ad hoc basis (see Levin, 1954).

We have so far taken for granted that the person responsible for making the diagnosis is different from the person responsible for eliciting information on the background factors. The two roles are not always capable of separation, but the results of a study in which the same person assumed both roles must always be suspect. An example of the bias which may arise is provided by a study of the psychiatric concomitants of systemic lupus erythematosus (SLE).

There has been an increasing number of reports that the frequency (on both an incidence and prevalence basis) of psychological disturbance among SLE patients is unusually high. As part of a study of this phenomenon (Ganz, Gurland, Deming, and Fisher, 1972), a psychologist interviewed, using a structured interview schedule, samples of SLE and rheumatoid arthritis (RA) patients coming to clinics at four New York City hospitals. RA patients were selected as a control group because of many similarities between the two diseases and because few reports exist of psychiatric complications in RA.

Sixty-eight SLE and 36 RA patients were interviewed, on the basis of which each patient's psychiatric symptomatology was characterized as severe (e.g., extreme anxiety or depression, and, occasionally, delusional thinking), moderate (e.g., slight degrees of worrying and depression), or none. Every attempt was made to keep the interviewer ignorant of the patients' diagnoses. Table 12.3 gives the results of the categorization of symptomatology elicited by the interview.

Table 12.3. Psychiatric symptomatology by diagnosis —results from interview

	Systemic Lupus Erythematosus		Rheumatoid Arthritis	
Symptoms	*N*	(%)	*N*	(%)
Severe	24	(35%)	9	(25%)
Moderate	19	(28%)	12	(33%)
None	25	(37%)	15	(42%)
Total	68	(100%)	36	(100%)

The proportions for the two diagnostic categories are seen to be similar. The chi square statistic, with 2 degrees of freedom, for comparing the two distributions is equal to 1.16, which is not significant at any reasonable level. Thus, on the basis of a structured interview conducted by an interviewer ignorant of the diagnosis, the conclusion would have to be that the degree of psychiatric symptomatology in SLE patients is not essentially different from that in RA patients.

As part of the study, the notes made by the physician on the same day as the interview were also examined. The physician was not told the results of the interview. With the same criteria as used before for characterizing the degree of psychiatric symptomatology, only now applied to the case notes, the data of Table 12.4 resulted. The assessment of the information recorded on the case notes was made by a person other than the interviewer.

The proportions for the two diagnostic categories are seen to be quite different. For example, nearly a third of the SLE patients as opposed to only 6% of the RA patients were characterized on the basis of case notes as having psychiatric symptoms of any kind. The value of chi square for the data of Table 12.4 is 8.27, indicating a difference significant at the .05 level.

Table 12.4. Psychiatric symptomatology by diagnosis —results from case notes

	Systemic Lupus Erythematosus		Rheumatoid Arthritis	
Symptoms	*N*	(%)	*N*	(%)
Severe	5	(7%)	0	(0%)
Moderate	15	(22%)	2	(6%)
None	48	(71%)	34	(94%)
Total	68	(100%)	36	(100%)

A possible explanation of the difference just found, when the results from the interview are taken into account, is that a self-perpetuating myth may be in operation. As more and more reports are published indicating a high frequency of psychological disturbance in SLE, more and more clinicians will be influenced to observe and record its presence. In itself, this is not a bad thing. But, if care in recording such disturbances in SLE is at the expense of equal care in other disorders, then the scientific value of these observations becomes highly questionable.

It therefore seems clear that an investigator should rely on a person ignorant of the status of the patient for the collection of background and other concomitant information. The desirability of a structured interview or questionnaire—where the questions to be asked and the probes to be made are set forth, and where the responses are precoded—may not be so clear.

Setting down the questions to be asked ensures that each interviewer covers the same ground and that each subject is asked the same questions. In this way differences in interviewing style and biases due to different kinds of patients being asked different questions are reduced. Having the responses precoded serves not only to have the data in a form suitable for keypunching and machine processing, but also to reduce the errors in having a person make sense of a verbal, sometimes anecdotal report.

Such structured interviews have met with great success in psychiatry (Spitzer, Fleiss, Endicott, and Cohen, 1967; Wing, et al., 1967; and Spitzer, Endicott, Fleiss, and Cohen, 1970) and in the study of respiratory diseases (Medical Research Council, 1966). The need for such procedures in psychiatry and bronchopulmonary medicine is due mainly to the variability among clinicians in the way they elicit information from patients, and, having elicited the data, in the way they interpret their findings.

Similar factors are present in almost every branch of medicine, and there is no compelling reason why the idea of the structured interview cannot be extended. Its applicability is clear in the interview for history, but rather subtle in the assessment of X-ray negatives, electrocardiogram (EKG) tracings, and the like. An important reason for diagnostic disagreement in heart disease, for example, is that different cardiologists interpret EKGs differently; even the same cardiologist may, on two occasions, interpret the same EKG differently. Surely many of these differences would be reduced if cardiologists were instructed to note, on a precoded form, just what the abnormalities were that they thought they detected in each wave of the EKG. The same idea can certainly be applied to the recording of lesions thought to be detected from an X-ray negative.

Having information recorded on such forms, in addition to increasing uniformity, serves yet another purpose. The data, provided they are suitably recorded, can be quantified and can thus serve to provide a more objective gradation of disease severity than that based on clinical judgment. The major difficulty in the use of such a form, assuming that someone has the fortitude and patience to develop, validate, and if necessary revise one, is that clinicians unfamiliar with it might resent having to use it. Considering, however, that the use of such techniques brings medical and epidemiological research closer to the ideal of all scientific endeavor—that all criteria be publicly specified and thus that every study be reproducible—not using them is virtually impossible to justify.

Problem 12.1. We considered, in Section 12.1, a means of correcting for the bias in an observed proportion. We now consider the comparison of two proportions, both of which are subject to error.

(*a*) The value of p_L, the rate of heavy smoking among lung cancer patients based on verbal report, was .44. Suppose that 200 women with cancer of the breast are interviewed, with 60 of them indicating that they are heavy smokers. What is the value of p_B, the rate of heavy smoking among breast cancer patients? What is the value of the odds ratio measuring the degree of association between type of cancer and heavy smoking?

(*b*) Based on the determination of smoking status by both response to interview and chemical test for 50 lung cancer patients, the rate of heavy smoking was adjusted to $P_L = .51$. Suppose that 50 breast cancer patients are likewise given a chemical test in addition to an interview, with the following results:

	Chemical Test		
Report	Heavy	Not Heavy	Total
Heavy	18	0	18
Not heavy	2	30	32

What are the values of the two correction factors, n_{00}/n_0 and n_{10}/n_1, which are to be applied to p_B? What value of P_B results from the adjustment

$$P_B = \frac{n_{00}}{n_0} p_B + \frac{n_{10}}{n_1}(1 - p_B)?$$

What is the value of the odds ratio associated with the adjusted rates? Does the association between heavy smoking and type of cancer now appear weaker or stronger than in (a)?

(c) Suppose instead that the cross-classification for the subsample of 50 breast cancer patients yielded the following results:

	Chemical Test		
Report	Heavy	Not Heavy	Total
Heavy	16	2	18
Not heavy	7	25	32

What are the values of the correction factors n_{00}/n_0 and n_{10}/n_1? What is the resulting value of P_B? What is the resulting value of the odds ratio? Does the association appear weaker or stronger than in (a)?

Problem 12.2. Suppose that the value of the correction factor n_{00}/n_0 was the same for the two kinds of patients and that the value of the correction factor n_{10}/n_1 was likewise the same. What is a simple expression for $P_L - P_B$ as a function of $p_L - p_B$ and of the difference $n_{00}/n_0 - n_{10}/n_1$?

REFERENCES

Barron, B. A. (1977). The effects of misclassification on the estimation of relative risk. *Biometrics*, **33**, 414–418.

Bryson, M. R. (1965). Errors of classification in a binomial population. *J. Am. Stat. Assoc.*, **60**, 217–224.

Chiacchierini, R. P. and Arnold, J. C. (1977). A two-sample test for independence in 2 × 2 contingency tables with both margins subject to misclassification. *J. Am. Stat. Assoc.*, **72**, 170–174.

Deming, W. E. (1977). An essay on screening, or on two-phase sampling, applied to surveys of a community. *Int. Stat. Rev.*, **45**, 29–37.

Densen, P. M., Davidow, B., Bass, H. E., and Jones, E. W. (1967). A chemical test for smoking exposure. *Arch. Environ. Health*, **14**, 865–874.

Ganz, V. H., Gurland, B. J., Deming, W. E., and Fisher, B. (1972). The study of the psychiatric symptoms of systemic lupus erythematosus: A biometric study. *Psychosom. Med.*, **34**, 199–206.

Harper, D. (1964). Misclassification in epidemiological surveys. *Am. J. Public Health*, **54**, 1882–1886.

Hochberg, Y. (1977). On the use of double sampling schemes in analyzing categorical data with misclassification errors. *J. Am. Stat. Assoc.*, **72**, 914–921.

Levin, M. L. (1954). Etiology of lung cancer: Present status. *N.Y. State J. Med.*, **54**, 769–777.

Medical Research Council (1966). Questionnaire on respiratory diseases. Dawlish, Devon, England: W. J. Holman, Ltd.

Press, S. J. (1968). Estimating from misclassified data. *J. Am. Stat. Assoc.*, **63**, 123–133.

Rubin, T., Rosenbaum, J., and Cobb, S. (1956). The use of interview data for the detection of association in field studies. *J. Chronic Dis.*, **4**, 253–266.

Spitzer, R. L., Endicott, J., Fleiss, J. L., and Cohen, J. (1970). Psychiatric Status Schedule: A technique for evaluating psychopathology and impairment in role functioning. *Arch. Gen. Psychiatry*, **23**, 41–55.

Spitzer, R. L., Fleiss, J. L., Endicott, J., and Cohen, J. (1967). Mental Status Schedule: Properties of factor analytically derived scales. *Arch. Gen. Psychiatry*, **16**, 479–493.

Tenenbein, A. (1970). A double sampling scheme for estimating from binomial data with misclassifications. *J. Am. Stat. Assoc.*, **65**, 1350–1361.

Tenenbein, A. (1971). A double sampling scheme for estimating from binomial data with misclassifications: Sample size determination. *Biometrics*, **27**, 935–944.

Wing, J. K., Birley, J. L. T., Cooper, J. E., Graham, P., and Isaacs, A. D. (1967). Reliability of a procedure for measuring and classifying "present psychiatric state." *Br. J. Psychiatry*, **113**, 499–515.

CHAPTER 13

The Measurement of Interrater Agreement

The statistical methods described in the preceding chapter for controlling for error are applicable only when the rates of misclassification are known from external sources or are estimable by applying a well defined standard classification procedure to a subsample of the group under study. For some variables of importance, however, no such standard is readily apparent.

To assess the extent to which a given characterization of a subject is reliable, it is clear that we must have a number of subjects classified more than once, for example by more than one rater. The degree of agreement among the raters provides no more than an upper bound on the degree of accuracy present in the ratings, however. If agreement among the raters is high, then there is a possibility, but by no means a guarantee, that the ratings do in fact reflect the dimension they are purported to reflect. If their agreement is low, on the other hand, then the usefulness of the ratings is severely limited, for it is meaningless to ask what is associated with the variable being rated when one cannot even trust those ratings to begin with.

In this chapter we consider the measurement of interrater agreement when the ratings are on categorical scales. Section 13.1 is devoted to the case of two raters per subject; Section 13.2, to the case of multiple ratings per subject. Applications to other problems are indicated in Section 13.3.

13.1. THE CASE OF TWO RATERS

Suppose that each of a sample of n subjects is rated independently by the same two raters, with the ratings being on a categorical scale consisting of k categories. Consider the hypothetical example of Table 13.1, in which each cell entry is the proportion of all subjects classified into one of $k = 3$ diagnostic categories by rater A and into another by rater B. Thus, for example, 5% of all subjects were diagnosed neurotic by rater A and psychotic by rater B.

Table 13.1. Diagnoses on n = 100 subjects by two raters

	Rater B			
Rater A	Psychotic	Neurotic	Organic	Total
Psychotic	.75	.01	.04	.80
Neurotic	.05	.04	.01	.10
Organic	0	0	.10	.10
Total	.80	.05	.15	1.00

Suppose it is desired to measure the degree of agreement on each category separately as well as across all categories. The analysis begins by collapsing the original $k \times k$ table into a 2×2 table in which all categories other than the one of current interest are combined into a single "all others" category. Table 13.2 presents the results in general as well as for neurosis from Table 13.1 in particular. It must be borne in mind that the entries a, b, c, and d in the general table refer to *proportions* of subjects, not to their numbers.

Table 13.2. Data for measuring agreement on a single category

	General Rater B				For Neurosis Rater B		
Rater A	Given Category	All Others	Total	Rater A	Neurosis	All Others	Total
Given category	a	b	p_1	Neurosis	.04	.06	.10
All others	c	d	q_1	All others	.01	.89	.90
Total	p_2	q_2	1	Total	.05	.95	1.00

The simplest and most frequently used index of agreement is the overall proportion of agreement, say

$$p_o = a + d. \tag{13.1}$$

p_o, or a simple variant of it such as $2p_o - 1$, has been proposed as the agreement index of choice by Holley and Guilford (1964) and by Maxwell

(1977). For neurosis, the overall proportion of agreement is

$$p_o = .04 + .89 = .93.$$

This value, along with the overall proportions of agreement for the other two categories, is given in the column labeled p_o in Table 13.3. The conclusion that might be drawn from these values is that agreement is,

Table 13.3. Values of several indices of agreement from data of Table 13.1

Category	p_o	p_s	λ_r	p_s'	A	κ
Psychotic	.90	.94	.88	.75	.84	.69
Neurotic	.93	.53	.06	.96	.75	.50
Organic	.95	.80	.60	.97	.89	.77

effectively, equally good on all three categories, with agreement on organic disorders being somewhat better than on neurosis, and agreement on neurosis being somewhat better than on psychosis.

Suppose the category under study is relatively rare, so that the proportion d, representing agreement on absence, is likely to be large and thus to inflate the value of p_o. A number of indices of agreement have been proposed that are based only on the proportions a, b, and c. Of all of them, only the so-called proportion of specific agreement, say

$$p_s = \frac{2a}{2a+b+c} = \frac{a}{\bar{p}}, \tag{13.2}$$

where $\bar{p} = (p_1 + p_2)/2$, has a sensible probabilistic interpretation. Let one of the two raters be selected at random, and let attention be focused on the subjects assigned to the category of interest. The quantity p_s is the conditional probability that the second rater will also make an assignment to that category, given that the randomly selected first rater did. This index was first proposed by Dice (1945) as a measure of similarity.

The proportion of specific agreement on neurosis is

$$p_s = \frac{2 \times .04}{2 \times .04 + .06 + .01} = .53,$$

and the values for all three categories are presented in the column headed p_s in Table 13.3. The conclusions based on p_s are rather different from those based on p_o. Agreement now seems best on psychosis, rather less good on organic disorders, and much poorer than either on neurosis.

Define $\bar{q} = 1 - \bar{p}$, or,

$$\bar{q} = \tfrac{1}{2}(q_1 + q_2) = d + \frac{b+c}{2}, \tag{13.3}$$

and suppose that $\bar{q} > \bar{p}$. Goodman and Kruskal (1954) proposed

$$\lambda_r = \frac{(a+d) - \bar{q}}{1 - \bar{q}} = \frac{2a - (b+c)}{2a + (b+c)} \tag{13.4}$$

as an index of agreement; it is motivated less by notions of agreement than by a consideration of the frequencies of correct predictions of a subject's category when predictions are made with and without knowledge of the joint ratings. λ_r assumes its maximum value of $+1$ when there is complete agreement, but assumes its minimum value of -1 whenever $a = 0$, irrespective of the value of d [not, as Goodman and Kruskal (1954, p. 758) imply, only when $a + d = 0$].

For neurosis,

$$\lambda_r = \frac{2 \times .04 - (.06 + .01)}{2 \times .04 + (.06 + .01)} = .06,$$

and the values of λ_r for all three categories are listed under the indicated column of Table 10.3. Because of the identity

$$\lambda_r = 2p_s - 1, \tag{13.5}$$

the categories are ordered on λ_r identically as on p_s.

The proportion of specific agreement ignores the proportion d. If, instead, we choose to ignore a, we would calculate the corresponding index, say

$$p_s' = \frac{d}{\bar{q}} = \frac{2d}{2d + b + c}, \tag{13.6}$$

where $\bar{q} = 1 - \bar{p}$. For neurosis

$$p_s' = \frac{2 \times .89}{2 \times .89 + .06 + .01} = .96,$$

and this value plus the other two are presented in the indicated column of Table 13.3. Yet a different picture emerges from these values than from earlier ones. Agreement (with respect to absence) on organic disorders and on neurosis seems to be equally good and apparently substantially better than on psychosis.

Rather than having to choose between p_s and p'_s, Rogot and Goldberg (1966) proposed simply taking their mean, say

$$A = \tfrac{1}{2}(p_s + p'_s) = \frac{a}{p_1 + p_2} + \frac{d}{q_1 + q_2}, \tag{13.7}$$

as an index of agreement. For neurosis,

$$A = \frac{.04}{.10 + .05} + \frac{.89}{.90 + .95} = .75.$$

As seen in the indicated column of Table 13.3, the index A orders the three categories in yet a new way: agreement on organic disorders is better than on psychosis, and agreement on organic disorders and on psychosis is better than on neurosis.

Yet other indices of agreement between two raters have been proposed (e.g., by Fleiss, 1965; Armitage, Blendis, and Smyllie, 1966; Rogot and Goldberg, 1966; and Bennett, 1972), but it should already be clear that there must be more to the measurement of interrater agreement than the arbitrary selection of an index of agreement.

The new dimension is provided by a realization that, except in the most extreme circumstances (either $p_1 = q_2 = 0$ or $p_2 = q_1 = 0$), some degree of agreement is to be expected by chance alone (see Table 13.4). For example, if rater A employs one set of criteria for distinguishing between the presence or absence of a condition, and if rater B employs an entirely different and independent set of criteria, then *all* the observed agreement is explainable by chance.

Table 13.4. Chance-expected proportions of joint judgments by two raters, for data of Table 13.2

	General Rater B				For Neurosis Rater B		
Rater A	Given Category	All Others	Total	Rater A	Neurosis	All Others	Total
Given category	p_1p_2	p_1q_2	p_1	Neurosis	.005	.095	.10
All others	q_1p_2	q_1q_2	q_1	All others	.045	.855	.90
Total	p_2	q_2	1	Total	.05	.95	1

Different opinions have been stated on the need to incorporate chance-expected agreement into the assessment of interrater reliability. Rogot and Goldberg (1966), for example, emphasize the importance of contrasting observed with expected agreement when comparisons are to be made between different pairs of raters or different kinds of subjects. Goodman and Kruskal (1954, p. 758), on the other hand, contend that chance-expected agreement need not cause much concern, that the observed degree of agreement may usually be assumed to be in excess of chance. (Even if one is willing to grant this assumption, one should nevertheless check whether the excess is trivially small or substantially large.)

Armitage, Blendis, and Smyllie (1966, p. 102) occupy a position between that of Rogot and Goldberg and that of Goodman and Kruskal. They appreciate the necessity for introducing chance-expected agreement whenever different sets of data are being compared, but claim that too much uncertainty exists as to how the correction for chance is to be incorporated into the measure of agreement.

There does exist, however, a natural means for correcting for chance. Consider any index that assumes the value 1 when there is complete agreement. Let I_o denote the observed value of the index (calculated from the proportions in Table 13.2) and let I_e denote the value expected on the basis of chance alone (calculated from the proportions in Table 13.4).

The obtained excess beyond chance is $I_o - I_e$, whereas the maximum possible excess is $1 - I_e$. The ratio of these two differences is denoted kappa,

$$\hat{\kappa} = \frac{I_o - I_e}{1 - I_e}. \tag{13.8}$$

Kappa is a measure of agreement with desirable properties. If there is complete agreement, $\hat{\kappa} = +1$. If observed agreement is greater than or equal to chance agreement, $\hat{\kappa} \geqq 0$, and if observed agreement is less than or equal to chance agreement, $\hat{\kappa} \leqq 0$. The minimum value of $\hat{\kappa}$ depends on the marginal proportions. If they are such that $I_e = .5$, then the minimum equals -1. Otherwise, the minimum is between -1 and 0.

It may be checked by simple algebra that, *for each of the indices of agreement defined above*, the same value of $\hat{\kappa}$ results after the chance-expected value is incorporated as in (13.8) (see Problem 13.1):

$$\hat{\kappa} = \frac{2(ad - bc)}{p_1 q_2 + p_2 q_1}. \tag{13.9}$$

An important unification of various approaches to the indexing of agreement is therefore achieved by introducing a correction for chance-expected agreement.

For neurosis,

$$\hat{\kappa}=\frac{2(.04\times.89-.06\times.01)}{.10\times.95+.05\times.90}=.50.$$

This value and the other two are presented in the final column of Table 13.3. They are close to those found by Spitzer and Fleiss (1974) in a review of the literature on the reliability of psychiatric diagnosis. Agreement is best on organic disorders, less good on psychosis, and poorest on neurosis.

The kappa statistic was first proposed by Cohen (1960). Variants of kappa have been proposed by Scott (1955) and by Maxwell and Pilliner (1968). All have interpretations as *intraclass correlation coefficients* (see Ebel, 1951). The intraclass correlation coefficient is a widely used measure of interrater reliability for the case of quantitative ratings. As shown by Fleiss (1975) and Krippendorff (1970), only kappa is identical (except for a term involving the factor $1/n$, where n is the number of subjects) to that version of the intraclass correlation coefficient due to Bartko (1966) in which a difference between the raters in their base rates (i.e., a difference between p_1 and p_2) is considered a source of unwanted variability.

Landis and Koch (1977a) have characterized different ranges of values for kappa with respect to the degree of agreement they suggest. For most purposes, values greater than .75 or so may be taken to represent excellent agreement beyond chance, values below .40 or so may be taken to represent poor agreement beyond chance, and values between .40 and .75 may be taken to represent fair to good agreement beyond chance.

Often, a composite measure of agreement across all categories is desired. An overall value of kappa may be defined as a weighted average of the individual kappa values, where the weights are the denominators of the individual kappas [i.e., the quantities $p_1q_2+p_2q_1$ in (13.9)]. An equivalent and more suggestive formula is based on arraying the data as in Table 13.5.

Table 13.5. Joint proportions of ratings by two raters on a scale with k categories

	Rater B				
Rater A	1	2	$\cdots$	k	Total
1	p_{11}	p_{12}	$\cdots$	p_{1k}	$p_{1.}$
2	p_{21}	p_{22}	$\cdots$	p_{2k}	$p_{2.}$
$\vdots$					
k	p_{k1}	p_{k2}	$\cdots$	p_{kk}	$p_{k.}$
Total	$p_{.1}$	$p_{.2}$	$\cdots$	$p_{.k}$	1

The overall proportion of observed agreement is, say,

$$p_o = \sum_{i=1}^{k} p_{ii}, \tag{13.10}$$

and the overall proportion of chance-expected agreement is, say,

$$p_e = \sum_{i=1}^{k} p_{i.}p_{.i}. \tag{13.11}$$

The overall value of kappa is then, say,

$$\hat{\kappa} = \frac{p_o - p_e}{1 - p_e}. \tag{13.12}$$

For the data of Table 13.1,

$$p_o = .75 + .04 + .10 = .89$$

and

$$p_e = .80 \times .80 + .10 \times .05 + .10 \times .15 = .66,$$

so that

$$\hat{\kappa} = \frac{.89 - .66}{1 - .66} = .68.$$

For testing the hypothesis that the ratings are independent (so that the underlying value of kappa is zero), Fleiss, Cohen, and Everitt (1969) showed that the appropriate standard error of kappa is estimated by

$$\text{s.e.}_0(\hat{\kappa}) = \frac{1}{(1-p_e)\sqrt{n}} \sqrt{p_e + p_e^2 - \sum_{i=1}^{k} p_{i.}p_{.i}(p_{i.} + p_{.i})}, \tag{13.13}$$

where p_e is defined in (13.11). The hypothesis may be tested against the alternative that agreement is better than chance would predict by referring the quantity

$$z = \frac{\hat{\kappa}}{\text{s.e.}_0(\hat{\kappa})} \tag{13.14}$$

to tables of the standard normal distribution and rejecting the hypothesis if z is sufficiently large (a one-sided test is more appropriate here than a two-sided test).

For the data at hand,

$$\text{s.e.}_0(\hat{\hat{\kappa}}) = \frac{1}{(1-.66)\sqrt{100}}\sqrt{.66 + .66^2 - 1.0285} = .076$$

and

$$z = \frac{.68}{.076} = 8.95.$$

The overall value of kappa is therefore statistically highly significant, and, by virtue of its magnitude, it indicates a good degree of agreement beyond chance.

Formulas (13.10)–(13.14) apply even when k, the number of categories, is equal to two. They may therefore be applied to the study of each category's reliability, as shown in Table 13.6 for the data of Table 13.1.

Table 13.6. Kappas for individual categories and across all categories of Table 13.1

Category	p_o	p_e	$\hat{\kappa}$	$\text{s.e.}_0(\hat{\kappa})$	z
Psychotic	.90	.68	.69	.100	6.90
Neurotic	.93	.86	.50	.093	5.38
Organic	.95	.78	.77	.097	7.94
Overall	.89	.66	.68	.076	8.95

Note that the overall value of kappa is equal to the sum of the individual $p_o - p_e$ differences (i.e., of the numerators of the individual kappas) divided by the sum of the individual $1 - p_e$ differences (i.e., of the denominators of the individual kappas),

$$\hat{\hat{\kappa}} = \frac{(.90-.68)+(.93-.86)+(.95-.78)}{(1-.68)+(1-.86)+(1-.78)} = \frac{.46}{.68} = .68,$$

confirming that $\hat{\hat{\kappa}}$ is a weighted average of the individual $\hat{\kappa}$'s.

For testing the hypothesis that the underlying value of kappa (either overall or for a single category) is equal to a prespecified value κ *other than zero*, Fleiss, Cohen, and Everitt (1969) showed that the appropriate standard error of $\hat{\kappa}$ is estimated by

$$\text{s.e.}(\hat{\kappa}) = \frac{\sqrt{A+B-C}}{(1-p_e)\sqrt{n}}, \tag{13.15}$$

where

$$A = \sum_{i=1}^{k} p_{ii}\left[1-(p_{i.}+p_{.i})(1-\hat{\kappa})\right]^2, \tag{13.16}$$

$$B = (1-\hat{\kappa})^2 \sum_{i \neq j}\sum p_{ij}(p_{.i}+p_{j.})^2, \tag{13.17}$$

and

$$C = \left[\hat{\kappa} - p_e(1-\hat{\kappa})\right]^2. \tag{13.18}$$

The hypothesis that κ is the underlying value would be rejected if the critical ratio

$$z = \frac{|\hat{\kappa}-\kappa|}{\text{s.e.}(\hat{\kappa})} \tag{13.19}$$

were found to be significantly large from tables of the normal distribution. An approximate $100(1-\alpha)\%$ confidence interval for κ is

$$\hat{\kappa} - c_{\alpha/2}\,\text{s.e.}(\hat{\kappa}) \leqq \kappa \leqq \hat{\kappa} + c_{\alpha/2}\,\text{s.e.}(\hat{\kappa}). \tag{13.20}$$

Consider testing the hypothesis that the overall value of kappa underlying the data in Table 13.1 is .80. The three quantities (13.16)–(13.18) needed to determine the standard error of $\hat{\hat{\kappa}}$ are

$$\begin{aligned} A &= .75\left[1-(.80+.80)(1-.68)\right]^2 \\ &\quad + .04\left[1-(.10+.05)(1-.68)\right]^2 \\ &\quad + .10\left[1-(.10+.15)(1-.68)\right]^2 \\ &= .2995, \\ B &= (1-.68)^2\left[.01(.80+.10)^2 + .04(.80+.10)^2\right. \\ &\quad + .05(.05+.80)^2 + .01(.05+.10)^2 \\ &\quad \left. + 0(.15+.80)^2 + 0(.15+.10)^2\right] \\ &= .0079, \end{aligned}$$

and

$$C = [.68 - .66(1 - .68)]^2 = .2198.$$

Thus

$$\text{s.e.}(\hat{\bar{\kappa}}) = \frac{\sqrt{.2995 + .0079 - .2198}}{(1 - .66)\sqrt{100}} = .087$$

and

$$z = \frac{|.68 - .80|}{.087} = 1.38,$$

so the hypothesis that $\bar{\kappa} = .80$ is not rejected.

Suppose one wishes to compare and combine $g(\geqq 2)$ independent estimates of kappa. The theory of Section 10.1 applies. Define, for the mth estimate, $V_m(\hat{\kappa}_m)$ to be the squared standard error of $\hat{\kappa}_m$, that is, the square of the expression in (13.15). The combined estimate of the supposed common value of kappa is, say,

$$\hat{\kappa}_{\text{overall}} = \frac{\displaystyle\sum_{m=1}^{g} \frac{\hat{\kappa}_m}{V_m(\hat{\kappa}_m)}}{\displaystyle\sum_{m=1}^{g} \frac{1}{V_m(\hat{\kappa}_m)}}. \tag{13.21}$$

To test the hypothesis that the g underlying values of kappa are equal, the value of

$$\chi^2_{\text{equal } \kappa\text{'s}} = \sum_{m=1}^{g} \frac{\left(\hat{\kappa}_m - \hat{\kappa}_{\text{overall}}\right)^2}{V_m(\hat{\kappa}_m)} \tag{13.22}$$

may be referred to tables of chi square with $g - 1$ degrees of freedom. The hypothesis is rejected if the value is significantly large. The limits of an approximate $100(1 - \alpha)\%$ confidence interval for the supposed common underlying value are given by

$$\hat{\kappa}_{\text{overall}} \pm c_{\alpha/2} \sqrt{\frac{1}{\displaystyle\sum_{m=1}^{g} \frac{1}{V_m(\hat{\kappa}_m)}}}. \tag{13.23}$$

Cohen (1968) (see also Spitzer, Cohen, Fleiss, and Endicott, 1967) generalized his kappa measure of interrater agreement to the case where the relative seriousness of each possible disagreement could be quantified. Suppose that, independently of the data actually collected, agreement weights, say w_{ij} ($i=1,\ldots,k$; $j=1,\ldots,k$), are assigned on rational or clinical grounds to the k^2 cells (see Cicchetti, 1976). The weights are restricted to lie in the interval $0 \leqq w_{ij} \leqq 1$ and to be such that

$$w_{ii} = 1 \tag{13.24}$$

(i.e., exact agreement is given maximal weight),

$$0 \leqq w_{ij} < 1 \quad \text{for} \quad i \neq j \tag{13.25}$$

(i.e., all disagreements are given less than maximal weight), and

$$w_{ij} = w_{ji} \tag{13.26}$$

(i.e., the two raters are considered symmetrically).

The observed weighted proportion of agreement is, say,

$$p_{o(w)} = \sum_{i=1}^{k} \sum_{j=1}^{k} w_{ij} p_{ij}, \tag{13.27}$$

where the proportions p_{ij} are arrayed as in Table 13.5, and the chance-expected weighted proportion of agreement is, say,

$$p_{e(w)} = \sum_{i=1}^{k} \sum_{j=1}^{k} w_{ij} p_{i.} p_{.j}. \tag{13.28}$$

Weighted kappa is then given by

$$\hat{\kappa}_w = \frac{p_{o(w)} - p_{e(w)}}{1 - p_{e(w)}}. \tag{13.29}$$

Note that, when $w_{ij} = 0$ for all $i \neq j$ (i.e., when all disagreements are considered as being equally serious), then weighted kappa becomes identical to the overall kappa given in (13.12).

The interpretation of the magnitude of weighted kappa is like that of unweighted kappa: $\hat{\kappa}_w \geqq .75$ or so signifies excellent agreement, for most purposes, and $\hat{\kappa}_w \leqq .40$ or so signifies poor agreement.

Suppose that the k categories are ordered and that the decision is made to apply a two-way analysis of variance to the data resulting from taking the numerals $1, 2, \ldots, k$ as bona fide measurements. Bartko (1966) gives a formula for the intraclass correlation coefficient derived from this analysis of variance, and Fleiss and Cohen (1973) have shown that, aside from a term involving the factor $1/n$, the intraclass correlation coefficient is identical to weighted kappa provided the weights are taken as

$$w_{ij} = 1 - \frac{(i-j)^2}{(k-1)^2}. \tag{13.30}$$

Independently of Cohen (1968), Cicchetti and Allison (1971) proposed a statistic for measuring interrater reliability that is formally identical to weighted kappa. They suggested that the weights be taken as

$$w_{ij} = 1 - \frac{|i-j|}{k-1}. \tag{13.31}$$

The sampling distribution of weighted kappa was derived by Fleiss, Cohen, and Everitt (1969) and confirmed by Cicchetti and Fleiss (1977), Landis and Koch (1977a), Fleiss and Cicchetti (1978), and Hubert (1978). For testing the hypothesis that the underlying value of weighted kappa is zero, the appropriate standard error of $\hat{\kappa}_w$ is

$$\text{s.e.}_0(\hat{\kappa}_w) = \frac{1}{(1-p_{e(w)})\sqrt{n}} \sqrt{\sum_{i=1}^{k} \sum_{j=1}^{k} p_{i.}p_{.j}\left[w_{ij} - (\bar{w}_{i.} + \bar{w}_{.j})\right]^2 - p_{e(w)}^2}, \tag{13.32}$$

where

$$\bar{w}_{i.} = \sum_{j=1}^{k} p_{.j} w_{ij} \tag{13.33}$$

and

$$\bar{w}_{.j} = \sum_{i=1}^{k} p_{i.} w_{ij}. \tag{13.34}$$

The hypothesis may be tested by referring the value of the critical ratio

$$z = \frac{\hat{\kappa}_w}{\text{s.e.}_0(\hat{\kappa}_w)} \tag{13.35}$$

to tables of the standard normal distribution.

For testing the hypothesis that the underlying value of weighted kappa is equal to a prespecified κ_w *other than zero*, the appropriate formula for the standard error of $\hat{\kappa}_w$ is

$$\text{s.e.}(\hat{\kappa}_w) = \frac{1}{(1-p_{e(w)})\sqrt{n}}$$

$$\times\sqrt{\sum_{i=1}^{k}\sum_{j=1}^{k} p_{ij}\left[w_{ij}-(\bar{w}_{i.}+\bar{w}_{.j})(1-\hat{\kappa}_w)\right]^2-\left[\hat{\kappa}_w-p_{e(w)}(1-\hat{\kappa}_w)\right]^2}\,. \tag{13.36}$$

The hypothesis may be tested by referring the value of the critical ratio

$$z=\frac{|\hat{\kappa}_w-\kappa_w|}{\text{s.e.}(\hat{\kappa}_w)} \tag{13.37}$$

to tables of the standard normal distribution and rejecting the hypothesis if the critical ratio is too large.

It may be shown (see problem 13.4) that the standard errors of unweighted kappa given in (13.13) and (13.15) are special cases of the standard errors of weighted kappa given in (13.32) and (13.36) when $w_{ii}=1$ for all i and $w_{ij}=0$ for all $i \neq j$.

Some attempts have been made to generalize kappa to the case where each subject is rated by each of the same set of more than two raters (Light, 1971; Landis and Koch, 1977a), but the few available results are not yet in a form suitable for general use. We consider in the next section the related but simpler problem of different raters for different subjects.

13.2. MULTIPLE RATINGS PER SUBJECT

Suppose that a sample of n subjects has been studied, with m_i being the number of ratings on the ith subject. The raters responsible for rating one subject are not assumed to be same as those responsible for rating another. Suppose, further, that $k=2$, that is, that the ratings consist of classifications into one of two categories; the case $k>2$ will be considered later in this section. Finally, let x_i denote the number of (arbitrarily defined) positive ratings on subject i, so that $m_i - x_i$ is the number of negative ratings on him.

Identities between intraclass correlation coefficients and kappa statistics will be exploited to derive a kappa statistic by starting with an analysis of variance applied to the data (forming a one-way layout) obtained by coding a positive rating as 1 and a negative rating as 0. This was precisely the approach taken by Landis and Koch (1977b), except that they took the number of degrees of freedom for the mean square between subjects to be $n-1$ instead of, as below, n.

Define the overall proportion of positive ratings to be

$$\bar{p} = \frac{\sum_{i=1}^{n} x_i}{n\bar{m}}, \tag{13.38}$$

where

$$\bar{m} = \frac{\sum_{i=1}^{n} m_i}{n}, \tag{13.39}$$

the mean number of ratings per subject. If the number of subjects is large (say, $n \geqq 20$), the mean square between subjects (BMS) is approximately equal to

$$\text{BMS} = \frac{1}{n} \sum_{i=1}^{n} \frac{(x_i - m_i\bar{p})^2}{m_i} \tag{13.40}$$

and the mean square within subjects (WMS) is equal to

$$\text{WMS} = \frac{1}{n(\bar{m}-1)} \sum_{i=1}^{n} \frac{x_i(m_i - x_i)}{m_i}. \tag{13.41}$$

Technically, the intraclass correlation coefficient should be estimated as

$$r = \frac{\text{BMS} - \text{WMS}}{\text{BMS} + (m_0 - 1)\,\text{WMS}}, \tag{13.42}$$

where

$$m_0 = \bar{m} - \frac{\sum_{i=1}^{n} (m_i - \bar{m})^2}{n(n-1)\bar{m}}. \tag{13.43}$$

If n is at all large, though, m_0 and $\bar{m}$ will be very close in magnitude. If m_0

is replaced by $\bar{m}$ in (13.42), the resulting expression for the intraclass correlation coefficient, and therefore for kappa, is

$$\hat{\kappa} = \frac{\text{BMS} - \text{WMS}}{\text{BMS} + (\bar{m} - 1)\,\text{WMS}}$$

$$= 1 - \frac{\sum_{i=1}^{n} \frac{x_i(m_i - x_i)}{m_i}}{n(\bar{m} - 1)\bar{p}\bar{q}}, \qquad (13.44)$$

where $\bar{q} = 1 - \bar{p}$.

$\hat{\kappa}$ has the following properties. If there is no subject-to-subject variation in the proportion of positive ratings (i.e., if $x_i/m_i = \bar{p}$ for all i, with $\bar{p}$ not equal to either 0 or 1), then there is more disagreement within subjects than between subjects. In this case $\hat{\kappa}$ may be seen to assume its minimum value of $-1/(\bar{m} - 1)$.

If the several proportions x_i/m_i vary exactly as binomial proportions with parameters m_i and a common probability $\bar{p}$, then there is as much similarity within subjects as between subjects. In this case, the value of $\hat{\kappa}$ is equal to 0.

If each proportion x_i/m_i assumes either the values 0 or 1, then there is perfect agreement within subjects. In this case, $\hat{\kappa}$ may be seen to assume the value 1.

Consider the hypothetical data of Table 13.7 on $n = 25$ subjects.

Table 13.7. Hypothetical ratings by different sets of raters on n = 25 subjects

Subject (i)	Number of Raters (m_i)	Number of Positive Ratings (x_i)	i	m_i	x_i
1	2	2	14	4	3
2	2	0	15	2	0
3	3	2	16	2	2
4	4	3	17	3	1
5	3	3	18	2	1
6	4	1	19	4	1
7	3	0	20	5	4
8	5	0	21	3	2
9	2	0	22	4	0
10	4	4	23	3	0
11	5	5	24	3	3
12	3	3	25	2	2
13	4	4	Total	81	46

For these data, the mean number of ratings per subject is

$$\overline{m} = \frac{81}{25} = 3.24,$$

the overall proportion of positive ratings is

$$\bar{p} = \frac{46}{25 \times 3.24} = .568,$$

and the value of $\Sigma x_i(m_i - x_i)/m_i$ is

$$\sum_{i=1}^{25} \frac{x_i(m_i - x_i)}{m_i} = 6.30.$$

The value of kappa in (13.44) for these ratings is therefore

$$\hat{\kappa} = 1 - \frac{6.30}{25(3.24 - 1) \times .568 \times .432}$$
$$= .54,$$

indicating only a modest degree of interrater agreement.

Fleiss and Cuzick (1979) derived the standard error of $\hat{\kappa}$ appropriate for testing the hypothesis that the underlying value of kappa is 0. Define $\overline{m}_H$ to be the *harmonic mean* of the number of ratings per subject, that is,

$$\overline{m}_H = \frac{n}{\sum_{i=1}^{n} \frac{1}{m_i}}. \quad (13.45)$$

The standard error of $\hat{\kappa}$ is approximately equal to

$$\text{s.e.}_0(\hat{\kappa}) = \frac{1}{(\overline{m} - 1)\sqrt{n\overline{m}_H}} \sqrt{2(\overline{m}_H - 1) + \frac{(\overline{m} - \overline{m}_H)(1 - 4\bar{p}\bar{q})}{\overline{m}\bar{p}\bar{q}}}, \quad (13.46)$$

and the hypothesis may be tested by referring the value of the critical ratio

$$z = \frac{\hat{\kappa}}{\text{s.e.}_0(\hat{\kappa})} \quad (13.47)$$

to tables of the standard normal distribution.

For the data of Table 13.7,

$$\overline{m}_H = \frac{25}{8.5167} = 2.935$$

and

$$\text{s.e.}_0(\hat{\kappa}) = \frac{1}{(3.24 - 1)\sqrt{25 \times 2.935}} \times \sqrt{2(2.935 - 1) + \frac{(3.24 - 2.935)(1 - 4 \times .568 \times .432)}{3.24 \times .568 \times .432}} = .103.$$

The value of the critical ratio in (13.47) is then

$$z = \frac{.54}{.103} = 5.24,$$

indicating that $\hat{\kappa}$ is significantly greater than zero.

Suppose, now, that the number of categories into which ratings are made is $k \geqq 2$. Denote by $\bar{p}_j$ the overall proportion of ratings in category j and by $\hat{\kappa}_j$ the value of kappa for category j, $j = 1, \ldots, k$. Landis and Koch (1977b) proposed taking the weighted average

$$\hat{\bar{\kappa}} = \frac{\sum_{j=1}^{k} \bar{p}_j \bar{q}_j \hat{\kappa}_j}{\sum_{j=1}^{k} \bar{p}_j \bar{q}_j} \tag{13.48}$$

as an overall measure of interrater agreement, where $\bar{q}_j = 1 - \bar{p}_j$. The standard error of $\hat{\bar{\kappa}}$ has yet to be derived, when the numbers of ratings per subject vary, to test the hypothesis that the underlying value is zero.

When, however, the number of ratings per subject is constant and equal to m, simple expressions for $\hat{\kappa}_j$, $\hat{\bar{\kappa}}$, and their standard errors are available. Define x_{ij} to be the number of ratings on subject $i (i = 1, \ldots, n)$ into category $j (j = 1, \ldots, k)$; note that

$$\sum_{j=1}^{k} x_{ij} = m \tag{13.49}$$

for all i. The value of $\hat{\kappa}_j$ is then

$$\hat{\kappa}_j = 1 - \frac{\sum_{i=1}^{n} x_{ij}(m - x_{ij})}{nm(m-1)\bar{p}_j\bar{q}_j}, \tag{13.50}$$

and the value of $\hat{\bar{\kappa}}$ is

$$\hat{\bar{\kappa}} = 1 - \frac{nm^2 - \sum_{i=1}^{n}\sum_{j=1}^{k} x_{ij}^2}{nm(m-1)\sum_{j=1}^{k}\bar{p}_j\bar{q}_j}. \tag{13.51}$$

Algebraically equivalent versions of these formulas were first presented by Fleiss (1971), who showed explicitly how they represent chance-corrected measures of agreement.

Table 13.8 presents hypothetical data representing, for each of $n = 10$ subjects, $m = 5$ ratings into one of $k = 3$ categories.

Table 13.8. Five ratings on each of ten subjects into one of three categories

	Number of ratings into category			
Subject	1	2	3	$\sum_{j=1}^{3} x_{ij}^2$
1	1	4	0	17
2	2	0	3	13
3	0	0	5	25
4	4	0	1	17
5	3	0	2	13
6	1	4	0	17
7	5	0	0	25
8	0	4	1	17
9	1	0	4	17
10	3	0	2	13
Total	20	12	18	174

The three overall proportions are $\bar{p}_1 = 20/50 = .40$, $\bar{p}_2 = 12/50 = .24$, and $\bar{p}_3 = 18/50 = .36$. For category 1, the numerator in expression (13.50) for $\hat{\kappa}_1$ is

$$\sum_{i=1}^{10} x_{i1}(5-x_{i1}) = 1\times(5-1)+2\times(5-2)+\cdots+3\times(5-3)=34,$$

and thus

$$\hat{\kappa}_1 = 1 - \frac{34}{10\times5\times4\times.40\times.60} = .29.$$

Similarly, $\hat{\kappa}_2 = .67$ and $\hat{\kappa}_3 = .35$. The overall value of $\hat{\bar{\kappa}}$ is, by (13.51),

$$\hat{\bar{\kappa}} = 1 - \frac{10\times25-174}{10\times5\times4\times(.40\times.60+.24\times.76+.36\times.64)} = .42.$$

Alternatively,

$$\hat{\bar{\kappa}} = \frac{(.40\times.60)\times.29+(.24\times.76)\times.67+(.36\times.64)\times.35}{.40\times.60+.24\times.76+.36\times.64} = .42.$$

When the numbers of ratings per subject are equal, Fleiss, Nee, and Landis (1979) derived and confirmed the following formulas for the approximate standard errors of $\hat{\bar{\kappa}}$ and $\hat{\kappa}_j$, each appropriate for testing the hypothesis that the underlying value is zero:

$$\text{s.e.}_0(\hat{\bar{\kappa}}) = \frac{\sqrt{2}}{\sum_{j=1}^{k}\bar{p}_j\bar{q}_j\sqrt{nm(m-1)}} \times\sqrt{\left(\sum_{j=1}^{k}\bar{p}_j\bar{q}_j\right)^2 - \sum_{j=1}^{k}\bar{p}_j\bar{q}_j(\bar{q}_j-\bar{p}_j)}, \tag{13.52}$$

and

$$\text{s.e.}_0(\hat{\kappa}_j) = \sqrt{\frac{2}{nm(m-1)}}. \tag{13.53}$$

Note that $\text{s.e.}_0(\hat{\kappa}_j)$ is independent of $\bar{p}_j$ and $\bar{q}_j$! Further, it is easily checked that formula (13.53) is a special case of (13.46) when the m_i's are all equal, because then $\bar{m} = \bar{m}_H = m$.

For the data of Table 13.8,

$$\sum_{j=1}^{3}\bar{p}_j\bar{q}_j = .40\times.60+.24\times.76+.36\times.64 = .6528$$

and

$$\sum_{j=1}^{3} \bar{p}_j\bar{q}_j(\bar{q}_j - \bar{p}_j) = .40 \times .60 \times (.60 - .40) + .24 \times .76 \times (.76 - .24)$$
$$+ .36 \times .64 \times (.64 - .36)$$
$$= .2074,$$

so that

$$\text{s.e.}_0(\hat{\bar{\kappa}}) = \frac{\sqrt{2}}{.6528\sqrt{10 \times 5 \times 4}} \sqrt{.6528^2 - .2074} = .072.$$

Because

$$z = \frac{\hat{\bar{\kappa}}}{\text{s.e.}_0(\hat{\bar{\kappa}})} = \frac{.42}{.072} = 5.83,$$

the overall value of kappa is significantly different from zero (although its magnitude indicates only mediocre reliability).

The approximate standard error of each $\hat{\kappa}_j$ is, by (13.53),

$$\text{s.e.}_0(\hat{\kappa}_j) = \sqrt{\frac{2}{10 \times 5 \times 4}} = .10.$$

Each individual kappa is significantly different ($p < .01$) from zero, but only $\hat{\kappa}_2$ approaches a value suggestive of fair reliability.

Landis and Koch (1977b) describe how standard errors of the $\hat{\kappa}_j$'s and of $\hat{\bar{\kappa}}$ appropriate to the case of nonzero underlying values can be calculated. The methods and results are too complicated for presentation here.

13.3. FURTHER APPLICATIONS

Even though the various kappa statistics were originally developed and were illustrated here for the measurement of interrater agreement, their applicability extends far beyond this specific problem. In fact, they are useful for measuring, on categorical data, such constructs as "similarity," "concordance," and "clustering." Some examples will be given.

1. In a study of the correlates or determinants of drug use among teenagers, it may be of interest to determine how concordant the attitudes

toward drug use are between each subject's same-sex parent and the subject's best friend. Either unweighted kappa or weighted kappa (Section 13.1) may be used, with rater A replaced by parent and rater B by best friend.

2. Suppose that m monitoring stations are set up in a city to measure levels of various pollutants and that, on each of n days, each station is characterized by whether or not the level of a specified pollutant (e.g., sulfur dioxide) exceeds an officially designated threshold. The version of kappa presented in Section 13.2 may be applied to describe how well (or poorly) the several stations agree.

3. Consider a study of the role of familial factors in the development of a condition such as adolescent hypertension. Suppose that n sibships are studied and that m_i is the number of siblings in the ith sibship. The version of kappa presented in Section 13.2 may be applied to describe the degree to which there is familial aggregation in the condition.

4. Many of the indices of agreement cited in Section 13.1 are used in numerical taxonomy (Sneath and Sokal, 1973) to describe the degree of similarity between different study units; in fact, p_s (13.2) was originally proposed for this purpose by Dice (1945). Suppose that two units (people, languages, or whatever) are being compared with respect to whether they possess or do not possess each of n dichotomous characteristics. The proportions $a-d$ in the left-hand part of Table 13.2 then refer to the proportion of all n characteristics that both units possess, the proportion that one possesses but the other does not, and so on. Corrections for chance-expected similarity in this kind of problem are as important as corrections for chance-expected agreement in the case of interrater reliability.

5. Studies in which several controls are matched with each case or each experimental unit were discussed in Section 8.3. If the several controls in each matched set were successfully matched, the responses by the controls from the same set should be more similar than the responses by controls from different sets. The version of kappa presented in Section 13.2 may be used to describe how successful the matching was.

Whether used to measure agreement, or, more generally, similarity, kappa in effect treats all the raters or units symmetrically. When one or more of the sources of ratings may be viewed as a standard, however (two of $m = 5$ raters, e.g., may be senior to the others or one of the air pollution monitoring stations in Example 2 may employ more precise measuring instruments than the others), kappa may no longer be appropriate, and the procedures described by Light (1971), Williams (1976), and Wackerley, McClave, and Rao (1978) should be employed instead.

Problem 13.1. Prove that, when each of the indices of agreement given by (13.1), (13.2), (13.4), (13.6), and (13.7) is corrected for chance-expected agreement using formula (13.8), the same formula for kappa (13.9) is obtained.

Problem 13.2. Prove that, when $k = 2$, the square of the critical ratio given in (13.14) is identical to the standard chi square statistic without the continuity correction.

Problem 13.3. Suppose that $g = 3$ independent reliability studies of a given kind of rating have been conducted, with results as follows:

Study 1 ($n = 20$)	Rater B		Study 2 ($n = 20$)	Rater D		Study 3 ($n = 30$)	Rater F	
Rater A	+	−	Rater C	+	−	Rater E	+	−
+	.60	.05	+	.75	.10	+	.50	.20
−	.20	.15	−	.05	.10	−	.10	.20

(*a*) What are the three values of kappa? What are their standard errors [see (13.15)]? What is the overall value of kappa [see (13.21)]?

(*b*) Are the three estimates of kappa significantly different? [Refer the value of the statistic in (13.22) to tables of chi square with 2 degrees of freedom.]

(*c*) Using (13.23), find an approximate 95% confidence interval for the common value of kappa.

Problem 13.4. Prove that, when $w_{ii} = 1$ for all i and $w_{ij} = 0$ for all $i \neq j$, the standard error formulas (13.13) and (13.32) are identical. Prove that, with this same system of agreement weights, the standard error formulas (13.15) and (13.36) are identical.

Problem 13.5. Prove that, when $k = 2$, formulas (13.52) and (13.53) are identical.

REFERENCES

Armitage, P., Blendis, L. M., and Smyllie, H. C. (1966). The measurement of observer disagreement in the recording of signs. *J. R. Stat. Soc., Ser. A*, **129**, 98–109.

Bartko, J. J. (1966). The intraclass correlation coefficient as a measure of reliability. *Psychol. Rep.*, **19**, 3–11,

Bennett, B. M. (1972). Measures for clinicians' disagreements over signs. *Biometrics*, **28**, 607–612.

Cicchetti, D. V. (1976). Assessing inter-rater reliability for rating scales: Resolving some basic issues. *Br. J. Psychiatry*, **129**, 452–456.

Cicchetti, D. V. and Allison, T. (1971). A new procedure for assessing reliability of scoring EEG sleep recordings. *Am. J. EEG Technol.*, **11**, 101–109.

Cicchetti, D. V. and Fleiss, J. L. (1977). Comparison of the null distributions of weighted kappa and the C ordinal statistic. *Appl. Psychol. Meas.*, **1**, 195–201.

Cohen, J. (1960). A coefficient of agreement for nominal scales. *Educ. Psychol. Meas.*, **20**, 37–46.

Cohen, J. (1968). Weighted kappa: Nominal scale agreement with provision for scaled disagreement or partial credit. *Psychol. Bull.*, **70**, 213–220.

Dice, L. R. (1945). Measures of the amount of ecologic association between species. *Ecology*, **26**, 297–302.

Ebel, R. L. (1951). Estimation of the reliability of ratings. *Psychometrika*, **16**, 407–424.

Fleiss, J. L. (1965). Estimating the accuracy of dichotomous judgments. *Psychometrika*, **30**, 469–479.

Fleiss, J. L. (1971). Measuring nominal scale agreement among many raters. *Psychol. Bull.*, **76**, 378–382.

Fleiss, J. L. (1975). Measuring agreement between two judges on the presence or absence of a trait. *Biometrics*, **31**, 651–659.

Fleiss, J. L. and Cicchetti, D. V. (1978). Inference about weighted kappa in the non-null case. *Appl. Psychol. Meas.*, **2**, 113–117.

Fleiss, J. L. and Cohen, J. (1973). The equivalence of weighted kappa and the intraclass correlation coefficient as measures of reliability. *Educ. Psychol. Meas.*, **33**, 613–619.

Fleiss, J. L., Cohen, J., and Everitt, B. S. (1969). Large sample standard errors of kappa and weighted kappa. *Psychol. Bull.*, **72**, 323–327.

Fleiss, J. L. and Cuzick, J. (1979). The reliability of dichotomous judgments: Unequal numbers of judges per subject. *Appl. Psychol. Meas.*, **3**, 537–542.

Fleiss, J. L., Nee, J. C. M., and Landis, J. R. (1979). The large sample variance of kappa in the case of different sets of raters. *Psychol. Bull.*, **86**, 974–977.

Goodman, L. A. and Kruskal, W. H. (1954). Measures of association for cross classifications. *J. Am. Stat. Assoc.*, **49**, 732–764.

Holley, J. W. and Guilford, J. P. (1964). A note on the *G* index of agreement. *Educ. Psychol. Meas.*, **32**, 281–288.

Hubert, L. J. (1978). A general formula for the variance of Cohen's weighted kappa. *Psychol. Bull.*, **85**, 183–184.

Krippendorff, K. (1970). Bivariate agreement coefficients for reliability of data. Pp. 139–150 in E. F. Borgatta (Ed.). *Sociological methodology 1970*. San Francisco: Jossey-Bass.

Landis, J. R. and Koch, G. G. (1977a). The measurement of observer agreement for categorical data. *Biometrics*, **33**, 159–174.

Landis, J. R. and Koch, G. G. (1977b). A one-way components of variance model for categorical data. *Biometrics*, **33**, 671–679.

Light, R. J. (1971). Measures of response agreement for qualitative data: Some generalizations and alternatives. *Psychol. Bull.*, **76**, 365–377.

Maxwell, A. E. (1977). Coefficients of agreement between observers and their interpretation. *Br. J. Psychiatry*, **130**, 79–83.

Maxwell, A. E. and Pilliner, A. E. G. (1968). Deriving coefficients of reliability and agreement for ratings. *Br. J. Math. Stat. Psychol.*, **21**, 105–116.

Rogot, E. and Goldberg, I. D. (1966). A proposed index for measuring agreement in test-retest studies. *J. Chronic Dis.*, **19**, 991–1006.

Scott, W. A. (1955). Reliability of content analysis: The case of nominal scale coding. *Public Opinion Quart.*, **19**, 321–325.

Sneath, P. H. A. and Sokal, R. R. (1973). *Numerical Taxonomy*. San Francisco: W. H. Freeman.

Spitzer, R. L., Cohen, J., Fleiss, J. L., and Endicott, J. (1967). Quantification of agreement in psychiatric diagnosis. *Arch. Gen. Psychiatry*, **17**, 83–87.

Spitzer, R. L. and Fleiss, J. L. (1974). A reanalysis of the reliability of psychiatric diagnosis. *Br. J. Psychiatry*, **125**, 341–347.

Wackerley, D. D., McClave, J. T., and Rao, P. V. (1978). Measuring nominal scale agreement betwen a judge and a known standard. *Psychometrika*, **43**, 213–223.

Williams, G. W. (1976). Comparing the joint agreement of several raters with another rater. *Biometrics*, **32**, 619–627.

CHAPTER 14

The Standardization of Rates

One of the most frequently occurring problems in epidemiology and vital statistics is the comparison of the rate for some event or characteristic across different populations or for the same population over time. If the populations were similar with respect to factors associated with the event under study—factors such as age, sex, race, or marital status—there would be no problem in comparing the overall rates (synonyms are total or crude rates) as they stand.

If the populations are not similarly constituted, however, the direct comparison of the overall rates may be misleading. Algebraically, the problem with such a comparison is as follows. Let $p_1, \ldots, p_I$ denote the proportions of all members of one of the populations being compared who fall into the various strata (age intervals, socioeconomic groups, etc.), there being I strata in all. Thus, $\sum p_i = 1$. If c_i denotes the rate specific to the ith stratum of this population, then the overall or crude rate for it is

$$c = \sum_{i=1}^{I} c_i p_i. \tag{14.1}$$

If the distribution for the second population across the I strata is represented by the proportions $P_1, \ldots, P_I$, so that $\sum P_i = 1$, and if C_i denotes the rate specific to the ith stratum in the second population, then the crude rate for this population is

$$C = \sum_{i=1}^{I} C_i P_i. \tag{14.2}$$

The difference between the two crude rates is

$$d = c - C, \tag{14.3}$$

and it is easy to check that

$$d = \sum_{i=1}^{I} \frac{p_i + P_i}{2}(c_i - C_i) + \sum_{i=1}^{I} \frac{c_i + C_i}{2}(p_i - P_i) \tag{14.4}$$

(see Kitagawa, 1955; Hemphill and Ament, 1970). Miettinen (1972) conducted a similar analysis for the ratio of two crude rates.

It is thus seen that the difference between two crude rates has two components. One of them,

$$d_1 = \sum_{i=1}^{I} \frac{p_i + P_i}{2}(c_i - C_i), \tag{14.5}$$

is a true summarization of the differences between the two schedules of specific rates, differences that are usually of major interest. However, the second component,

$$d_2 = \sum_{i=1}^{I} \frac{c_i + C_i}{2}(p_i - P_i), \tag{14.6}$$

is a summarization of the differences between the two sets of population distributions, differences that are of little if any interest.

A number of conclusions can be drawn from the representation (14.4):

1. If the two population distributions are equal, that is, if $p_1 = P_1, \ldots, p_I = P_I$, then $d_2 = 0$ and the difference between the crude rates indeed summarizes the differences between the schedules of specific rates.

2. If the two schedules of specific rates are equal, that is, if $c_1 = C_1, \ldots, c_I = C_I$, then $d_1 = 0$ and the difference between the crude rates measures only the difference in population distributions across the strata, a difference usually of no importance.

3. For any given value of d_1, that is, for a given summarization of the differences between two schedules of specific rates, the apparent difference between the two populations measured by d may be unaltered, increased, decreased, or even changed in sign depending on the differences between the two population distributions. The effect of d_2 is an additive one if the first population has a larger proportion of its members in the strata where the rates are high than does the second—that is, if $p_i > P_i$ in those strata where $(c_i + C_i)/2$ is large—and the effect of d_2 is a subtractive one if the converse holds. It is these kinds of effects that explain why, when each of one population's specific rates is greater than another's, the first population's crude rate may nevertheless be lower than the second's.

Section 14.1 presents some reasons for standardization, and some warnings against its uncritical use. Section 14.2 describes the indirect method of standardization, and Section 14.3 illustrates how it can give misleading results. Section 14.4 describes the direct method of standardization, Section 14.5 presents some other methods of standardization, and Section 14.6 discusses techniques for standardizing on two correlated dimensions.

14.1. REASONS FOR AND WARNINGS AGAINST STANDARDIZATION

It is to prevent anomalies of the sort described in (2) and (3) above that resort is made to standardization (synonymously, adjustment) in the comparison of two or more schedules of specific rates. Standardization should never, however, substitute for a comparison of the specific rates themselves. It is these that characterize the experience (morbidity, mortality, or whatever the rate refers to) of the populations being studied.

Woolsey has pointed out that "The specific rates are essential (because) it is only through the analysis of specific rates that an accurate and detailed study can be made of the variation (of the phenomenon under study) among population classes" [1959, p. 60]. Elveback (1966), too, stresses the importance of studying the specific rates and strongly criticizes the calculation of adjusted rates.

One criticism of the adjustment of rates is that if the specific rates vary in different ways across the various strata, then no single method of standardization will indicate that these differences exist. Standardization will, on the contrary, tend to mask these differences. As one example (see Kitagawa, 1966), there is the contrast between the age-specific death rates for white males resident in metropolitan counties of the United States in 1960 and those for white males resident in nonmetropolitan counties. Up to age 40, the rates in metropolitan counties are lower than in nonmetropolitan counties; after age 40, the reverse is true. No single summary comparison will reveal this information. On the contrary, at least two summary comparisons are needed.

Another example is provided by data reported by El-Badry (1969). He points out that in Ceylon, India, and Pakistan, mortality among males occurs at a lower rate than among females in many age categories. Single summary indexes for males and females might mask this phenomenon and thus fail to reveal data suggestive of further research. Doll and Cook (1967), in addition, cite the inadequacy of a single index for summarizing age- and sex-specific incidence rates.

Bearing in mind that there is no substitute for examining the specific rates themselves, we may consider some of the reasons for standardization.

1. A single summary index for a population is more easily compared with other summary indices than are entire schedules of specific rates.

2. If some strata are comprised of small numbers of people, the associated specific rates may be too imprecise and unreliable for use in detailed comparisons.

3. For small populations, or for some groups of especial interest, specific rates may not exist. This may be the case for selected occupational groups and for populations from geographic areas especially demarcated for a single study. In such cases, only the total number of events (e.g., deaths) may be available and not their subdivision by strata.

Other reasons for standardization are given by Woolsey (1959), Kalton (1968), and Cochran (1968), who in addition studied the effects of varying the number of strata, I. Mausner and Bahn (1974, p. 138) give an elegant summary of the advantages and disadvantages of analyzing crude, specific, and adjusted rates.

14.2. INDIRECT STANDARDIZATION

The second and third reasons just given for standardization, the unreliability and possibly even the unavailability of some specific rates, lead to perhaps the most frequently adopted method of standardization, the so-called indirect method. The ingredients necessary for its implementation are:

1. The crude rate for the population being studied, say c.

2. The distribution across the various strata for that population, say $p_1, \ldots, p_I$.

3. The schedule of specific rates for a selected standard population, say $c_{S1}, \ldots, c_{SI}$.

4. The crude rate for the standard population, say c_S.

The first calculation in indirect standardization is of the overall rate that would obtain if the schedule of specific rates for the standard population were applied to the given population. It is

$$c' = \sum_{i=1}^{I} c_{Si} p_i. \tag{14.7}$$

The indirect adjusted rate is then

$$c_{\text{indirect}} = c_S \frac{c}{c'}; \tag{14.8}$$

that is, the crude rate for the standard population, c_S, multiplied by the ratio of the actual crude rate for the given population, c, to the crude rate, c', that would exist if the given population were subject to the standard population's schedule of rates.

As an example, consider the following data from Stark and Mantel (1966). In the state of Michigan, from 1950 to 1964, 731,177 infants were the first-born to their mothers; of these, 412 were mongoloid, giving a crude rate of $c = 56.3$ mongoloids per 100,000 first-born live births. In the same 15-year interval, 442,811 infants were the fifth-born or more to their mothers; of these, 740 were mongoloid, giving a crude rate of $C = 167.1$ mongoloids per 100,000 fifth-born or more live births.

The comparison as it stands is not a fair one, because maternal age is known to be associated with both birth order and mongolism. Some method has therefore to be applied to adjust for possible differences between the first-born and later-born in maternal age distributions. Table 14.1 illustrates indirect adjustment.

Table 14.1. An example of indirect standardization

		Birth Order			
	Specific Rates for all Michigan	First		Fifth or More	
Maternal Age	per 100,000—c_{Si}	p_i	$c_{Si}p_i$	P_i	$c_{Si}P_i$
Under 20	42.5	.315	13.4	.001	0.0
20–24	42.5	.451	19.2	.069	2.9
25–29	52.3	.157	8.2	.279	14.6
30–34	87.7	.054	4.7	.339	29.7
35–39	264.0	.019	5.0	.235	62.0
40 and over	864.4	.004	3.5	.078	67.4
Sum			54.0 ($= c'$)		176.6 ($= C'$)

The selected standard population was all the live births in Michigan during the years 1950–1964. The crude rate of mongolism for the state as a whole was $c_S = 89.5$ per 100,000 live births, and the rates specific for maternal age are given in the column headed c_{Si} in the table. For the infants born first and born fifth or more, the maternal age distributions are given in the columns headed p_i and P_i. The results of applying formula (14.7) are shown in the bottom row of the table.

To review, we are given the crude rates

$$c = 56.3 \text{ mongoloids per 100,000 first-borns} \tag{14.9}$$

and

$$C = 167.1 \text{ mongoloids per 100,000 fifth- or later-borns.} \tag{14.10}$$

By applying the rates of mongolism specific to maternal age for the state of Michigan as a whole, we would have expected

$$c' = 54.0 \text{ mongoloids per 100,000 first-borns,}$$

and

$$C' = 176.6 \text{ mongoloids per 100,000 fifth- or later-borns.}$$

Given the crude rate for the entire state,

$$c_S = 89.5 \text{ mongoloids per 100,000 live births,}$$

we find, by (14.8), the indirect adjusted rates

$$\begin{aligned} c_{\text{indirect}} &= 89.5 \times \frac{56.3}{54.0} \\ &= 93.3 \text{ mongoloids per 100,000 first-borns} \end{aligned} \tag{14.11}$$

and

$$\begin{aligned} C_{\text{indirect}} &= 89.5 \times \frac{167.1}{176.6} \\ &= 84.7 \text{ mongoloids per 100,000 fifth- or later-borns.} \end{aligned} \tag{14.12}$$

By just comparing the crude rates given in (14.9) and (14.10), we would conclude that there was a threefold increase in the risk of mongolism from first-borns to infants born fifth or more. By comparing the adjusted rates given by (14.11) and (14.12), on the other hand, we would conclude that there was no effective difference in the risk of mongolism.

It seems that the apparent greater risk of mongolism for later births, suggested by a comparison of the crude rates in (14.9) and (14.10), is a reflection of differences in maternal age distribution. Proportionately more mothers of later-born infants are in the older age categories, where the specific rates are higher, than are mothers of first-born infants. After adjustment for differences in the maternal age distributions, it appears that, if anything, the rate of mongolism in later-born infants is somewhat less than the rate in first-born infants.

14.3. A FEATURE OF INDIRECT STANDARDIZATION

Consider the data of Table 14.2, giving hypothetical sex-specific mortality rates (per 1000) in each of two groups. The two sets of sex-specific rates

Table 14.2. Sex-specific mortality rates in two groups

	Group 1		Group 2	
Sex	p_i	Rate/1000	P_i	Rate/1000
Male	.60	2.0	.80	2.0
Female	.40	1.0	.20	1.0

are equal, but the unequal sex distributions in the two groups yield unequal overall rates. For group 1, the crude rate is

$$c = 2.0 \times .60 + 1.0 \times .40 = 1.6 \text{ deaths/1000}, \qquad (14.13)$$

and for group 2 the crude rate is

$$C = 2.0 \times .80 + 1.0 \times .20 = 1.8 \text{ deaths/1000}. \qquad (14.14)$$

Suppose, now, that the two sets of sex-specific rates could have been obtained only with the greatest difficulty, so that the only data actually available were the two sex distributions and the two crude rates. Indirect adjustment would therefore have to be resorted to. For the population chosen as the standard, suppose that the crude rate is

$$c_S = 1.5 \text{ deaths/1000}, \qquad (14.15)$$

and that the sex-specific rates are

$$c_{S1} = 2.2 \text{ deaths/1000 males} \qquad (14.16)$$

and

$$c_{S2} = 0.9 \text{ deaths/1000 females}. \qquad (14.17)$$

The expected crude rate in group 1 is

$$c' = 2.2 \times .60 + 0.9 \times .40 = 1.68 \text{ deaths/1000}, \qquad (14.18)$$

yielding an indirect adjusted rate of

$$c_{\text{indirect}} = 1.5 \times \frac{1.6}{1.68} = 1.43 \text{ deaths}/1000. \tag{14.19}$$

The expected crude rate in group 2 is

$$C' = 2.2 \times .80 + 0.9 \times .20 = 1.94 \text{ deaths}/1000, \tag{14.20}$$

so that, for group 2,

$$C_{\text{indirect}} = 1.5 \times \frac{1.8}{1.94} = 1.39 \text{ deaths}/1000. \tag{14.21}$$

The two adjusted rates given by (14.19) and (14.21) are more nearly equal than the two unadjusted rates given by (14.13) and (14.14), reflecting more accurately the equality of sex-specific rates indicated in Table 14.2. It is a bit disquieting, however, that equality of the two schedules of specific rates has not been reflected by precise equality of the two indirect adjusted rates. This is a feature of indirect adjustment that does not characterize direct adjustment, the method to be considered next. Neither in this example nor in most other instances is the distortion great, however. Furthermore, such distortion will not occur if the standard population is the composite of the two populations studied.

It is clear that indirect standardization does not completely adjust for differences in population composition. Thus when attempting to explain variability across groups of indirect adjusted rates, one should bear in mind that, whereas variation of schedules of specific rates accounts for most of it, variation in population composition may still account for some of it. Additional criticisms of indirect adjustment are given by Yule (1934) and Kilpatrick (1963). Breslow and Day (1975), however, present a particular mathematical model for the specific rates (each specific rate is assumed to be the product of two terms, one descriptive of the stratum and the other of the population) under which indirect standardization is appropriate.

14.4. DIRECT STANDARDIZATION

The method of standardization used second most frequently is the so-called direct method. Direct standardization may be applied only when the schedule of specific rates for a given population is available. The data necessary for its implementation are:

1. The schedule of specific rates for the population being studied, say $c_1, \ldots, c_I$.

2. The distribution across the various strata for a selected standard population, say $p_{S1}, \ldots, p_{SI}$.

The direct adjusted rate is then simply

$$c_{\text{direct}} = \sum_{i=1}^{I} c_i p_{Si}. \tag{14.22}$$

The term direct refers to working directly with the specific rates of the population being studied, in distinction to what was done in the method previously presented.

As an example, let us consider the same event, mongolism, studied in Section 14.2. Table 14.3 gives the maternal age distribution for all infants born in the state of Michigan from 1950 to 1964 (the standard distribution), and the rates of mongolism specific to maternal age for first-borns and for infants born fifth or more (data from Stark and Mantel, 1966).

Table 14.3. An example of direct standardization (rates per 100,000)

		Birth Order			
	Distribution for all of	First		Fifth or More	
Maternal Age	Michigan—p_{Si}	c_i	$c_i p_{Si}$	C_i	$C_i p_{Si}$
Under 20	.113	46.5	5.3	0	0.0
20–24	.330	42.8	14.1	26.1	8.6
25–29	.278	52.2	14.5	51.0	14.2
30–34	.173	101.3	17.5	74.7	12.9
35–39	.084	274.5	23.1	251.7	21.1
40 and over	.022	819.1	18.0	857.8	18.9
Sum			92.5 ($= c_{\text{direct}}$)		75.7 ($= C_{\text{direct}}$)

The conclusion drawn from a comparison of these direct adjusted rates is the same as that drawn before from the indirect adjusted rates: the rates of mongolism are about the same for infants born first and for infants born fifth or more. Such concordance between the conclusions drawn from comparisons of indirect and direct adjusted rates is usually, although not invariably, the case.

Given the consistent contrast between the specific rates shown in Table 14.3—in five of the six maternal age categories, the rate of mongolism among first-borns was slightly greater than the rate among infants born fifth or more—there was clearly no compelling reason for any standardization at all. The specific rates spoke for themselves. The only legitimate reason for standardization in such a case is the one cited first at the conclusion of Section 14.1, namely, the greater simplicity of working with a single summary index than of working with an entire schedule of specific rates.

There is one decided advantage to direct over indirect standardization. If, stratum by stratum, the specific rate in one group is equal to the specific rate in a second group, then, no matter which population is chosen as standard, the direct adjusted rates will be equal. Consider, for example, the specific rates presented in Table 14.2. The direct adjusted rate for group 1 is

$$c_{\text{direct}} = 2.0 \times p_{S1} + 1.0 \times p_{S2}, \tag{14.23}$$

and that for group 2 is obviously the same.

Direct standardization has a more general property. Consistent inequalities among specific rates, stratum by stratum, yield direct adjusted rates bearing the same inequalities. Thus if each specific rate in group 1 is greater than the corresponding rate in group 2, the direct adjusted rate for group 1 will be greater than that for group 2, no matter what the composition of the standard population.

These features of direct adjusted rates are actually quite trivial, for the circumstances leading to them are fully described by the specific rates themselves. Here the adjusted rates serve merely as convenient summarizations.

An important point to bear in mind is that an adjusted rate, no matter which method of adjustment is used, has meaning only when compared with a similarly adjusted rate. Its magnitude means little in and of itself. For the rates of Table 14.2, for example, the direct adjusted rate varies from 1.25 through 1.50 to 1.75 as the standard sex distribution varies from (.25, .75) through (.50, .50) to (.75, .25). No matter which standard is used, the direct adjusted rates for the two groups will be identical. The magnitude of the rate, however, is seen to depend strongly on the composition of the standard population. Spiegelman and Marks (1966) have shown that, in the direct standardization of mortality rates, the choice of a standard population generally has little effect on the differences between adjusted rates, and tends to affect only their individual magnitudes.

When the specific rates in the groups being compared do not bear consistent relations across the strata, then any kind of overall standardization is questionable. Problem 14.1 is concerned with the comparison of two groups for which the specific rates of one are higher than those for the other in some strata but are lower in other strata. It is shown that, depending on the strata in which the standard population is concentrated, either of the two groups can end up with the larger adjusted rate. The standard population may even be chosen to give equal adjusted rates. In addition, the phenomenon of a crossover in specific rates is lost by the usual methods of standardization.

A compromise calls for the calculation of a number of adjusted rates, one for each contiguous set of strata in which the specific rates bear consistent relations over the groups being compared. This device is illustrated in Problem 14.1, where two such sets of strata may be constituted. The division of the strata into such sets is not always easy, and sometimes may have to be forced. Nevertheless, working with a number of adjusted rates is preferable to working with one overall adjusted rate that may be more of a distortion than a summarization.

14.5. SOME OTHER SUMMARY INDICES

Woolsey (1959) and Kitagawa (1964) have reviewed a number of approaches to the standardization of rates. Formulas for determining standard errors have been given by Chiang (1961) and Keyfitz (1966).

Variations are sometimes encountered of the two kinds of adjusted rates considered so far. With respect to indirect standardization, the quantity

$$\text{SMR} = \frac{c}{c'} = \frac{c_{\text{indirect}}}{c_S} \tag{14.24}$$

may be used, where c_S is the rate for the standard population. This quantity, called the *standardized mortality ratio* (or the *standard mortality figure*) when the event studied is mortality, is merely the ratio of the actual to the expected crude rate. The SMR can be calculated for deaths from all causes or for deaths from a specific cause. Kupper, et al. (1978) have shown how, under some modest assumptions, inferences about the SMR for a specific cause of death can be based on an analysis of proportional mortality rates, that is, the ratios of numbers of deaths from a specific cause to the number of deaths from all causes. Gail (1978) presents methods for analyzing variations in the SMR across different populations.

The corresponding ratio for direct adjustment,

$$\text{CMF} = \frac{c_{\text{direct}}}{c_S}, \tag{14.25}$$

is dubbed the *comparative mortality figure* when applied to mortality.

Some other methods of adjustment exist but are less frequently used than the ones so far studied. One is a simple average of the crude and the direct adjusted rates,

$$\mathrm{CMR} = \tfrac{1}{2}(c + c_{\mathrm{direct}}) = \sum_{i=1}^{I} \tfrac{1}{2}(p_{Si} + p_i)c_i, \tag{14.26}$$

and is referred to as the *comparative mortality rate* when mortality is studied. Its infrequent use is testimony to its virtual uninterpretability.

Two indices are available for use with age-specific rates that in effect give equal weight to each year of age. Let n_i denote the number of years in the ith age interval, so that if, for example, the first age interval is 0–4 years, then $n_1 = 5$. The first such index (see Yule, 1934) is

$$\mathrm{EADR} = \frac{\sum_{i=1}^{I} n_i c_i}{\sum_{i=1}^{I} n_i}, \tag{14.27}$$

named the *equivalent average death rate* when applied to mortality. This index may be viewed as a direct adjusted rate where each year of age is assumed to have the same number of people.

The second such index (see Yerushalmy, 1951, and Elveback, 1966) is

$$\mathrm{MI} = \frac{\sum_{i=1}^{I} n_i \frac{c_i}{c_{Si}}}{\sum_{i=1}^{I} n_i}, \tag{14.28}$$

named the *mortality index* when applied to mortality. The mortality index is a simple average of the ratios of specific rates, with weights given by the numbers of years in the various age intervals. The usefulness of the latter two indices is limited because of the questionable validity of assigning equal importance to each year of age.

An index similar to (14.28) is

$$\mathrm{RMI} = \sum_{i=1}^{I} p_i \frac{c_i}{c_{Si}}, \tag{14.29}$$

named the *relative mortality index* when applied to mortality. The relative mortality index is also an average of the ratios of the actual specific rates to the standard population's specific rates, but is weighted by the given population's actual age distribution.

An equivalent expression is

$$\text{RMI} = \frac{\sum_{i=1}^{I} \frac{e_i}{c_{Si}}}{N}, \tag{14.30}$$

where e_i is the observed *number* of deaths (in general, of events) in the ith stratum and N is the total number of people in the given population. It is thus seen that one only needs, for the given population, its total size and the distribution over its strata of its total number of events in order to calculate the relative mortality index. An implication is that the relative mortality index may be calculated in years between censuses when, say, the age distribution of the population is not available but when the age distribution of deaths may be determined from registration data.

14.6. ADJUSTMENT FOR TWO FACTORS

Table 14.4 presents data on the incidence of mongolism specific both to birth order and to maternal age. The two methods to be described, the first

Table 14.4. Distribution of discovered mongoloids and of total live births by maternal age and birth order, Michigan, 1950–1964 (Values in cells are number of mongoloids found per number of live births)[a]

	Birth Order					
Maternal Age	1	2	3	4	5+	Total
Under 20	107/230,061	25/72,202	3/15,050	1/2293	0/327	136/319,933
20–24	141/329,449	150/326,701	71/175,702	26/68,800	8/30,666	396/931,318
25–29	60/114,920	110/208,667	114/207,081	64/132,424	63/123,419	411/786,511
30–34	40/39,487	84/83,228	103/117,300	89/98,301	112/149,919	428/488,235
35–39	39/14,208	82/28,466	108/45,026	137/46,075	262/104,088	628/237,863
40 and over	25/3052	39/5375	75/8660	96/9834	295/34,392	530/61,313
Total	412/731,177	490/724,639	474/568,819	413/357,727	740/442,811	2529/2,825,173

[a] Data from Stark and Mantel (1966).

based on direct and the second on indirect standardization, are useful when two factors are both associated with some disorder and with each other, and when one wishes to identify and measure their separate effects.

Simultaneous Direct Adjustment. Table 14.5 presents the various specific rates, the crude rates, and the direct adjusted rates.

Table 14.5. Incidence rates of discovered mongolism (number of mongoloids found per 100,000 live births) by maternal age and birth order

Maternal Age	Birth Order 1	2	3	4	5+	Crude Rate	Adjusted Rate[a]
Under 20	46.5	34.6	19.9	43.6	0	42.5	30.4
20–24	42.8	45.9	40.4	37.8	26.1	42.5	39.9
25–29	52.2	52.7	55.1	48.3	51.0	52.3	52.2
30–34	101.3	100.9	87.8	90.5	74.7	87.7	92.9
35–39	274.5	288.1	239.9	297.3	251.7	264.0	270.3
40 and over	819.1	725.6	866.1	976.2	857.8	864.4	830.4
Crude rate	56.3	67.6	83.3	115.5	167.1	89.5	
Adjusted rate[b]	92.3	91.2	85.1	92.7	75.5		88.0[c]

[a] The last column contains rates specific for maternal age and directly adjusted for birth order, with the standard birth-order distribution being that of the total sample.

[b] The last row contains rates specific for birth order and directly adjusted for maternal age, with the standard maternal age distribution being that of the total sample.

[c] Whenever the two standard distributions are those of the total sample, the overall rates based on the two series of directly adjusted rates will be equal to each other, but not necessarily to the overall crude rate.

In this case the specific rates speak for themselves. Within none of the maternal age categories is there any appreciable variation in the rates of mongolism specific to birth order. The increasing gradient with birth order of the crude rates is therefore likely to be only a reflection of the association between birth order and maternal age, and not of any direct relationship between birth order and the incidence of mongolism.

Maternal age, on the other hand, is seen to be strongly associated with the incidence of mongolism. Within each birth-order category, there is a clear increase in the incidence rate with increasing maternal age.

The direct adjusted rates shown in the last row and last column of Table 14.5 serve only to summarize the information provided by the 30 specific rates: little effect of birth order but a strong effect of maternal age on the incidence of mongolism. In fact, direct adjustment is appropriate only because of the consistency found both within maternal age categories (rather little variability among the birth-order-specific rates) and within birth-order categories (a clear gradient with increasing maternal age).

Simultaneous Indirect Adjustment. If all the specific rates are available, and all are based on samples of sufficient size to possess adequate precision, then simultaneous direct adjustment, the method just described, may be applied. If the specific rates are not available or are based on relatively small sample sizes, then a method due to Mantel and Stark (1968), based on indirect adjustment, may be applied.

It might have happened that the data required to calculate rates specific for maternal age and birth order simultaneously did not exist. In fact, the only data available might have been those in Table 14.6, together with the crude rates for birth order and for maternal age.

Table 14.6. Distribution of total live births by maternal age and birth order, overall crude rates, and indirectly adjusted rates

	Birth Order					Crude	Adjusted
Maternal Age	1	2	3	4	5+	Rate	Rate[a]
Under 20	230,061	72,202	15,050	2,293	327	42.5	62.7
20–24	329,449	326,701	175,702	68,800	30,666	42.5	51.9
25–29	114,920	208,667	207,081	132,424	123,419	52.3	49.9
30–34	39,487	83,228	117,300	98,301	149,919	87.7	70.9
35–39	14,208	28,466	45,026	46,075	104,088	264.0	192.6
40 and over	3,052	5,375	8,660	9,834	34,392	864.4	582.9
Crude rate	56.3	67.6	83.3	115.5	167.1	89.5	79.2[c]
Adjusted rate[b]	93.0	92.7	87.3	94.3	84.8	90.7[c]	

[a]The last column contains rates specific for maternal age and indirectly adjusted for birth order, with the standard set of rates specific for birth order being that of the total sample. Thus, for example,

$$c_{20\text{–}24(\text{indirect})} = 51.9 = 89.5 \times \frac{42.5 \times 9.31318}{56.3 \times 3.29449 + \cdots + 167.1 \times .30666}.$$

[b]The last row contains rates specific for birth order and indirectly adjusted for maternal age, with the standard set of rates specific for maternal age being that of the total sample. Thus, for example,

$$c_{2(\text{indirect})} = 92.7 = 89.5 \times \frac{67.6 \times 7.24639}{42.5 \times .72202 + \cdots + 864.4 \times .05375}.$$

[c]The overall rates based on the two series of indirectly adjusted rates will almost never equal each other, nor will either of them equal the overall crude rate.

In contrast to the summarizing role played by the direct adjusted rates, the indirect adjusted rates, presented in the last row and last column of Table 14.6, must almost of necessity be calculated when the schedules of specific rates do not exist. They must be interpreted with a great deal of caution, however.

Two awkward features of the indirectly adjusted rates can be seen. For one thing, they do not yield equal overall rates. For another, the indirectly adjusted rates for some of the maternal age categories are totally out of the range of the specific rates (compare, e.g., the indirect adjusted rate of 62.7 for the first maternal age category with the specific rates ranging from 0 to 46.5, as seen in Table 14.5).

As a means of correcting these and possibly other anomalies, Mantel and Stark (1968) recommend the following procedure.

1. Beginning with the schedule of crude rates specific to maternal age, obtain the indirect adjusted rates specific for birth order. These have already been calculated and appear in the final row of Table 14.6.

2. Using *these latter rates* as the standard schedule, calculate the indirect adjusted rates specific for maternal age. Multiply the obtained rates by the ratio of the expected total crude rate (90.7) to the actual crude rate (89.5):

Maternal age	Under 20	20–24	25–29	30–34	35–39	40 and Over
Adjusted rate	41.6	42.0	52.4	89.0	270.3	892.9

For example, 41.6 is calculated as

$$41.6 = \frac{42.5}{92.7} \times 89.5 \times \frac{90.7}{89.5} = \frac{42.5 \times 90.7}{92.7},$$

where 42.5 is the crude rate in the first maternal age category, 90.7 is the overall expected crude rate, and

$$92.7 = \frac{93.0 \times 2.30061 + 92.7 \times .72202 + \cdots + 84.8 \times .00327}{3.19933}.$$

3. Continue in the same manner, each time using the previous set of indirect adjusted rates for one of the two variables to generate a new set for the other variable. Multiply each rate in the new set by the ratio of the expected total crude rate based on the previous set to the actual total crude rate.

4. Stop the process when successive sets of rates are unchanged. Here, about four cycles were necessary.

5. The sets of rates we end up with are

Maternal age	Under 20	20–24	25–29	30–34	35–39	40 and Over
Adjusted rate	41.0	41.8	52.6	89.7	273.3	904.5

for maternal age and

Birth order	1	2	3	4	5+
Adjusted rate	95.2	93.7	87.3	93.6	83.6

for birth order.

6. These two schedules of rates have the property that either, when used as the standard, implies the other. Furthermore, each of these rates happens for these data to lie within the range of the specific rates. This need not always occur with the Mantel-Stark adjustment method. Finally, their adjusted rates both yield the same expected total rate, 91.2. This final property would not have held had we not, at every step, multiplied the generated set of rates by the ratio of the total expected rate based on the preceding set of rates to the total observed rate.

The inferences from these adjusted rates are the same as were drawn previously: a strong effect of maternal age and little if any effect of birth order. Recall, however, that here we have made no use at all of the specific rates.

The Mantel-Stark procedure has the property that it will yield the same results no matter what set of rates one starts with, although the number of steps will vary with the starting set. An alternative approach to the problem of separating the effects of two correlated factors is to use log linear or logistic regression models (Berry, 1970; Breslow and Day, 1975; Gail, 1978). The required arithmetic, however, is more complex than for the Mantel-Stark method.

Problem 14.1. The following data are from Table 2 of Discher and Feinberg (1969).

Age-specific rates of abnormal lung functioning in males employed in manufacturing or service industries

	Manufacturing		Services	
Age Interval	Number	% Abnormal	Number	% Abnormal
20–29	403	2.2	256	4.8
30–39	688	3.2	525	3.2
40–49	683	2.2	599	2.8
50–59	539	6.9	453	6.6
60+	133	12.8	155	9.0

(*a*) Comparing the age-specific rates for the two kinds of industries, what conclusions would you be willing to draw?

(*b*) What do you think would be gained by calculating an age-adjusted rate for each kind of industry and then comparing the adjusted rates? What do you think might be lost?

(*c*) Consider the three following standard age distributions.

	Standard		
Age Interval	1	2	3
20–29	.25	.05	.07
30–39	.25	.05	.75
40–49	.30	.10	.06
50–59	.10	.40	.06
60 +	.10	.40	.06

1. The first standard distribution is concentrated below age 49. What are the values of the two direct adjusted rates using this standard? What is the direction and magnitude of difference between them?

2. The second standard distribution is concentrated above age 50. What are the values of the two direct adjusted rates using this standard? What is the direction and magnitude of difference between them?

3. The third standard distribution is concentrated in the age interval 30–39. What are the values of the two direct adjusted rates using this standard? What is the direction and magnitude of difference between them?

(*d*) Using the age distribution of the total sample as a standard, calculate two direct adjusted rates for both kinds of industries, one for the age interval 20–49 and the other for the interval 50 and over. Do the comparisons now seem fair to the data?

REFERENCES

Berry, G. (1970). Parametric analysis of disease incidences in multiway tables. *Biometrics*, **26**, 572–579.

Breslow, N. E. and Day, N. E. (1975). Indirect standardization and multiplicative models for rates, with reference to the age adjustment of cancer incidence and relative frequency data. *J. Chronic Dis.*, **28**, 289–303.

Chiang, C. L. (1961). Standard error of the age-adjusted death rate. U.S. Department of Health, Education and Welfare: Vital Statistics—Special Reports, **47**, 271–285.

Cochran, W. G. (1968). The effectiveness of adjustment by subclassification in removing bias in observational studies. *Biometrics*, **24**, 295–313.

Discher, D. P. and Feinberg, H. C. (1969). Screening for chronic pulmonary disease: Survey of 10,000 industrial workers. *Am. J. Public Health*, **59**, 1857–1867.

Doll, R. and Cook, P. (1967). Summarizing indices for comparison of cancer incidence data. *Int. J. Cancer*, **2**, 269–279.

El-Badry, M. A. (1969). Higher female than male mortality in some countries of south Asia: A digest. *J. Am. Stat. Assoc.*, **64**, 1234–1244.

Elveback, L. R. (1966). Discussion of "Indices of mortality and tests of their statistical significance." *Hum. Biol.*, **38**, 322–324.

Gail, M. (1978). The analysis of heterogeneity for indirect standardized mortality ratios. *J. R. Stat. Soc., Ser. A*, **141**, 224–234.

Hemphill, F. M. and Ament, R. P. (1970). Quantitative assessment of subcategory contributions to observed change in relative frequencies. Paper read at annual meeting of American Public Health Association, Houston.

Kalton, G. (1968). Standardization: A technique to control for extraneous variables. *Appl. Stat.*, **17**, 118–136.

Keyfitz, N. (1966). Sampling variance of standardized mortality rates. *Hum. Biol.*, **38**, 309–317.

Kilpatrick, S. J. (1963). Mortality comparisons in socio-economic groups. *Appl. Stat.*, **12**, 65–86.

Kitagawa, E. M. (1955). Components of a difference between two rates. *J. Am. Stat. Assoc.*, **50**, 1168–1194.

Kitagawa, E. M. (1964). Standardized comparisons in population research. *Demography*, **1**, 296–315.

Kitagawa, E. M. (1966). Theoretical considerations in the selection of a mortality index, and some empirical comparisons. *Hum. Biol.*, **38**, 293–308.

Kupper, L. L., McMichael, A. J., Symons, M. J., and Most, B. M. (1978). On the utility of proportional mortality analysis. *J. Chronic Dis.*, **31**, 15–22.

Mantel, N. and Stark, C. R. (1968). Computation of indirect-adjusted rates in the presence of confounding. *Biometrics*, **24**, 997–1005.

Mausner, J. S. and Bahn, A. K. (1974). *Epidemiology: An introductory text*. Philadelphia: W. B. Saunders.

Miettinen, O. S. (1972). Components of the crude risk ratio. *Am. J. Epidemiol.*, **96**, 168–172.

Spiegelman, M. and Marks, H. H. (1966). Empirical testing of standards for the age adjustment of death rates by the direct method. *Hum. Biol.*, **38**, 280–292.

Stark, C. R. and Mantel, N. (1966). Effects of maternal age and birth order on the risk of mongolism and leukemia. *J. Natl. Cancer Inst.*, **37**, 687–698.

Woolsey, T. D. (1959). Adjusted death rates and other indices of mortality. Chapter 4 in F. E. Linder and R. D. Grove. *Vital statistics rates in the United States, 1900–1940*. Washington, D.C.: U.S. Government Printing Office.

Yerushalmy, J. (1951). A mortality index for use in place of the age-adjusted death rate. *Am. J. Public Health*, **41**, 907–922.

Yule, G. U. (1934). On some points relating to vital statistics, more especially statistics of occupation mortality. *J. R. Stat. Soc.*, **97**, 1–72.

Appendix

Table A.1. Critical values of the chi square distribution

DEGREES OF FREEDOM	ALPHA .10	.05	.025	.01	.005	.001
1	2.71	3.84	5.02	6.63	7.88	10.83
2	4.61	5.99	7.38	9.21	10.60	13.82
3	6.25	7.81	9.35	11.34	12.84	16.27
4	7.78	9.49	11.14	13.28	14.86	18.47
5	9.24	11.07	12.83	15.09	16.75	20.52
6	10.64	12.59	14.45	16.81	18.55	22.46
7	12.02	14.07	16.01	18.48	20.28	24.32
8	13.36	15.51	17.53	20.09	21.96	26.12
9	14.68	16.92	19.02	21.67	23.59	27.88
10	15.99	18.31	20.48	23.21	25.19	29.59
11	17.28	19.68	21.92	24.72	26.76	31.26
12	18.55	21.03	23.34	26.22	28.30	32.91
13	19.81	22.36	24.74	27.69	29.82	34.53
14	21.06	23.68	26.12	29.14	31.32	36.12
15	22.31	25.00	27.49	30.58	32.80	37.70
16	23.54	26.30	28.85	32.00	34.27	39.25
17	24.77	27.59	30.19	33.41	35.72	40.79
18	25.99	28.87	31.53	34.81	37.16	42.31
19	27.20	30.14	32.85	36.19	38.58	43.82
20	28.41	31.41	34.17	37.57	40.00	45.32
25	34.38	37.65	40.65	44.31	46.93	52.62
30	40.26	43.77	46.98	50.89	53.67	59.70
40	51.80	55.76	59.34	63.69	66.77	73.40
60	74.40	79.08	83.30	88.38	91.95	99.61
100	118.50	124.34	129.56	135.81	140.17	149.45

ABRIDGED FROM TABLE 8 OF ''BIOMETRIKA TABLES FOR STATISTICIANS, VOL. I, 2ND EDITION,'' EDITED BY E.S. PEARSON AND H.O. HARTLEY. CAMBRIDGE UNIVERSITY PRESS, CAMBRIDGE, ENGLAND, 1958.

Table A.2. Critical values of the normal distribution

P = AREA IN THE TAILS OF THE NORMAL CURVE BELOW -Z AND ABOVE +Z. IN A TEST OF SIGNIFICANCE, P IS THE SIGNIFICANCE LEVEL ASSOCIATED WITH THE OBTAINED VALUE OF Z.

THE TOTAL AREA UNDER THE NORMAL CURVE TO THE RIGHT OF Z IS 1 - P/2 IF Z IS NEGATIVE, AND IS P/2 IF Z IS POSITIVE.

SUPPOSE ONE MUST FIND THAT VALUE OF Z SUCH THAT THE TOTAL AREA UNDER THE NORMAL CURVE TO THE RIGHT OF Z IS 1 - B. IF 1 - B IS GREATER THAN 0.50, TAKE THE VALUE OF Z CORRESPONDING TO P = 2B, AND AFFIX A MINUS SIGN TO IT. IF 1 - B IS LESS THAN 0.50, TAKE THE VALUE OF Z CORRESPONDING TO P = 2(1 - B).

Z	P	Z	P	Z	P
0.0	1.0000	1.2	0.2301	2.4	0.0164
0.1	0.9203	1.282	0.20	2.5	0.0124
0.126	0.90	1.3	0.1936	2.576	0.01
0.2	0.8415	1.4	0.1615	2.6	0.0093
0.3	0.7642	1.440	0.15	2.7	0.0069
0.385	0.70	1.5	0.1336	2.8	0.0051
0.4	0.6892	1.6	0.1096	2.813	0.005
0.5	0.6171	1.645	0.10	2.9	0.0037
0.524	0.60	1.7	0.0891	3.0	0.0027
0.6	0.5485	1.8	0.0719	3.090	0.002
0.674	0.50	1.9	0.0574	3.1	0.0019
0.7	0.4839	1.960	0.05	3.2	0.0014
0.8	0.4237	2.0	0.0455	3.3	0.0010
0.842	0.40	2.1	0.0357	3.4	0.0007
0.9	0.3681	2.2	0.0278	3.5	0.0005
1.0	0.3173	2.242	0.025	3.6	0.0003
1.036	0.30	2.3	0.0214	3.7	0.0002
1.1	0.2713	2.326	0.02	3.8	0.0001

ADAPTED FROM TABLES 1 AND 4 OF ''BIOMETRIKA TABLES FOR STATISTICIANS, VOL. I, 2ND EDITION,'' EDITED BY E.S. PEARSON AND H.O. HARTLEY. CAMBRIDGE UNIVERSITY PRESS, CAMBRIDGE, ENGLAND, 1958.

Table A.3. Sample sizes per group for a two-tailed test on proportions. $P_1 = .05$

P2	ALPHA	POWER								
		0.99	0.95	0.90	0.85	0.80	0.75	0.70	0.65	0.50
0.10	0.01	1368	1025	863	762	686	624	572	525	407
	0.02	1235	911	760	665	595	538	489	446	339
	0.05	1054	758	621	536	474	423	381	344	252
	0.10	910	637	513	437	381	336	299	267	188
	0.20	758	512	402	336	288	250	219	192	128
0.15	0.01	447	337	285	253	228	209	192	177	139
	0.02	404	300	252	221	199	180	165	151	117
	0.05	345	251	207	179	160	143	130	118	88
	0.10	299	212	172	147	130	115	103	93	67
	0.20	250	171	136	115	99	87	77	68	47
0.20	0.01	241	183	155	138	125	115	106	98	77
	0.02	218	163	137	121	109	99	91	84	65
	0.05	187	136	113	99	88	79	72	66	50
	0.10	162	115	94	81	72	64	58	52	38
	0.20	135	94	75	64	55	49	44	39	28
0.25	0.01	157	120	102	91	83	76	70	65	52
	0.02	142	107	90	80	72	66	61	56	44
	0.05	122	90	75	65	58	53	48	44	34
	0.10	106	76	62	54	48	43	39	35	26
	0.20	88	62	50	42	37	33	29	26	19
0.30	0.01	113	87	74	66	60	55	51	48	38
	0.02	102	77	66	58	53	48	44	41	33
	0.05	88	65	54	48	43	39	35	32	25
	0.10	76	55	45	39	35	32	29	26	20
	0.20	64	45	36	31	27	24	22	20	14
0.35	0.01	86	66	57	51	46	43	40	37	30
	0.02	78	59	50	45	41	37	34	32	25
	0.05	67	50	42	37	33	30	28	25	20
	0.10	58	42	35	30	27	25	22	20	16
	0.20	48	34	28	24	21	19	17	16	12
0.40	0.01	68	53	45	41	37	34	32	30	24
	0.02	62	47	40	36	33	30	28	26	21
	0.05	53	39	33	29	27	24	22	21	16
	0.10	46	34	28	24	22	20	18	17	13
	0.20	38	27	22	19	17	15	14	13	10
0.45	0.01	55	43	37	33	31	28	26	25	20
	0.02	50	38	33	29	27	25	23	21	17
	0.05	43	32	27	24	22	20	18	17	14
	0.10	37	27	23	20	18	16	15	14	11
	0.20	31	22	18	16	14	13	12	11	8
0.50	0.01	46	36	31	28	26	24	22	21	17
	0.02	41	32	27	25	23	21	19	18	15
	0.05	35	27	23	20	18	17	16	14	12
	0.10	31	23	19	17	15	14	13	12	9
	0.20	26	19	15	13	12	11	10	9	7
0.55	0.01	38	30	26	24	22	20	19	18	15
	0.02	35	27	23	21	19	18	17	16	13
	0.05	30	23	19	17	16	14	13	12	10
	0.10	26	19	16	14	13	12	11	10	8
	0.20	21	16	13	11	10	9	9	8	6

Table A.3. Sample sizes per group for a two-tailed test on proportions. $P_1 = .05$

P2	ALPHA	POWER 0.99	0.95	0.90	0.85	0.80	0.75	0.70	0.65	0.50
0.60	0.01	32	26	22	20	19	18	16	16	13
	0.02	29	23	20	18	17	15	14	13	11
	0.05	25	19	16	15	14	12	12	11	9
	0.10	22	16	14	12	11	10	9	9	7
	0.20	18	13	11	10	9	8	7	7	5
0.65	0.01	28	22	19	18	16	15	14	14	11
	0.02	25	20	17	16	14	13	13	12	10
	0.05	21	16	14	13	12	11	10	9	8
	0.10	18	14	12	11	10	9	8	8	6
	0.20	15	11	10	9	8	7	7	6	5
0.70	0.01	23	19	17	15	14	13	13	12	10
	0.02	21	17	15	14	13	12	11	10	9
	0.05	18	14	12	11	10	10	9	8	7
	0.10	16	12	10	9	9	8	7	7	6
	0.20	13	10	8	7	7	6	6	5	4
0.75	0.01	20	16	15	13	13	12	11	11	9
	0.02	18	15	13	12	11	10	10	9	8
	0.05	15	12	11	10	9	8	8	7	6
	0.10	13	10	9	8	7	7	7	6	5
	0.20	11	8	7	6	6	5	5	5	4
0.80	0.01	17	14	13	12	11	10	10	9	8
	0.02	15	13	11	10	10	9	9	8	7
	0.05	13	10	9	8	8	7	7	7	6
	0.10	11	9	8	7	7	6	6	5	5
	0.20	9	7	6	6	5	5	5	4	4
0.85	0.01	15	12	11	10	10	9	9	8	7
	0.02	13	11	10	9	8	8	8	7	6
	0.05	11	9	8	7	7	7	6	6	5
	0.10	9	8	7	6	6	5	5	5	4
	0.20	8	6	5	5	5	4	4	4	3
0.90	0.01	12	10	9	9	8	8	8	7	7
	0.02	11	9	8	8	7	7	7	7	6
	0.05	9	8	7	6	6	6	6	5	5
	0.10	8	6	6	5	5	5	5	4	4
	0.20	6	5	5	4	4	4	4	3	3
0.95	0.01	10	9	8	8	7	7	7	7	6
	0.02	9	8	7	7	7	6	6	6	5
	0.05	8	6	6	6	5	5	5	5	4
	0.10	6	5	5	5	4	4	4	4	4
	0.20	5	4	4	4	4	3	3	3	3

Table A.3. Sample sizes per group for a two-tailed test on proportions. $P_1 = .10$

		POWER								
P2	ALPHA	0.99	0.95	0.90	0.85	0.80	0.75	0.70	0.65	0.50
0.15	0.01	2137	1595	1340	1179	1060	963	880	806	620
	0.02	1928	1416	1176	1027	916	826	749	682	513
	0.05	1642	1174	957	823	725	646	579	520	375
	0.10	1415	984	787	667	579	509	450	399	275
	0.20	1175	787	613	508	433	373	324	281	182
0.20	0.01	627	471	397	351	316	288	264	243	189
	0.02	566	419	349	306	274	248	226	206	157
	0.05	483	348	286	247	219	196	176	159	117
	0.10	417	293	236	201	176	156	139	124	88
	0.20	347	236	185	155	133	116	102	89	60
0.25	0.01	316	238	202	179	162	147	136	125	98
	0.02	285	212	178	156	140	127	116	107	82
	0.05	244	177	146	127	113	101	91	83	62
	0.10	211	149	121	104	91	81	73	65	47
	0.20	176	120	96	81	70	61	54	48	33
0.30	0.01	196	149	126	112	102	93	86	79	63
	0.02	178	133	112	98	89	81	74	68	53
	0.05	152	111	92	80	71	64	58	53	40
	0.10	131	94	76	66	58	52	47	42	31
	0.20	110	76	61	51	45	39	35	31	22
0.35	0.01	136	104	88	79	72	66	61	56	45
	0.02	123	93	78	69	62	57	52	48	38
	0.05	105	77	64	56	50	45	41	38	29
	0.10	91	65	54	46	41	37	33	30	22
	0.20	76	53	43	36	32	28	25	22	16
0.40	0.01	101	77	66	59	54	49	46	42	34
	0.02	91	69	58	52	47	43	39	36	29
	0.05	78	58	48	42	38	34	31	29	22
	0.10	68	49	40	35	31	28	25	23	17
	0.20	56	40	32	27	24	21	19	17	13
0.45	0.01	78	60	51	46	42	39	36	33	27
	0.02	71	54	46	41	37	34	31	29	23
	0.05	60	45	38	33	30	27	25	23	18
	0.10	52	38	31	27	24	22	20	18	14
	0.20	44	31	25	22	19	17	15	14	10
0.50	0.01	62	48	41	37	34	31	29	27	22
	0.02	56	43	37	33	30	27	25	23	19
	0.05	48	36	30	27	24	22	20	19	15
	0.10	42	30	25	22	20	18	16	15	12
	0.20	35	25	20	18	16	14	13	11	9
0.55	0.01	51	39	34	31	28	26	24	23	19
	0.02	46	35	30	27	25	23	21	20	16
	0.05	39	29	25	22	20	18	17	16	12
	0.10	34	25	21	18	16	15	14	13	10
	0.20	28	20	17	15	13	12	11	10	7
0.60	0.01	42	33	28	26	24	22	20	19	16
	0.02	38	29	25	23	21	19	18	17	14
	0.05	32	24	21	18	17	15	14	13	11
	0.10	28	21	17	15	14	13	12	11	8
	0.20	23	17	14	12	11	10	9	8	6

Table A.3. Sample sizes per group for a two-tailed test on proportions. $P_1 = .10$

		POWER								
P2	ALPHA	0.99	0.95	0.90	0.85	0.80	0.75	0.70	0.65	0.50
0.65	0.01	35	27	24	22	20	19	17	16	14
	0.02	31	24	21	19	18	16	15	14	12
	0.05	27	20	18	16	14	13	12	11	9
	0.10	23	17	15	13	12	11	10	9	7
	0.20	19	14	12	10	9	8	8	7	6
0.70	0.01	29	23	20	19	17	16	15	14	12
	0.02	26	21	18	16	15	14	13	12	10
	0.05	22	17	15	13	12	11	11	10	8
	0.10	19	15	13	11	10	9	9	8	7
	0.20	16	12	10	9	8	7	7	6	5
0.75	0.01	25	20	17	16	15	14	13	12	11
	0.02	22	18	15	14	13	12	11	11	9
	0.05	19	15	13	12	11	10	9	9	7
	0.10	16	13	11	10	9	8	8	7	6
	0.20	14	10	9	8	7	6	6	6	4
0.80	0.01	21	17	15	14	13	12	11	11	9
	0.02	19	15	13	12	11	11	10	9	8
	0.05	16	13	11	10	9	9	8	8	6
	0.10	14	11	9	8	8	7	7	6	5
	0.20	11	9	7	7	6	6	5	5	4
0.85	0.01	18	14	13	12	11	11	10	10	8
	0.02	16	13	11	11	10	9	9	8	7
	0.05	13	11	9	9	8	8	7	7	6
	0.10	11	9	8	7	7	6	6	6	5
	0.20	10	7	6	6	5	5	5	4	4
0.90	0.01	15	12	11	10	10	9	9	8	7
	0.02	13	11	10	9	9	8	8	7	6
	0.05	11	9	8	7	7	7	6	6	5
	0.10	10	8	7	6	6	5	5	5	4
	0.20	8	6	6	5	5	4	4	4	3
0.95	0.01	12	10	9	9	8	8	8	7	7
	0.02	11	9	8	8	7	7	7	7	6
	0.05	9	8	7	6	6	6	6	5	5
	0.10	8	6	6	5	5	5	5	4	4
	0.20	6	5	5	4	4	4	4	3	3

Table A.3. Sample sizes per group for a two-tailed test on proportions. $P_1 = .15$

P2	ALPHA	POWER 0.99	0.95	0.90	0.85	0.80	0.75	0.70	0.65	0.50
0.20	0.01	2810	2094	1756	1545	1388	1259	1149	1052	806
	0.02	2534	1858	1541	1343	1198	1078	977	888	664
	0.05	2157	1538	1252	1075	945	840	751	674	483
	0.10	1856	1287	1027	868	753	660	582	515	351
	0.20	1539	1027	797	659	559	480	415	360	228
0.25	0.01	783	586	494	435	392	357	326	300	232
	0.02	707	521	434	380	340	307	279	254	193
	0.05	603	433	354	305	270	241	216	195	142
	0.10	520	363	292	248	216	191	169	151	106
	0.20	432	291	228	190	163	141	123	108	71
0.30	0.01	380	286	241	213	193	176	161	148	116
	0.02	343	254	213	187	167	152	138	126	97
	0.05	293	212	174	151	134	120	108	98	72
	0.10	253	178	144	123	108	95	85	76	54
	0.20	211	143	113	95	82	71	[illegible]	55	38
0.35	0.01	229	173	147	130	118	108	99	91	72
	0.02	207	154	130	114	102	93	85	78	60
	0.05	177	129	106	92	82	74	67	61	45
	0.10	153	109	88	76	67	59	53	48	35
	0.20	128	88	70	59	51	45	39	35	24
0.40	0.01	155	118	100	89	81	74	68	63	50
	0.02	140	105	89	78	70	64	59	54	42
	0.05	120	88	73	63	57	51	46	42	32
	0.10	104	74	60	52	46	41	37	33	25
	0.20	87	60	48	41	35	31	28	25	18
0.45	0.01	113	86	73	65	60	55	50	47	37
	0.02	102	77	65	57	52	47	44	40	32
	0.05	87	64	53	47	42	38	34	32	24
	0.10	75	54	44	39	34	31	28	25	19
	0.20	63	44	35	30	26	23	21	19	14
0.50	0.01	86	66	56	50	46	42	39	36	29
	0.02	78	59	50	44	40	37	34	31	25
	0.05	66	49	41	36	32	29	27	25	19
	0.10	57	42	34	30	26	24	22	20	15
	0.20	48	34	27	23	21	18	16	15	11
0.55	0.01	67	52	45	40	37	34	31	29	24
	0.02	61	46	39	35	32	29	27	25	20
	0.05	52	39	33	29	26	24	22	20	16
	0.10	45	33	27	24	21	19	17	16	12
	0.20	38	27	22	19	17	15	13	12	9
0.60	0.01	54	42	36	33	30	28	26	24	20
	0.02	49	37	32	29	26	24	22	21	17
	0.05	42	31	26	23	21	19	18	16	13
	0.10	36	27	22	19	17	16	14	13	10
	0.20	30	22	18	15	14	12	11	10	8
0.65	0.01	44	34	30	27	25	23	21	20	16
	0.02	40	31	26	24	22	20	19	17	14
	0.05	34	26	22	19	18	16	15	14	11
	0.10	29	22	18	16	15	13	12	11	9
	0.20	25	18	15	13	11	10	9	9	7

Table A.3. Sample sizes per group for a two-tailed test on proportions. $P_1 = .15$

P2	ALPHA	POWER 0.99	0.95	0.90	0.85	0.80	0.75	0.70	0.65	0.50
0.70	0.01	36	29	25	23	21	19	18	17	14
	0.02	33	26	22	20	18	17	16	15	12
	0.05	28	21	18	16	15	14	13	12	9
	0.10	24	18	15	14	12	11	10	10	8
	0.20	20	15	12	11	10	9	8	7	6
0.75	0.01	30	24	21	19	18	16	15	15	12
	0.02	27	21	19	17	16	14	13	13	11
	0.05	23	18	15	14	13	12	11	10	8
	0.10	20	15	13	11	10	10	9	8	7
	0.20	17	12	10	9	8	8	7	6	5
0.80	0.01	25	20	18	16	15	14	13	13	11
	0.02	23	18	16	14	13	12	12	11	9
	0.05	19	15	13	12	11	10	9	9	7
	0.10	17	13	11	10	9	8	8	7	6
	0.20	14	10	9	8	7	7	6	6	4
0.85	0.01	21	17	15	14	13	12	12	11	9
	0.02	19	15	13	12	11	11	10	10	8
	0.05	16	13	11	10	9	9	8	8	6
	0.10	14	11	9	8	8	7	7	6	5
	0.20	12	9	8	7	6	6	5	5	4
0.90	0.01	18	14	13	12	11	11	10	10	8
	0.02	16	13	11	11	10	9	9	8	7
	0.05	13	11	9	9	8	8	7	7	6
	0.10	11	9	8	7	7	6	6	6	5
	0.20	10	7	6	6	5	5	5	4	4
0.95	0.01	15	12	11	10	10	9	9	8	7
	0.02	13	11	10	9	8	8	8	7	6
	0.05	11	9	8	7	7	7	6	6	5
	0.10	9	8	7	6	6	5	5	5	4
	0.20	8	6	5	5	5	4	4	4	3

Table A.3. Sample sizes per group for a two-tailed test on proportions. $P_1 = .20$

					POWER					
P2	ALPHA	0.99	0.95	0.90	0.85	0.80	0.75	0.70	0.65	0.50
0.25	0.01	3386	2522	2114	1858	1668	1512	1379	1262	965
	0.02	3053	2236	1853	1615	1438	1294	1172	1064	794
	0.05	2597	1850	1504	1290	1134	1007	900	806	575
	0.10	2235	1547	1233	1041	901	789	695	614	417
	0.20	1852	1232	955	788	668	572	494	426	268
0.30	0.01	915	685	576	507	456	415	379	348	268
	0.02	826	608	506	442	395	356	324	295	222
	0.05	704	504	412	355	313	279	250	225	163
	0.10	607	423	339	288	250	220	195	174	121
	0.20	504	339	265	220	188	162	141	123	80
0.35	0.01	433	325	274	242	219	199	183	168	131
	0.02	391	289	242	212	190	172	156	143	109
	0.05	334	240	197	171	151	135	122	110	81
	0.10	288	202	163	139	122	107	96	85	61
	0.20	240	162	128	107	92	80	70	62	41
0.40	0.01	256	193	164	145	131	120	110	101	79
	0.02	232	172	144	127	114	103	94	86	66
	0.05	198	143	118	102	91	82	74	67	50
	0.10	171	121	98	84	74	65	58	52	38
	0.20	142	97	77	65	56	49	43	38	26
0.45	0.01	171	129	110	98	88	81	74	69	54
	0.02	154	115	97	85	77	70	64	59	46
	0.05	132	96	79	69	62	56	50	46	35
	0.10	114	81	66	57	50	45	40	36	26
	0.20	95	65	52	44	38	34	30	27	19
0.50	0.01	122	93	79	71	64	59	54	50	40
	0.02	110	83	70	62	56	51	47	43	34
	0.05	94	69	57	50	45	41	37	34	26
	0.10	82	58	48	41	37	33	29	27	20
	0.20	68	47	38	32	28	25	22	20	14
0.55	0.01	92	70	60	54	49	45	41	38	31
	0.02	83	63	53	47	43	39	36	33	26
	0.05	71	52	44	38	34	31	28	26	20
	0.10	61	44	36	32	28	25	23	21	16
	0.20	51	36	29	25	22	19	17	15	11
0.60	0.01	71	55	47	42	38	35	33	31	25
	0.02	64	49	42	37	34	31	28	26	21
	0.05	55	41	34	30	27	25	23	21	16
	0.10	47	35	29	25	22	20	18	17	13
	0.20	40	28	23	20	17	15	14	13	9
0.65	0.01	56	44	38	34	31	29	27	25	20
	0.02	51	39	33	30	27	25	23	21	17
	0.05	44	33	27	24	22	20	18	17	13
	0.10	38	28	23	20	18	16	15	14	11
	0.20	31	22	18	16	14	13	11	10	8
0.70	0.01	46	36	31	28	25	24	22	21	17
	0.02	41	32	27	24	22	21	19	18	14
	0.05	35	26	22	20	18	17	15	14	11
	0.10	30	22	19	17	15	14	12	11	9
	0.20	25	18	15	13	12	11	10	9	7

Table A.3. Sample sizes per group for a two-tailed test on proportions. $P_1 = .20$

		POWER								
P2	ALPHA	0.99	0.95	0.90	0.85	0.80	0.75	0.70	0.65	0.50
0.75	0.01	37	29	25	23	21	20	18	17	14
	0.02	34	26	23	20	19	17	16	15	12
	0.05	29	22	19	17	15	14	13	12	10
	0.10	25	18	16	14	12	11	10	10	8
	0.20	21	15	12	11	10	9	8	7	6
0.80	0.01	31	24	21	19	18	17	16	15	12
	0.02	28	22	19	17	16	15	14	13	11
	0.05	23	18	16	14	13	12	11	10	8
	0.10	20	15	13	12	11	10	9	8	7
	0.20	17	12	10	9	8	8	7	6	5
0.85	0.01	25	20	18	16	15	14	13	13	11
	0.02	23	18	16	14	13	12	12	11	9
	0.05	19	15	13	12	11	10	9	9	7
	0.10	17	13	11	10	9	8	8	7	6
	0.20	14	10	9	8	7	7	6	6	4
0.90	0.01	21	17	15	14	13	12	11	11	9
	0.02	19	15	13	12	11	11	10	9	8
	0.05	16	13	11	10	9	9	8	8	6
	0.10	14	11	9	8	8	7	7	6	5
	0.20	11	9	7	7	6	6	5	5	4
0.95	0.01	17	14	13	12	11	10	10	9	8
	0.02	15	13	11	10	10	9	9	8	7
	0.05	13	10	9	8	8	7	7	7	6
	0.10	11	9	8	7	7	6	6	5	5
	0.20	9	7	6	6	5	5	5	4	4

Table A.3. Sample sizes per group for a two-tailed test on proportions. $P_1 = .25$

		POWER								
P2	ALPHA	0.99	0.95	0.90	0.85	0.80	0.75	0.70	0.65	0.50
0.30	0.01	3867	2878	2411	2119	1902	1723	1572	1438	1098
	0.02	3486	2552	2114	1841	1639	1474	1334	1211	902
	0.05	2965	2110	1714	1470	1291	1145	1023	916	652
	0.10	2550	1764	1404	1185	1025	897	789	696	471
	0.20	2112	1404	1087	895	758	649	559	482	301
0.35	0.01	1023	765	643	566	509	462	423	387	298
	0.02	923	679	564	493	440	397	360	328	247
	0.05	786	563	459	395	348	310	278	250	181
	0.10	678	472	378	320	278	245	217	192	133
	0.20	562	377	294	244	208	179	156	136	88
0.40	0.01	476	357	301	266	240	218	200	183	142
	0.02	430	317	265	232	207	188	171	156	118
	0.05	366	264	216	187	165	147	133	120	88
	0.10	316	221	178	152	133	117	104	93	65
	0.20	263	178	140	117	100	87	76	67	44
0.45	0.01	278	209	177	156	141	129	118	109	85
	0.02	251	186	156	137	123	111	101	93	71
	0.05	214	155	127	110	98	88	79	72	53
	0.10	185	130	105	90	79	70	63	56	40
	0.20	154	105	83	70	60	52	46	41	28
0.50	0.01	182	138	117	104	94	86	79	73	57
	0.02	165	123	103	91	82	74	68	62	48
	0.05	141	102	85	74	65	59	53	48	36
	0.10	121	86	70	60	53	47	42	38	28
	0.20	101	70	55	47	41	36	31	28	20
0.55	0.01	129	98	83	74	67	62	57	53	42
	0.02	117	87	74	65	59	53	49	45	35
	0.05	99	73	60	53	47	43	39	35	27
	0.10	86	61	50	43	38	34	31	28	21
	0.20	72	50	40	34	29	26	23	21	15
0.60	0.01	96	73	62	56	51	47	43	40	32
	0.02	86	65	55	49	44	40	37	34	27
	0.05	74	54	45	40	36	32	29	27	21
	0.10	64	46	38	33	29	26	24	21	16
	0.20	53	37	30	26	22	20	18	16	12
0.65	0.01	73	56	48	43	39	36	34	31	25
	0.02	66	50	43	38	34	32	29	27	21
	0.05	57	42	35	31	28	25	23	21	17
	0.10	49	36	29	26	23	21	19	17	13
	0.20	41	29	23	20	18	16	14	13	9
0.70	0.01	58	45	38	34	32	29	27	25	21
	0.02	52	40	34	30	28	25	23	22	17
	0.05	44	33	28	25	22	20	19	17	14
	0.10	38	28	23	20	18	17	15	14	11
	0.20	32	23	19	16	14	13	12	10	8
0.75	0.01	46	36	31	28	26	24	22	21	17
	0.02	42	32	27	25	22	21	19	18	15
	0.05	35	27	23	20	18	17	15	14	11
	0.10	31	23	19	17	15	14	12	11	9
	0.20	26	18	15	13	12	11	10	9	7

Table A.3. Sample sizes per group for a two-tailed test on proportions. $P_1 = .25$

		POWER								
P2	ALPHA	0.99	0.95	0.90	0.85	0.80	0.75	0.70	0.65	0.50
0.80	0.01	37	29	25	23	21	20	18	17	14
	0.02	34	26	23	20	19	17	16	15	12
	0.05	29	22	19	17	15	14	13	12	10
	0.10	25	18	16	14	12	11	10	10	8
	0.20	21	15	12	11	10	9	8	7	6
0.85	0.01	30	24	21	19	18	16	15	15	12
	0.02	27	21	19	17	16	14	13	13	11
	0.05	23	18	15	14	13	12	11	10	8
	0.10	20	15	13	11	10	10	9	8	7
	0.20	17	12	10	9	8	8	7	6	5
0.90	0.01	25	20	17	16	15	14	13	12	11
	0.02	22	18	15	14	13	12	11	11	9
	0.05	19	15	13	12	11	10	9	9	7
	0.10	16	13	11	10	9	8	8	7	6
	0.20	14	10	9	8	7	6	6	6	4
0.95	0.01	20	16	15	13	13	12	11	11	9
	0.02	18	15	13	12	11	10	10	9	8
	0.05	15	12	11	10	9	8	8	7	6
	0.10	13	10	9	8	7	7	7	6	5
	0.20	11	8	7	6	6	5	5	5	4

Table A.3. Sample sizes per group for a two-tailed test on proportions. $P_1 = .30$

					POWER					
P2	ALPHA	0.99	0.95	0.90	0.85	0.80	0.75	0.70	0.65	0.50
0.35	0.01	4251	3163	2650	2328	2089	1892	1726	1578	1204
	0.02	3832	2804	2322	2022	1800	1618	1464	1329	989
	0.05	3259	2318	1882	1613	1416	1256	1122	1004	714
	0.10	2803	1937	1541	1300	1124	983	865	762	514
	0.20	2320	1541	1192	981	830	710	611	527	327
0.40	0.01	1108	827	695	612	550	499	456	418	322
	0.02	999	734	610	532	475	428	389	354	266
	0.05	851	608	496	427	376	334	300	269	194
	0.10	733	510	408	345	300	264	233	207	142
	0.20	608	407	317	263	224	193	167	145	94
0.45	0.01	508	381	321	283	255	232	213	195	151
	0.02	459	338	282	247	221	200	182	166	126
	0.05	391	281	230	199	175	157	141	127	93
	0.10	337	236	190	161	141	124	110	98	69
	0.20	280	189	148	124	106	92	80	70	47
0.50	0.01	293	220	186	165	149	135	124	114	89
	0.02	264	196	164	144	129	117	106	97	75
	0.05	225	163	134	116	103	92	83	75	56
	0.10	194	137	111	95	83	73	66	59	42
	0.20	162	110	87	73	63	55	48	42	29
0.55	0.01	190	144	122	108	98	89	82	76	60
	0.02	172	128	107	94	85	77	71	65	50
	0.05	147	106	88	76	68	61	55	50	38
	0.10	127	90	73	63	55	49	44	39	29
	0.20	105	72	57	48	42	37	33	29	20
0.60	0.01	133	101	86	76	69	63	59	54	43
	0.02	120	90	76	67	60	55	50	46	36
	0.05	103	75	62	54	48	44	40	36	27
	0.10	89	63	52	45	39	35	32	29	21
	0.20	74	51	41	35	30	27	24	21	15
0.65	0.01	98	75	64	57	52	47	44	41	32
	0.02	88	66	56	50	45	41	38	35	27
	0.05	75	55	46	40	36	33	30	27	21
	0.10	65	47	38	33	30	27	24	22	16
	0.20	54	38	31	26	23	20	18	16	12
0.70	0.01	74	57	49	44	40	37	34	32	25
	0.02	67	51	43	38	35	32	29	27	22
	0.05	57	42	36	31	28	26	23	21	17
	0.10	49	36	30	26	23	21	19	17	13
	0.20	41	29	24	20	18	16	14	13	9
0.75	0.01	58	45	38	34	32	29	27	25	21
	0.02	52	40	34	30	28	25	23	22	17
	0.05	44	33	28	25	22	20	19	17	14
	0.10	38	28	23	20	18	17	15	14	11
	0.20	32	23	19	16	14	13	12	10	8
0.80	0.01	46	36	31	28	25	24	22	21	17
	0.02	41	32	27	24	22	21	19	18	14
	0.05	35	26	22	20	18	17	15	14	11
	0.10	30	22	19	17	15	14	12	11	9
	0.20	25	18	15	13	12	11	10	9	7

Table A.3. Sample sizes per group for a two-tailed test on proportions. $P_1 = .30$

					POWER					
P2	ALPHA	0.99	0.95	0.90	0.85	0.80	0.75	0.70	0.65	0.50
0.85	0.01	36	29	25	23	21	19	18	17	14
	0.02	33	26	22	20	18	17	16	15	12
	0.05	28	21	18	16	15	14	13	12	9
	0.10	24	18	15	14	12	11	10	10	8
	0.20	20	15	12	11	10	9	8	7	6
0.90	0.01	29	23	20	19	17	16	15	14	12
	0.02	26	21	18	16	15	14	13	12	10
	0.05	22	17	15	13	12	11	11	10	8
	0.10	19	15	13	11	10	9	9	8	7
	0.20	16	12	10	9	8	7	7	6	5
0.95	0.01	23	19	17	15	14	13	13	12	10
	0.02	21	17	15	14	13	12	11	10	9
	0.05	18	14	12	11	10	10	9	8	7
	0.10	16	12	10	9	9	8	7	7	6
	0.20	13	10	8	7	7	6	6	5	4

Table A.3. Sample sizes per group for a two-tailed test on proportions. $P_1 = .35$

		POWER								
P2	ALPHA	0.99	0.95	0.90	0.85	0.80	0.75	0.70	0.65	0.50
0.40	0.01	4540	3377	2828	2484	2229	2019	1841	1683	1284
	0.02	4092	2993	2478	2157	1920	1726	1562	1417	1054
	0.05	3479	2474	2008	1721	1511	1340	1196	1070	760
	0.10	2992	2067	1644	1386	1198	1047	921	812	547
	0.20	2476	1644	1271	1046	885	756	650	560	347
0.45	0.01	1168	872	732	644	579	526	480	440	338
	0.02	1053	774	642	561	500	451	409	372	279
	0.05	897	641	522	449	395	352	315	283	204
	0.10	772	537	429	363	316	277	245	217	149
	0.20	640	429	334	276	235	202	176	152	98
0.50	0.01	530	397	334	295	265	241	221	203	157
	0.02	478	352	294	257	230	208	189	172	131
	0.05	407	293	239	207	182	163	146	132	96
	0.10	351	246	197	168	146	129	115	102	71
	0.20	292	197	154	128	110	95	83	73	48
0.55	0.01	302	227	192	169	153	139	128	118	92
	0.02	272	202	169	148	133	120	110	100	77
	0.05	232	168	138	119	106	95	85	77	57
	0.10	200	141	114	97	85	75	67	60	43
	0.20	167	113	89	75	65	56	49	43	29
0.60	0.01	194	147	124	110	100	91	84	77	61
	0.02	175	130	109	96	87	79	72	66	51
	0.05	149	109	90	78	69	62	56	51	38
	0.10	129	91	74	64	56	50	45	40	29
	0.20	108	74	59	49	43	37	33	29	20
0.65	0.01	134	102	87	77	70	64	59	55	43
	0.02	121	91	76	68	61	55	51	47	36
	0.05	104	76	63	55	49	44	40	36	28
	0.10	89	64	52	45	40	35	32	29	21
	0.20	75	52	41	35	30	27	24	21	15
0.70	0.01	98	75	64	57	52	47	44	41	32
	0.02	88	66	56	50	45	41	38	35	27
	0.05	75	55	46	40	36	33	30	27	21
	0.10	65	47	38	33	30	27	24	22	16
	0.20	54	38	31	26	23	20	18	16	12
0.75	0.01	73	56	48	43	39	36	34	31	25
	0.02	66	50	43	38	34	32	29	27	21
	0.05	57	42	35	31	28	25	23	21	17
	0.10	49	36	29	26	23	21	19	17	13
	0.20	41	29	23	20	18	16	14	13	9
0.80	0.01	56	44	38	34	31	29	27	25	20
	0.02	51	39	33	30	27	25	23	21	17
	0.05	44	33	27	24	22	20	18	17	13
	0.10	38	28	23	20	18	16	15	14	11
	0.20	31	22	18	16	14	13	11	10	8
0.85	0.01	44	34	30	27	25	23	21	20	16
	0.02	40	31	26	24	22	20	19	17	14
	0.05	34	26	22	19	18	16	15	14	11
	0.10	29	22	18	16	15	13	12	11	9
	0.20	25	18	15	13	11	10	9	9	7

		POWER								
P2	ALPHA	0.99	0.95	0.90	0.85	0.80	0.75	0.70	0.65	0.50
0.90	0.01	35	27	24	22	20	19	17	16	14
	0.02	31	24	21	19	18	16	15	14	12
	0.05	27	20	18	16	14	13	12	11	9
	0.10	23	17	15	13	12	11	10	9	7
	0.20	19	14	12	10	9	8	8	7	6
0.95	0.01	28	22	19	18	16	15	14	14	11
	0.02	25	20	17	16	14	13	13	12	10
	0.05	21	16	14	13	12	11	10	9	8
	0.10	18	14	12	11	10	9	8	8	6
	0.20	15	11	10	9	8	7	7	6	5

Table A.3. Sample sizes per group for a two-tailed test on proportions. P_1 = .40

					POWER					
P2	ALPHA	0.99	0.95	0.90	0.85	0.80	0.75	0.70	0.65	0.50
0.45	0.01	4732	3520	2947	2589	2322	2104	1918	1753	1337
	0.02	4265	3119	2582	2248	2001	1798	1627	1476	1097
	0.05	3626	2578	2093	1793	1573	1395	1245	1114	791
	0.10	3118	2153	1713	1444	1248	1090	959	845	568
	0.20	2581	1712	1323	1089	921	787	677	582	360
0.50	0.01	1204	898	754	664	597	542	495	453	348
	0.02	1086	797	662	578	515	464	421	383	287
	0.05	924	660	538	463	407	362	324	291	210
	0.10	796	553	442	374	325	285	252	223	153
	0.20	660	441	343	284	242	208	180	157	100
0.55	0.01	540	405	341	301	271	246	225	207	160
	0.02	488	359	299	262	234	212	192	175	133
	0.05	415	298	244	211	186	166	149	134	98
	0.10	358	250	201	171	149	131	117	104	73
	0.20	298	201	157	131	112	97	85	74	49
0.60	0.01	305	229	193	171	154	141	129	119	93
	0.02	275	204	170	149	134	121	111	101	77
	0.05	235	169	139	120	107	96	86	78	58
	0.10	202	142	115	98	86	76	68	61	43
	0.20	168	114	90	76	65	57	50	44	30
0.65	0.01	194	147	124	110	100	91	84	77	61
	0.02	175	130	109	96	87	79	72	66	51
	0.05	149	109	90	78	69	62	56	51	38
	0.10	129	91	74	64	56	50	45	40	29
	0.20	108	74	59	49	43	37	33	29	20
0.70	0.01	133	101	86	76	69	63	59	54	43
	0.02	120	90	76	67	60	55	50	46	36
	0.05	103	75	62	54	48	44	40	36	27
	0.10	89	63	52	45	39	35	32	29	21
	0.20	74	51	41	35	30	27	24	21	15
0.75	0.01	96	73	62	56	51	47	43	40	32
	0.02	86	65	55	49	44	40	37	34	27
	0.05	74	54	45	40	36	32	29	27	21
	0.10	64	46	38	33	29	26	24	21	16
	0.20	53	37	30	26	22	20	18	16	12
0.80	0.01	71	55	47	42	38	35	33	31	25
	0.02	64	49	42	37	34	31	28	26	21
	0.05	55	41	34	30	27	25	23	21	16
	0.10	47	35	29	25	22	20	18	17	13
	0.20	40	28	23	20	17	15	14	13	9
0.85	0.01	54	42	36	33	30	28	26	24	20
	0.02	49	37	32	29	26	24	22	21	17
	0.05	42	31	26	23	21	19	18	16	13
	0.10	36	27	22	19	17	16	14	13	10
	0.20	30	22	18	15	14	12	11	10	8
0.90	0.01	42	33	28	26	24	22	20	19	16
	0.02	38	29	25	23	21	19	18	17	14
	0.05	32	24	21	18	17	15	14	13	11
	0.10	28	21	17	15	14	13	12	11	8
	0.20	23	17	14	12	11	10	9	8	6
					POWER					
P2	ALPHA	0.99	0.95	0.90	0.85	0.80	0.75	0.70	0.65	0.50
0.95	0.01	32	26	22	20	19	18	16	16	13
	0.02	29	23	20	18	17	15	14	13	11
	0.05	25	19	16	15	14	12	12	11	9
	0.10	22	16	14	12	11	10	9	9	7
	0.20	18	13	11	10	9	8	7	7	5

Table A.3. Sample sizes per group for a two-tailed test on proportions. $P_1 = .45$

P2	ALPHA	POWER 0.99	0.95	0.90	0.85	0.80	0.75	0.70	0.65	0.50
0.50	0.01	4828	3591	3007	2641	2369	2146	1956	1788	1364
	0.02	4352	3182	2635	2293	2041	1834	1659	1505	1119
	0.05	3700	2630	2135	1829	1605	1423	1270	1136	806
	0.10	3181	2197	1747	1472	1273	1112	978	861	579
	0.20	2633	1747	1350	1110	939	802	690	593	367
0.55	0.01	1216	907	762	670	603	547	499	458	352
	0.02	1097	805	668	583	520	469	425	387	290
	0.05	933	667	543	467	411	366	328	294	212
	0.10	804	558	446	378	328	288	254	225	155
	0.20	667	446	347	287	244	210	182	158	101
0.60	0.01	540	405	341	301	271	246	225	207	160
	0.02	488	359	299	262	234	212	192	175	133
	0.05	415	298	244	211	186	166	149	134	98
	0.10	358	250	201	171	149	131	117	104	73
	0.20	298	201	157	131	112	97	85	74	49
0.65	0.01	302	227	192	169	153	139	128	118	92
	0.02	272	202	169	148	133	120	110	100	77
	0.05	232	168	138	119	106	95	85	77	57
	0.10	200	141	114	97	85	75	67	60	43
	0.20	167	113	89	75	65	56	49	43	29
0.70	0.01	190	144	122	108	98	89	82	76	60
	0.02	172	128	107	94	85	77	71	65	50
	0.05	147	106	88	76	68	61	55	50	38
	0.10	127	90	73	63	55	49	44	39	29
	0.20	105	72	57	48	42	37	33	29	20
0.75	0.01	129	98	83	74	67	62	57	53	42
	0.02	117	87	74	65	59	53	49	45	35
	0.05	99	73	60	53	47	43	39	35	27
	0.10	86	61	50	43	38	34	31	28	21
	0.20	72	50	40	34	29	26	23	21	15
0.80	0.01	92	70	60	54	49	45	41	38	31
	0.02	83	63	53	47	43	39	36	33	26
	0.05	71	52	44	38	34	31	28	26	20
	0.10	61	44	36	32	28	25	23	21	16
	0.20	51	36	29	25	22	19	17	15	11
0.85	0.01	67	52	45	40	37	34	31	29	24
	0.02	61	46	39	35	32	29	27	25	20
	0.05	52	39	33	29	26	24	22	20	16
	0.10	45	33	27	24	21	19	17	16	12
	0.20	38	27	22	19	17	15	13	12	9
0.90	0.01	51	39	34	31	28	26	24	23	19
	0.02	46	35	30	27	25	23	21	20	16
	0.05	39	29	25	22	20	18	17	16	12
	0.10	34	25	21	18	16	15	14	13	10
	0.20	28	20	17	15	13	12	11	10	7
0.95	0.01	38	30	26	24	22	20	19	18	15
	0.02	35	27	23	21	19	18	17	16	13
	0.05	30	23	19	17	16	14	13	12	10
	0.10	26	19	16	14	13	12	11	10	8
	0.20	21	16	13	11	10	9	9	8	6

Table A.3. Sample sizes per group for a two-tailed test on proportions. $P_1 = .50$

		POWER								
P2	ALPHA	0.99	0.95	0.90	0.85	0.80	0.75	0.70	0.65	0.50
0.55	0.01	4828	3591	3007	2641	2369	2146	1956	1788	1364
	0.02	4352	3182	2635	2293	2041	1834	1659	1505	1119
	0.05	3700	2630	2135	1829	1605	1423	1270	1136	806
	0.10	3181	2197	1747	1472	1273	1112	978	861	579
	0.20	2633	1747	1350	1110	939	802	690	593	367
0.60	0.01	1204	898	754	664	597	542	495	453	348
	0.02	1086	797	662	578	515	464	421	383	287
	0.05	924	660	538	463	407	362	324	291	210
	0.10	796	553	442	374	325	285	252	223	153
	0.20	660	441	343	284	242	208	180	157	100
0.65	0.01	530	397	334	295	265	241	221	203	157
	0.02	478	352	294	257	230	208	189	172	131
	0.05	407	293	239	207	182	163	146	132	96
	0.10	351	246	197	168	146	129	115	102	71
	0.20	292	197	154	128	110	95	83	73	48
0.70	0.01	293	220	186	165	149	135	124	114	89
	0.02	264	196	164	144	129	117	106	97	75
	0.05	225	163	134	116	103	92	83	75	56
	0.10	194	137	111	95	83	73	66	59	42
	0.20	162	110	87	73	63	55	48	42	29
0.75	0.01	182	138	117	104	94	86	79	73	57
	0.02	165	123	103	91	82	74	68	62	48
	0.05	141	102	85	74	65	59	53	48	36
	0.10	121	86	70	60	53	47	42	38	28
	0.20	101	70	55	47	41	36	31	28	20
0.80	0.01	122	93	79	71	64	59	54	50	40
	0.02	110	83	70	62	56	51	47	43	34
	0.05	94	69	57	50	45	41	37	34	26
	0.10	82	58	48	41	37	33	29	27	20
	0.20	68	47	38	32	28	25	22	20	14
0.85	0.01	86	66	56	50	46	42	39	36	29
	0.02	78	59	50	44	40	37	34	31	25
	0.05	66	49	41	36	32	29	27	25	19
	0.10	57	42	34	30	26	24	22	20	15
	0.20	48	34	27	23	21	18	16	15	11
0.90	0.01	62	48	41	37	34	31	29	27	22
	0.02	56	43	37	33	30	27	25	23	19
	0.05	48	36	30	27	24	22	20	19	15
	0.10	42	30	25	22	20	18	16	15	12
	0.20	35	25	20	18	16	14	13	11	9
0.95	0.01	46	36	31	28	26	24	22	21	17
	0.02	41	32	27	25	23	21	19	18	15
	0.05	35	27	23	20	18	17	16	14	12
	0.10	31	23	19	17	15	14	13	12	9
	0.20	26	19	15	13	12	11	10	9	7

Table A.3. Sample sizes per group for a two-tailed test on proportions. $P_1 = .55$

P2	ALPHA	POWER 0.99	0.95	0.90	0.85	0.80	0.75	0.70	0.65	0.50
0.60	0.01	4732	3520	2947	2589	2322	2104	1918	1753	1337
	0.02	4265	3119	2582	2248	2001	1798	1627	1476	1097
	0.05	3626	2578	2093	1793	1573	1395	1245	1114	791
	0.10	3118	2153	1713	1444	1248	1090	959	845	568
	0.20	2581	1712	1323	1089	921	787	677	582	360
0.65	0.01	1168	872	732	644	579	526	480	440	338
	0.02	1053	774	642	561	500	451	409	372	279
	0.05	897	641	522	449	395	352	315	283	204
	0.10	772	537	429	363	316	277	245	217	149
	0.20	640	429	334	276	235	202	176	152	98
0.70	0.01	508	381	321	283	255	232	213	195	151
	0.02	459	338	282	247	221	200	182	166	126
	0.05	391	281	230	199	175	157	141	127	93
	0.10	337	236	190	161	141	124	110	98	69
	0.20	280	189	148	124	106	92	80	70	47
0.75	0.01	278	209	177	156	141	129	118	109	85
	0.02	251	186	156	137	123	111	101	93	71
	0.05	214	155	127	110	98	88	79	72	53
	0.10	185	130	105	90	79	70	63	56	40
	0.20	154	105	83	70	60	52	46	41	28
0.80	0.01	171	129	110	98	88	81	74	69	54
	0.02	154	115	97	85	77	70	64	59	46
	0.05	132	96	79	69	62	56	50	46	35
	0.10	114	81	66	57	50	45	40	36	26
	0.20	95	65	52	44	38	34	30	27	19
0.85	0.01	113	86	73	65	60	55	50	47	37
	0.02	102	77	65	57	52	47	44	40	32
	0.05	87	64	53	47	42	38	34	32	24
	0.10	75	54	44	39	34	31	28	25	19
	0.20	63	44	35	30	26	23	21	19	14
0.90	0.01	78	60	51	46	42	39	36	33	27
	0.02	71	54	46	41	37	34	31	29	23
	0.05	60	45	38	33	30	27	25	23	18
	0.10	52	38	31	27	24	22	20	18	14
	0.20	44	31	25	22	19	17	15	14	10
0.95	0.01	55	43	37	33	31	28	26	25	20
	0.02	50	38	33	29	27	25	23	21	17
	0.05	43	32	27	24	22	20	18	17	14
	0.10	37	27	23	20	18	16	15	14	11
	0.20	31	22	18	16	14	13	12	11	8

Table A.3. Sample sizes per group for a two-tailed test on proportions. $P_1 = .60$

P2	ALPHA	POWER 0.99	0.95	0.90	0.85	0.80	0.75	0.70	0.65	0.50
0.65	0.01	4540	3377	2828	2484	2229	2019	1841	1683	1284
	0.02	4092	2993	2478	2157	1920	1726	1562	1417	1054
	0.05	3479	2474	2008	1721	1511	1340	1196	1070	760
	0.10	2992	2067	1644	1386	1198	1047	921	812	547
	0.20	2476	1644	1271	1046	885	756	650	560	347
0.70	0.01	1108	827	695	612	550	499	456	418	322
	0.02	999	734	610	532	475	428	389	354	266
	0.05	851	608	496	427	376	334	300	269	194
	0.10	733	510	408	345	300	264	233	207	142
	0.20	608	407	317	263	224	193	167	145	94
0.75	0.01	476	357	301	266	240	218	200	183	142
	0.02	430	317	265	232	207	188	171	156	118
	0.05	366	264	216	187	165	147	133	120	88
	0.10	316	221	178	152	133	117	104	93	65
	0.20	263	178	140	117	100	87	76	67	44
0.80	0.01	256	193	164	145	131	120	110	101	79
	0.02	232	172	144	127	114	103	94	86	66
	0.05	198	143	118	102	91	82	74	67	50
	0.10	171	121	98	84	74	65	58	52	38
	0.20	142	97	77	65	56	49	43	38	26
0.85	0.01	155	118	100	89	81	74	68	63	50
	0.02	140	105	89	78	70	64	59	54	42
	0.05	120	88	73	63	57	51	46	42	32
	0.10	104	74	60	52	46	41	37	33	25
	0.20	87	60	48	41	35	31	28	25	18
0.90	0.01	101	77	66	59	54	49	46	42	34
	0.02	91	69	58	52	47	43	39	36	29
	0.05	78	58	48	42	38	34	31	29	22
	0.10	68	49	40	35	31	28	25	23	17
	0.20	56	40	32	27	24	21	19	17	13
0.95	0.01	68	53	45	41	37	34	32	30	24
	0.02	62	47	40	36	33	30	28	26	21
	0.05	53	39	33	29	27	24	22	21	16
	0.10	46	34	28	24	22	20	18	17	13
	0.20	38	27	22	19	17	15	14	13	10

Table A.3. Sample sizes per group for a two-tailed test on proportions. $P_1 = .65$

P2	ALPHA	POWER 0.99	0.95	0.90	0.85	0.80	0.75	0.70	0.65	0.50
0.70	0.01	4251	3163	2650	2328	2089	1892	1726	1578	1204
	0.02	3833	2804	2322	2022	1800	1618	1464	1329	989
	0.05	3259	2318	1882	1613	1416	1256	1122	1004	714
	0.10	2803	1937	1541	1300	1124	983	865	762	514
	0.20	2320	1541	1192	981	830	710	611	527	327
0.75	0.01	1023	765	643	566	509	462	423	387	298
	0.02	923	679	564	493	440	397	360	328	247
	0.05	786	563	459	395	348	310	278	250	181
	0.10	678	472	378	320	278	245	217	192	133
	0.20	562	377	294	244	208	179	156	136	88
0.80	0.01	433	325	274	242	219	199	183	168	131
	0.02	391	289	242	212	190	172	156	143	109
	0.05	334	240	197	171	151	135	122	110	81
	0.10	288	202	163	139	122	107	96	85	61
	0.20	240	162	128	107	92	80	70	62	41
0.85	0.01	229	173	147	130	118	108	99	91	72
	0.02	207	154	130	114	102	93	85	78	60
	0.05	177	129	106	92	82	74	67	61	45
	0.10	153	109	88	76	67	59	53	48	35
	0.20	128	88	70	59	51	45	39	35	24
0.90	0.01	136	104	88	79	72	66	61	56	45
	0.02	123	93	78	69	62	57	52	48	38
	0.05	105	77	64	56	50	45	41	38	29
	0.10	91	65	54	46	41	37	33	30	22
	0.20	76	53	43	36	32	28	25	22	16
0.95	0.01	86	66	57	51	46	43	40	37	30
	0.02	78	59	50	45	41	37	34	32	25
	0.05	67	50	42	37	33	30	28	25	20
	0.10	58	42	35	30	27	25	22	20	16
	0.20	48	34	28	24	21	19	17	16	12

Table A.3. Sample sizes per group for a two-tailed test on proportions. $P_1 = .70$

P2	ALPHA	POWER 0.99	0.95	0.90	0.85	0.80	0.75	0.70	0.65	0.50
0.75	0.01	3867	2878	2411	2119	1902	1723	1572	1438	1098
	0.02	3486	2552	2114	1841	1639	1474	1334	1211	902
	0.05	2965	2110	1714	1470	1291	1145	1023	916	652
	0.10	2550	1764	1404	1185	1025	897	789	696	471
	0.20	2112	1404	1087	895	758	649	559	482	301
0.80	0.01	915	685	576	507	456	415	379	348	268
	0.02	826	608	506	442	395	356	324	295	222
	0.05	704	504	412	355	313	279	250	225	163
	0.10	607	423	339	288	250	220	195	174	121
	0.20	504	339	265	220	188	162	141	123	80
0.85	0.01	380	286	241	213	193	176	161	148	116
	0.02	343	254	213	187	167	152	138	126	97
	0.05	293	212	174	151	134	120	108	98	72
	0.10	253	178	144	123	108	95	85	76	54
	0.20	211	143	113	95	82	71	63	55	38
0.90	0.01	196	149	126	112	102	93	86	79	63
	0.02	178	133	112	98	89	81	74	68	53
	0.05	152	111	92	80	71	64	58	53	40
	0.10	131	94	76	66	58	52	47	42	31
	0.20	110	76	61	51	45	39	35	31	22
0.95	0.01	113	87	74	66	60	55	51	48	38
	0.02	102	77	66	58	53	48	44	41	33
	0.05	88	65	54	48	43	39	35	32	25
	0.10	76	55	45	39	35	32	29	26	20
	0.20	64	45	36	31	27	24	22	20	14

Table A.3. Sample sizes per group for a two-tailed test on proportions. $P_1 = .75$

		POWER								
P2	ALPHA	0.99	0.95	0.90	0.85	0.80	0.75	0.70	0.65	0.50
0.80	0.01	3386	2522	2114	1858	1668	1512	1379	1262	965
	0.02	3053	2236	1853	1615	1438	1294	1172	1064	794
	0.05	2597	1850	1504	1290	1134	1007	900	806	575
	0.10	2235	1547	1233	1041	901	789	695	614	417
	0.20	1852	1232	955	788	668	572	494	426	268
0.85	0.01	783	586	494	435	392	357	326	300	232
	0.02	707	521	434	380	340	307	279	254	193
	0.05	603	433	354	305	270	241	216	195	142
	0.10	520	363	292	248	216	191	169	151	106
	0.20	432	291	228	190	163	141	123	108	71
0.90	0.01	316	238	202	179	162	147	136	125	98
	0.02	285	212	178	156	140	127	116	107	82
	0.05	244	177	146	127	113	101	91	83	62
	0.10	211	149	121	104	91	81	73	65	47
	0.20	176	120	96	81	70	61	54	48	33
0.95	0.01	157	120	102	91	83	76	70	65	52
	0.02	142	107	90	80	72	66	61	56	44
	0.05	122	90	75	65	58	53	48	44	34
	0.10	106	76	62	54	48	43	39	35	26
	0.20	88	62	50	42	37	33	29	26	19

Table A.3. Sample sizes per group for a two-tailed test on proportions. $P_1 = .80$

		POWER								
P2	ALPHA	0.99	0.95	0.90	0.85	0.80	0.75	0.70	0.65	0.50
0.85	0.01	2810	2094	1756	1545	1388	1259	1149	1052	806
	0.02	2534	1858	1541	1343	1198	1078	977	888	664
	0.05	2157	1538	1252	1075	945	840	751	674	483
	0.10	1856	1287	1027	868	753	660	582	515	351
	0.20	1539	1027	797	659	559	480	415	360	228
0.90	0.01	627	471	397	351	316	288	264	243	189
	0.02	566	419	349	306	274	248	226	206	157
	0.05	483	348	286	247	219	196	176	159	117
	0.10	417	293	236	201	176	156	139	124	88
	0.20	347	236	185	155	133	116	102	89	60
0.95	0.01	241	183	155	138	125	115	106	98	77
	0.02	218	163	137	121	109	99	91	84	65
	0.05	187	136	113	99	88	79	72	66	50
	0.10	162	115	94	81	72	64	58	52	38
	0.20	135	94	75	64	55	49	44	39	28

Table A.3. Sample sizes per group for a two-tailed test on proportions. $P_1 = .85$

		POWER								
P2	ALPHA	0.99	0.95	0.90	0.85	0.80	0.75	0.70	0.65	0.50
0.90	0.01	2137	1595	1340	1179	1060	963	880	806	620
	0.02	1928	1416	1176	1027	916	826	749	682	513
	0.05	1642	1174	957	823	725	646	579	520	375
	0.10	1415	984	787	667	579	509	450	399	275
	0.20	1175	787	613	508	433	373	324	281	182
0.95	0.01	447	337	285	253	228	209	192	177	139
	0.02	404	300	252	221	199	180	165	151	117
	0.05	345	251	207	179	160	143	130	118	88
	0.10	299	212	172	147	130	115	103	93	67
	0.20	250	171	136	115	99	87	77	68	47

Table A.3. Sample sizes per group for a two-tailed test on proportions. $P_1 = .90$

		POWER								
P2	ALPHA	0.99	0.95	0.90	0.85	0.80	0.75	0.70	0.65	0.50
0.95	0.01	1368	1025	863	762	686	624	572	525	407
	0.02	1235	911	760	665	595	538	489	446	339
	0.05	1054	758	621	536	474	423	381	344	252
	0.10	910	637	513	437	381	336	299	267	188
	0.20	758	512	402	336	288	250	219	192	128

Table A.4. 20,000 random digits

08939	53632	41345	65379	20165	32576	13967	90616	17995	92422
92578	23668	08801	39792	59541	99117	58830	60923	36068	68101
83994	91054	90377	22776	23263	34593	98191	77811	83144	98563
43080	71414	40760	01831	44145	48387	93018	22618	98547	87716
39372	46789	26381	37186	85684	79426	05395	17538	56671	82181
83046	58644	04452	98912	53406	30224	00687	32099	86414	29590
99808	32539	96961	88917	60847	64826	41332	64557	15354	11111
28478	70870	68912	75644	33648	21097	23745	52593	01849	37760
09916	19651	28659	95093	12626	19919	05879	56003	83100	94572
19537	66067	20569	28808	87722	67059	12851	73573	25776	92500
23013	05574	26320	07754	09642	88068	41626	57139	68199	94938
55838	80585	80967	60540	34528	62310	63106	17843	39104	74036
92279	87344	93556	75233	09394	79265	91047	32891	77925	71530
27850	23332	89336	26026	52130	78544	02090	05645	15060	39550
01760	54605	11794	79312	69728	04554	99775	57659	47981	68954
81889	70751	87501	88247	41966	57574	67745	88304	20118	25964
74722	14654	15425	60665	25162	04987	03467	75915	24282	62456
56196	75068	44643	92240	51651	79743	13598	63901	61020	91003
96842	62021	00543	45073	65545	87612	35765	26079	34589	72821
25619	98328	59393	71401	93871	20611	78830	87477	15390	05044
91746	05084	04781	82933	54564	80986	94843	40178	87483	63288
92384	84706	76778	98313	98875	08427	60687	88272	83448	06237
86390	62208	95735	14535	25591	22730	06059	31786	36181	31016
60458	83606	57510	92609	38061	94881	26736	06489	98303	31419
03783	39922	05489	73630	92379	91602	18193	84741	44704	05558
31011	36035	37113	98362	56149	51634	04468	62096	32361	35301
20555	05621	48728	41776	12101	96615	70781	55151	93876	66892
56466	36766	12400	43510	49456	05140	85736	68155	37306	10438
26875	67304	61950	65962	38223	35676	70043	99178	64677	95457
90648	84770	92791	93814	27760	22232	83545	01183	55188	20482
26197	72840	01264	52019	00739	36259	10905	39097	36437	66743
72522	34445	53975	13840	97262	59007	78685	41044	38103	59216
12370	41270	36290	46307	51230	90614	82613	80148	37371	02895
81028	60112	31415	47478	02131	85480	93699	92876	13958	47867
61573	38634	77650	18189	10283	97999	95442	90657	84963	93863
98511	46300	91199	30492	62159	98525	31710	03540	35844	83200
76606	10834	75548	55779	54744	26450	66001	57949	53685	00567
20237	16311	15733	47599	43998	35594	17577	85113	52487	48900
21022	86025	26951	87480	82317	06580	98627	32536	07573	52612
47512	11564	41777	46581	03492	01722	78900	57901	37307	02727
80598	59041	28861	41793	91007	69907	00376	73086	35132	53014
01892	34226	88327	21926	36607	22307	04376	25491	13563	51955
89657	70349	15176	57916	10911	44218	67108	04678	24097	02476
97983	65616	11841	80504	76452	34176	16986	94328	13091	29592
59727	92033	14654	59622	25844	18460	78162	02832	13528	55683
12340	72894	26303	01771	73895	27432	99536	50328	06141	83886
48049	33318	67463	04914	22316	89663	37132	15825	60759	22131
85953	16537	25639	05004	99269	50577	10036	05022	39800	93605
03426	78111	37828	23967	03350	04397	96227	37787	60680	23993
97837	71085	45973	36073	02680	91425	24425	23725	22521	21601

Table A.4. (Continued)

26916	67086	60270	57846	04646	07258	01734	45079	54869	23505
47205	71678	05222	86233	70398	46287	44139	48247	92230	19157
84869	36794	56943	10512	50582	08884	98068	08447	68071	32397
81740	98868	57546	55461	14850	89946	06024	26626	05543	93616
11808	28306	63559	26600	87569	86007	27922	93468	09509	15841
09464	14219	00130	72813	35704	58905	32091	62397	85560	51783
40656	77886	01411	07490	32240	26028	66002	61762	76551	03442
31693	59176	69817	86317	89547	60424	56618	95888	65770	31622
97799	02197	32987	78146	71992	28633	23868	85504	98216	19756
34590	29732	67082	34899	05654	19830	68088	30054	67535	34721
00504	90537	38681	17248	55362	76935	63352	87699	56022	46835
76814	39363	44851	14836	85357	78617	03482	13336	48678	72047
94171	16606	52092	63096	09752	90644	56092	20751	19678	31311
10758	82747	99662	53243	22501	55820	32406	92052	60659	35477
66933	82305	91425	07804	24003	73777	26634	95806	35126	48503
74883	12771	02671	01090	82498	85176	68569	44827	51844	07616
79102	06066	24478	92267	33300	69392	16652	75381	02415	36065
94649	43308	08005	58253	77473	40559	46096	11540	54375	22388
44952	68217	04728	14414	43931	33854	07744	41771	80933	09655
29531	90289	75949	43091	75005	62207	98196	29316	92128	88918
04355	25867	16008	63243	35388	43138	40330	53741	59469	73144
03640	63541	48488	19060	77959	96217	75666	88042	47261	91184
21749	09836	63276	91133	77308	43654	66146	03991	28629	35848
57425	21919	14688	90852	12918	59833	42736	17916	22868	75963
49962	16108	46986	36939	98761	60113	71822	89915	93090	49299
57985	04214	03417	82576	64699	45011	87770	21525	39212	41547
96374	04318	58540	12375	47382	72917	11063	10129	61201	76044
05457	48338	40916	10453	94473	72759	86299	62959	01064	39749
28918	15769	34348	64162	75841	77582	82921	99286	49425	02973
73010	11300	10710	62560	78969	10771	53899	26454	73627	03681
59435	23480	05967	24479	93169	38697	93658	13676	39128	11680
17929	02455	53366	98097	08284	66830	26423	57062	04563	13822
41862	26768	83848	62175	18414	50906	39708	80097	23206	18358
01294	42540	43590	78681	79771	70501	05062	95860	29602	14866
22775	02858	12165	47273	74148	19427	49227	20518	80065	95722
53747	60983	82171	44180	25536	55599	03762	22186	99253	95841
81766	28025	32247	41257	57319	72602	19740	65016	12435	89463
24862	44004	40269	45574	11018	55941	36479	15404	64110	46027
99169	80770	92093	25630	24942	18977	89382	65496	88534	41734
53600	45992	93546	47348	42169	26882	81774	48703	56244	99137
88627	39523	39496	81268	32137	61411	79234	22696	23073	34171
97367	76657	83638	11912	18723	05129	62265	27431	04195	78294
32005	87382	36246	31037	60009	80722	44244	38968	35608	62938
57154	76478	44478	78561	71064	19331	76406	84452	19058	54278
54146	36375	30932	58210	70875	01355	70257	09341	23730	58309
36283	92917	30953	49460	18185	63965	20121	45041	89156	29563
74973	83767	27843	13152	28328	51597	54624	63371	88603	61277
70237	69924	87413	95159	84237	48986	35781	73808	20817	60630
76426	12882	89455	20792	19655	30803	07915	70264	50346	69701
65088	63220	93521	92145	11180	37773	26018	16150	62735	31062

Table A.4. (Continued)

21755	50969	10016	01373	18088	96168	14217	19786	90759	66476
82024	93860	24943	04919	05019	85844	69890	46740	51431	87922
64649	96595	97725	16988	22404	81529	87537	91453	60886	42239
05455	52581	66391	25111	53143	92863	78886	37547	15306	53911
85711	29066	02999	56394	11372	60689	61784	24499	90934	25106
32230	67428	14496	80119	50249	80419	30275	57878	74784	27806
93773	12383	30343	70604	50537	67783	51863	01132	40022	29939
19436	47161	08039	23786	70362	08094	15302	18963	76059	85683
29564	06230	71308	71770	88850	87166	23344	55564	23287	39647
28294	12945	23018	21604	22457	40306	39721	75568	95922	95419
09211	96490	96042	07837	82647	25343	08236	21325	53823	31010
01652	30822	70058	42947	27160	76437	14177	97132	55193	56972
45091	57793	40937	25483	84462	77419	04356	29363	36969	57549
12567	57462	31667	72844	52056	56741	71936	20944	78241	80949
81524	60599	29872	33841	34193	00587	95783	69415	54442	01910
21482	11696	76840	55775	43085	56535	51444	99849	36099	17950
82810	35306	66543	81499	90106	07145	31914	27172	75808	10295
79498	84331	90497	84000	89528	81166	81247	56983	10673	51195
11109	05896	35392	59285	37186	89548	02607	09712	34804	21413
15244	98745	55271	42923	60096	74268	04743	60039	17547	64932
31666	05605	48629	41332	10329	89982	46927	71723	07996	02466
42826	34764	23143	25983	47607	51791	82282	27570	24876	01128
82881	87130	76850	76921	69879	26981	32973	55008	33291	04669
28391	28322	14413	31579	59754	74317	08112	79815	05879	16938
48719	39869	00739	45610	67010	91567	46312	53765	05780	50798
52763	59397	81517	54521	93475	70156	79661	46562	62420	55458
70967	26680	21377	88141	36450	25424	24495	18149	88435	67268
07692	40737	75193	84524	30406	21722	56673	44542	57189	42256
38832	52688	66638	25632	54050	93604	75178	08625	75145	73248
63182	19854	50484	24217	90941	27692	47680	36849	91973	20190
93388	78611	31175	79544	96694	64262	15325	13587	43599	39302
43423	06816	50091	39199	84373	53446	61320	86900	69517	35003
84358	91122	87506	04936	42059	07924	69016	42775	35505	28060
48808	40305	02561	52614	92636	82287	60001	19417	76491	84195
16750	42742	05696	49496	45709	28786	61339	08953	01668	29427
61552	04467	72828	96765	25138	79942	53404	00946	25034	41690
24829	73764	19122	50857	33043	40546	45884	10391	49390	02819
72401	09034	02594	34257	82193	84846	69338	52408	90406	70765
10932	28706	73841	84692	43581	99260	03325	26610	29737	38927
65930	45238	78052	61167	64536	36708	39425	06176	82227	37781
86639	79801	99050	76091	65094	05740	48597	39918	02130	53520
01947	29996	62454	04755	66442	55854	37146	20187	86811	39179
53770	70012	36138	86720	95077	89978	84171	95222	13796	25774
30475	50884	31026	28195	89935	85855	05715	61588	18092	54261
62739	88081	63832	90260	67072	90095	36914	10629	31549	93630
77844	24386	45720	54845	17591	11938	13307	72402	14648	36263
85747	88110	56936	48625	79327	38333	16052	51315	63422	18693
41361	79928	07316	24546	81431	67669	73127	29744	20315	54192
42317	66171	16169	01470	63300	66571	29722	79191	07644	33148
30088	82194	22366	94453	65418	42430	65820	13046	85896	97958

Table A.4. (Continued)

14679	06451	06588	81467	29514	43874	97618	90837	66459	81223
89204	04220	22479	96891	79032	02169	01727	90834	29465	59280
90491	09422	42113	67877	28245	58572	97229	56225	62283	67668
45465	63773	73453	66756	39456	35932	60485	39521	92761	00876
70407	14550	53641	90888	62562	57373	81180	70722	93714	25780
11254	35507	11749	50931	99843	40677	45472	06738	31180	26435
77992	30030	78254	99249	91207	54003	03149	32871	68881	17444
84733	51402	45811	12247	61182	56872	90592	31698	74858	63867
09476	82052	07023	03671	58821	19882	89900	60456	53637	77719
83577	01987	59725	40464	05258	67372	56907	35085	98351	74336
60855	96100	28905	08526	91010	69331	96888	57348	72097	16448
49813	50842	56729	82794	86204	72075	56610	90983	36404	99578
69832	22786	08230	44306	91759	51802	84976	43633	97016	89399
63245	74878	80059	06446	18147	50861	41846	93322	83316	59991
10105	20707	33291	13385	80687	63653	55732	61518	67517	51439
00977	42906	48644	05707	96448	27318	95873	29376	13401	16019
75501	08838	73515	61548	07645	91940	82831	69523	71658	04241
80525	44978	45143	66055	90038	45735	64065	72771	60040	05302
91446	00959	82739	87803	82164	02753	77038	37448	31342	51018
24031	85017	93826	97797	42345	73895	62266	08538	90827	68122
91285	34606	86074	04984	55238	86574	22300	54630	27078	70794
81062	49226	65798	51749	18313	44134	99983	02693	16310	73623
94945	07061	87861	63618	90956	73170	20417	63972	00075	24697
47508	62774	49067	29437	35021	45184	55918	28848	69823	59308
27440	30866	38231	51593	65475	28517	81818	34260	69189	06796
18072	47268	20958	00335	13388	77308	43355	64352	95916	96329
14722	21374	95744	82090	30844	46254	99924	83255	00209	51956
09853	91508	60367	38631	88481	83202	02881	68468	84877	93045
94373	28832	23632	82309	81160	46174	46608	64077	64988	13237
94526	48016	37362	92027	15906	67185	59957	55076	90844	49373
78640	27483	57136	95467	58574	92235	26245	27355	27918	21312
76610	67852	17624	95070	11800	15172	84826	72409	71016	74412
07321	74221	79429	08584	36645	42613	25865	71671	97244	38423
55337	86213	19237	39510	63919	27925	92274	02322	83458	79852
27987	49253	43638	18546	18533	44280	98884	94782	78736	19373
07610	71301	59319	14196	51309	80085	18726	91587	80992	61663
41047	91314	78463	48950	87529	84618	19947	58119	69184	92035
29552	48994	27991	27000	10082	17489	14188	27732	38692	82566
47166	39897	14884	30234	47445	12559	48351	77069	27255	69909
74159	15772	27203	21269	82781	05262	86541	71890	52467	67787
34518	97021	71462	55582	90333	41760	37564	49543	59181	09194
22533	52449	11893	99313	88836	39198	35665	61203	46236	26586
43385	21033	35736	25113	29055	48307	28348	35323	56803	22908
26224	51169	70990	65422	53620	32916	14914	93239	25205	12074
45601	64938	79216	90857	32195	75453	62960	98682	25443	64524
58151	68188	85768	00919	33595	93299	57440	04948	12724	31809
76340	71754	43465	15001	98816	57885	57961	26798	39138	93645
09625	14943	03037	83731	75049	96716	04849	58650	08254	21673
55746	39415	34777	53925	10558	78021	73106	36443	60700	36211
70966	99897	60680	65010	43936	78523	75711	47553	03918	95528

Table A.4. (Continued)

37906	67684	64945	80515	98580	66844	13841	81451	64133	51736
33214	33660	03036	15271	64298	48348	11405	33299	97141	83158
24673	99616	75635	57261	62852	61758	04883	73473	96889	20071
48424	09907	23621	52564	02788	43652	36819	28046	36901	68989
81824	70042	83835	72632	81280	33986	72398	28508	19464	60208
86072	74558	72701	65183	36787	34078	73381	33580	41052	44089
95065	73583	85914	53236	46186	97991	72273	51718	59845	93607
23036	95754	67198	41400	43613	89077	41942	49956	22261	83956
03385	64705	57768	64255	65622	15430	01983	73023	20295	64558
04689	47109	40454	18734	48321	21315	93002	66173	60023	64576
47246	02296	88556	10674	67034	06143	33545	45982	73986	30075
14570	16742	69321	65249	67601	18364	01776	45378	56279	29273
69122	51274	85540	51772	40770	78669	05078	22446	88971	31817
90157	54589	16114	05382	87269	75173	65610	17102	12127	18837
03879	53737	87508	41417	60925	92795	08442	15497	17004	62546
22236	70503	22890	02810	10852	39816	41230	89031	63119	77428
96497	82130	24298	06620	21037	66636	10484	63177	84701	39611
75355	95630	95584	12836	16760	85032	59349	90801	10663	46768
84645	86953	59912	76888	22124	40748	45370	05479	24547	97967
66878	19564	15482	16818	61566	18034	54106	62327	87010	61112
83586	51499	56724	76848	50567	11768	15509	87138	83245	15223
42133	15790	15539	51120	66864	41105	44854	99171	91340	55499
10936	66119	98287	94653	83328	48089	38583	98693	44910	81224
83154	67905	59041	43101	27232	75480	07791	67427	34439	99786
88765	34520	08226	38677	58026	00065	78152	68330	06605	24659
88509	09116	58112	66633	76782	60996	74937	00650	29469	70960
60539	24597	02726	94980	45345	17247	95374	17014	43723	09207
61732	87526	69563	29647	51811	44040	97946	91308	66335	76237
60406	76296	14122	98066	61291	85153	59295	89395	02711	11708
45845	69699	05584	06215	87034	66264	14275	89270	07145	39440
72327	79003	23073	27408	56790	94062	53259	72995	58638	94869
41471	95004	96783	25663	82927	66588	53186	19823	40260	63150
16553	30969	36833	42272	22134	52349	14887	18182	75107	87002
46044	93244	45068	31210	81646	08985	19099	33722	30442	79154
00942	93267	51118	67305	43765	56843	47168	71421	04014	81289
51611	78061	03866	20642	89054	48544	89778	01766	02593	99660
74620	50774	33064	41415	50910	32721	38133	34307	62641	67084
38731	28630	23199	81522	80336	48317	66878	66416	96592	81803
21487	92695	62781	42431	89552	55433	26629	60878	25607	05738
03962	72124	97090	33420	26704	59442	16319	62932	30719	17924
31070	25105	70997	00037	61248	67153	51682	05715	69150	63466
58445	79472	50823	89685	80704	77058	36012	22548	11179	64137
84211	28033	10300	09500	64296	00270	22957	35313	05259	13737
35087	86887	05537	51242	57611	84488	88429	70179	25215	19675
91108	69575	97474	58666	74734	20403	49813	15245	43150	21696
41824	55683	57679	44922	50418	98213	35513	29157	55322	69516
19196	89530	64420	80747	04696	01452	66444	85596	15581	23122
98495	82872	10780	18824	15923	26125	13436	45495	52042	02791
48367	65082	55187	45381	75601	45172	90624	37190	07520	10413
94793	75039	97093	07207	69403	51555	84698	44192	02867	19475

Table A.4. (Continued)

91046	70999	06579	00484	43692	57795	53540	01085	24645	38105
06824	97998	26567	77415	25388	55594	05067	30055	34726	37855
14595	46870	49867	77370	15416	96161	38219	63861	39195	60419
09751	14737	00661	71330	22026	90185	42872	45571	87570	15285
03574	83877	71094	71667	90150	95901	64049	21185	50667	13340
24033	24135	28513	53860	66540	14499	88130	98287	96551	94717
99341	43590	67468	12497	67768	94135	54895	82151	98848	53733
32760	12964	82939	80956	39285	07097	89021	70247	20290	89516
54491	21296	37353	32456	58553	59215	28308	36913	66700	67981
07583	33673	33787	99662	93888	66367	53206	21934	52744	19057
39647	66364	41356	50860	32955	39987	43329	00091	10580	62660
80732	20453	96128	92692	90996	11743	51493	03267	56168	07600
40090	72135	71999	82780	48686	47093	44383	42461	55315	49745
00632	56081	30796	80045	03103	98211	61335	84114	52662	58949
79731	47847	69497	86357	92669	78799	38769	23422	91608	38848
08614	02049	34765	90145	27982	56589	87690	16839	11823	19391
09935	85090	21118	60897	75323	03859	45248	36755	13296	48981
74216	04471	58876	13015	48205	72092	98703	43385	71982	41419
00677	31291	81644	08248	14690	13905	51216	82150	31951	52357
26580	88272	90406	41986	74259	13682	13754	59388	32537	60731
71552	82733	52426	69959	47917	57872	15979	75022	06320	62721
19442	52160	37978	58426	08753	26686	81330	47810	54884	99014
00122	09604	56525	52716	07567	70954	57616	07110	24118	80713
67216	76877	56318	46014	69224	01212	84256	94624	09438	05009
45110	25576	47459	54570	00287	10586	60938	70350	73650	08754
89671	59232	48355	57037	07029	28837	09759	99012	06245	46358
21941	14417	89031	04431	25308	11970	44044	56533	42796	47979
02705	84418	82163	33219	59841	60076	21881	90604	46695	64728
68115	26133	43759	27355	70302	75614	20963	45249	82820	89682
92707	49102	60251	19581	75228	75131	73738	66245	33821	06721
35939	55142	07399	48110	22069	99420	97897	92602	74539	13811
12019	47810	78689	41839	42832	80434	97117	58792	78698	43063
50092	12981	27059	45518	29575	67789	40553	33217	34323	06982
32986	35078	13588	65822	72642	43450	06917	50448	40435	88575
67536	08040	40407	70084	56838	10269	50074	08019	97450	12526
98106	75894	72414	51431	56860	78280	57941	43121	37253	35425
77276	44827	73475	37404	63144	42226	85056	30301	16297	25073
03764	96921	47652	13621	52855	94537	91525	98316	66168	12161
77455	55276	34556	09855	48125	00049	67169	02566	10876	42160
55073	50995	10314	02925	24721	22000	09511	59060	68761	81026
67299	74565	41692	79065	99163	83388	07859	96658	09217	85375
29298	07412	80784	17996	80921	23560	13066	66357	80544	86050
91407	73988	21260	61665	78651	16919	93657	09667	15090	03532
85379	80488	14514	62693	45528	08934	43846	82672	01415	64436
73877	63341	15148	20821	88589	44145	67572	08120	40576	70372
57044	08909	40062	60187	00559	61672	65001	34955	24717	33706
79784	75348	34030	26050	50030	65726	44081	72952	40980	89306
67019	98358	86971	36607	36901	91939	19522	89680	62379	67151
41496	44616	94227	63819	34868	34838	95217	57756	89580	17676
99834	39917	40995	86715	51336	27577	03438	72434	03659	70046

Table A.4. (Continued)

87340	93621	75668	11412	87464	22069	45241	72818	29736	23057
70713	16765	64165	34106	27152	55954	91352	44525	44980	69154
10102	38228	38442	86601	73625	62341	11417	07431	41828	84087
28072	11642	17204	98450	35856	29088	51820	49122	28351	28007
12884	25234	35447	85574	94423	96364	28378	02991	62542	48328
27094	27605	21783	82254	97475	44559	90078	39433	25896	00475
69789	14454	58626	21664	02351	19124	93582	89374	94004	59659
11915	34559	00117	89667	36950	14694	55609	01408	07963	35106
38964	17831	56305	77347	57337	47900	71366	97092	40254	67697
43892	54073	29411	89805	74131	36540	52057	83475	32340	42757
65481	08075	59117	82023	60088	22314	93095	57736	08564	31758
13466	94974	48647	37115	84859	75118	86979	45810	92049	40000
11558	09341	52027	28088	00284	48909	90897	05195	13099	31837
73126	52226	55216	61265	70639	72451	98949	41632	59251	80815
51628	82433	29946	37772	57118	02761	02502	90157	18424	99129
28958	81585	28890	39072	74422	94883	99498	43036	62730	89054
69755	17039	74438	93269	89674	98623	84667	20393	60353	78578
28290	62543	20640	60954	79960	31174	67404	23853	36480	04202
96886	43500	89025	42647	54653	44096	72699	39322	81643	35955
80940	62044	43802	04416	32270	53875	32820	12044	76883	52900
25451	76608	30584	94033	88940	87339	23568	55358	20032	21972
51540	11490	05081	27069	16685	56486	88750	24122	45978	58768
38809	03936	74336	10590	94518	71798	80119	34531	86118	05922
60470	09522	12897	91680	34003	78900	67368	94108	58328	02998
93030	31200	49927	18762	63223	10480	93871	68905	68583	91355
30880	63084	00584	35748	09225	33618	18680	09517	88981	48227
88533	97152	86114	42311	78843	92251	43919	33256	04265	26280
19299	79270	01925	98119	71388	45254	29028	66878	40017	38202
69054	70503	01526	74633	34061	32661	89417	42554	50565	20407
67352	20450	16530	15129	41996	15818	16940	59277	03202	85715
85468	41376	79039	01850	99744	81812	93169	22705	97709	81907
12057	35180	02566	98772	69539	28280	43827	08444	56220	61323
61951	19798	61229	89189	84075	01746	53801	07092	58342	86217
92224	77389	34320	09414	47608	00915	77016	53859	30015	95353
01982	53711	04423	43139	19021	25871	84038	71386	71973	89367
88443	26351	62114	35523	54106	04932	42631	11399	84712	05686
71700	79026	28855	61893	11658	12915	72563	19139	61766	98345
34173	19937	12062	92936	49051	57880	05821	14005	31644	63815
98089	14200	02395	86565	97833	07913	66981	30666	81164	10981
35414	13649	63167	56163	68467	05339	15824	46897	38963	11698
19522	11849	95389	65695	35664	22725	15372	87703	87868	37883
36480	77931	39265	34212	51883	03389	53384	89802	58350	41882
26138	79888	44088	45532	76396	48583	03931	86340	82657	18881
69367	46269	53313	03455	40910	14365	18002	78724	10321	53409
27566	84710	60165	98597	50090	13172	28217	50750	50546	46520
24208	26566	41518	10015	86420	28387	92538	99744	65618	96011
85497	48885	23833	03027	03664	94744	35480	60186	41792	09076
78331	88300	24816	54195	01822	23175	22655	27351	60205	15074
48594	55901	98052	85207	28770	05754	75593	01764	30247	65604
21400	37964	35184	69430	99920	74647	48604	19798	81347	09895

Table A.4. (Continued)

27247	74427	01337	38176	17021	58543	98070	61531	86549	65516
14156	95285	44308	08275	50878	30794	26858	84002	62289	17713
45673	14618	76646	28312	80058	25534	32686	66307	03660	25200
18256	82731	32083	47921	98771	61337	79083	22463	23032	36026
08862	28939	93878	02811	71963	06475	91182	88814	12247	74153
34694	40786	32471	27744	74229	95669	05957	74718	94691	95680
21859	70031	23453	10432	51516	15202	27567	28584	23404	83160
88328	81523	94184	31389	40680	61577	03337	65833	64959	97256
98901	18100	18490	48034	21791	98440	94518	81149	36224	87001
95993	92944	93729	25874	11685	37237	18258	74414	82159	23219
99886	90343	43080	45388	84604	99135	33371	08008	47709	14181
55707	06608	38286	70240	76861	29006	82287	32662	55387	38363
31696	44902	84149	00774	47296	76811	35197	19884	02527	36204
94480	41042	95933	06218	73908	87481	59713	70946	88258	91030
51854	91851	84421	79859	19367	97465	10491	85755	20110	48867
12205	33427	90714	43441	44220	74345	48089	19423	83736	27611
12035	23714	33964	90354	36447	05492	04934	80168	36601	98093
59152	72070	00051	51676	09597	92497	68607	79172	57563	32830
78914	78012	57844	44952	49118	90138	98763	81334	99134	62792
84548	42156	91999	72590	07545	91955	83829	75373	97778	08310
69856	44345	37359	25052	14076	58986	27231	32509	49973	07253
93760	97283	39850	63554	22674	64053	80248	05012	07838	01917
40959	28502	02379	57758	25133	30976	59655	79147	37982	15567
51562	69272	51570	85970	51690	36403	53207	91615	70820	00385
64929	86104	32268	18664	21571	61451	74567	94342	94943	20582
69002	28768	51591	50633	39475	81155	31652	59516	72230	97730
36308	38281	02907	72917	11332	11740	68452	05049	14222	39893
11359	09112	52439	32620	23768	49023	80229	40159	18894	51933
41544	81869	17314	67064	46558	75767	35582	31585	69270	31356
64702	06004	53711	68222	25934	41607	16231	22923	91454	42414
31398	06654	57345	84180	88969	76195	56445	52913	09476	80635
98527	65438	05890	46391	25333	34481	78886	62987	67946	40786
33203	32139	94004	74772	02591	42598	32264	10207	70859	33289
62005	72426	76511	07230	54774	63575	88484	58723	55982	07384
40465	76337	93835	75973	11321	84168	03118	61194	39104	83876
51313	52998	56165	60010	54571	87333	32852	11115	71019	26079
17303	69107	58913	31510	58840	69451	87141	97787	02447	34599
85562	01985	41848	33221	22690	37149	18677	77720	98226	89880
55239	22513	37930	24957	08372	25614	78337	39488	31896	35980
28818	49087	35155	69148	98487	68591	25162	33651	75444	49804
19825	70717	45870	38769	19778	69745	40464	15079	26297	22066
95718	75713	92817	75483	17542	25903	97538	52096	34734	39538
24624	91901	29784	51594	41504	84679	34535	45096	59754	52661
78177	95113	67082	46472	75090	32286	17907	16865	40027	88378
70022	24726	18150	86370	54867	51867	17396	37574	68877	75099
30696	08284	73441	66084	35537	18463	90940	79472	58367	34949
84393	91813	91338	21704	08183	53759	48903	09586	17386	18038
51751	48159	23192	05719	25583	02027	81915	73246	02237	54207
05106	42768	10657	79027	78247	58237	45199	47058	75553	29795
98790	24578	58358	28940	48422	30065	44593	96972	80495	10221

Table A.5. Natural logarithm of x—ln(x)

LN(10)=2.30259.

IF X IS GREATER THAN 10, EXPRESS X AS THE PRODUCT OF A NUMBER, Y, BETWEEN 1 AND 10 TIMES 10 RAISED TO A CERTAIN POWER. IF N IS THE POWER OF 10, THEN LN(X) = LN(Y) + N*LN(10). FOR EXAMPLE, 974 = 9.74*(10**2), SO THAT Y = 9.74 AND N= 2. THEREFORE, LN(974) = LN(9.74) + 2*LN(10) = 2.27624 + 2*2.30259 = 6.88142.

IF X IS LESS THAN 1, EXPRESS X AS THE RATIO OF A NUMBER, Y, BETWEEN 1 AND 10 DIVIDED BY 10 RAISED TO A CERTAIN POWER. IF N IS THE POWER OF 10, THEN LN(X) = LN(Y) - N*LN(10). FOR EXAMPLE, 0.974 = 9.74/(10**1), SO THAT Y= 9.74 AND N=1. THEREFORE, LN(0.974) = LN(9.74) - LN(10) = 2.27624 - 2.30259 = -0.02635.

X	0.00	0.01	0.02	0.03	0.04	0.05	0.06	0.07	0.08	0.09
1.0	0.00000	0.00995	0.01980	0.02956	0.03922	0.04879	0.05827	0.06766	0.07696	0.08618
1.1	0.09531	0.10436	0.11333	0.12222	0.13103	0.13976	0.14842	0.15700	0.16551	0.17395
1.2	0.18232	0.19062	0.19885	0.20701	0.21511	0.22314	0.23111	0.23902	0.24686	0.25464
1.3	0.26236	0.27003	0.27763	0.28518	0.29267	0.30010	0.30748	0.31481	0.32208	0.32930
1.4	0.33647	0.34359	0.35066	0.35767	0.36464	0.37156	0.37844	0.38526	0.39204	0.39878
1.5	0.40546	0.41211	0.41871	0.42527	0.43178	0.43825	0.44468	0.45107	0.45742	0.46373
1.6	0.47000	0.47623	0.48243	0.48858	0.49470	0.50077	0.50682	0.51282	0.51879	0.52473
1.7	0.53063	0.53649	0.54232	0.54812	0.55388	0.55961	0.56531	0.57098	0.57661	0.58221
1.8	0.58779	0.59333	0.59884	0.60432	0.60977	0.61518	0.62058	0.62594	0.63127	0.63658
1.9	0.64185	0.64710	0.65232	0.65752	0.66269	0.66783	0.67294	0.67803	0.68310	0.68813
2.0	0.69315	0.69813	0.70310	0.70804	0.71295	0.71784	0.72271	0.72755	0.73237	0.73716
2.1	0.74194	0.74669	0.75142	0.75612	0.76081	0.76547	0.77011	0.77473	0.77932	0.78390
2.2	0.78846	0.79299	0.79751	0.80200	0.80648	0.81093	0.81536	0.81978	0.82417	0.82855
2.3	0.83291	0.83725	0.84157	0.84587	0.85015	0.85441	0.85866	0.86289	0.86710	0.87129
2.4	0.87547	0.87963	0.88377	0.88789	0.89200	0.89609	0.90016	0.90422	0.90826	0.91228
2.5	0.91629	0.92028	0.92426	0.92822	0.93216	0.93609	0.94001	0.94391	0.94779	0.95166
2.6	0.95551	0.95935	0.96317	0.96698	0.97078	0.97456	0.97833	0.98208	0.98582	0.98954
2.7	0.99325	0.99695	1.00063	1.00430	1.00796	1.01160	1.01523	1.01885	1.02245	1.02604
2.8	1.02962	1.03318	1.03674	1.04028	1.04380	1.04732	1.05082	1.05431	1.05779	1.06126
2.9	1.06471	1.06815	1.07158	1.07500	1.07841	1.08180	1.08519	1.08856	1.09192	1.09527
3.0	1.09861	1.10194	1.10526	1.10856	1.11186	1.11514	1.11841	1.12168	1.12493	1.12817
3.1	1.13140	1.13462	1.13783	1.14103	1.14422	1.14740	1.15057	1.15373	1.15688	1.16002
3.2	1.16315	1.16627	1.16938	1.17248	1.17557	1.17865	1.18173	1.18479	1.18784	1.19089
3.3	1.19392	1.19695	1.19996	1.20297	1.20597	1.20896	1.21194	1.21491	1.21787	1.22083
3.4	1.22377	1.22671	1.22964	1.23256	1.23547	1.23837	1.24127	1.24415	1.24703	1.24990
3.5	1.25276	1.25562	1.25846	1.26130	1.26413	1.26695	1.26976	1.27256	1.27536	1.27815
3.6	1.28093	1.28371	1.28647	1.28923	1.29198	1.29473	1.29746	1.30019	1.30291	1.30563
3.7	1.30833	1.31103	1.31372	1.31641	1.31908	1.32175	1.32442	1.32707	1.32972	1.33237
3.8	1.33500	1.33763	1.34025	1.34286	1.34547	1.34807	1.35067	1.35325	1.35583	1.35841
3.9	1.36098	1.36354	1.36609	1.36864	1.37118	1.37371	1.37624	1.37877	1.38128	1.38379
4.0	1.38629	1.38879	1.39128	1.39377	1.39624	1.39872	1.40118	1.40364	1.40610	1.40854
4.1	1.41099	1.41342	1.41585	1.41828	1.42070	1.42311	1.42551	1.42792	1.43031	1.43270
4.2	1.43508	1.43746	1.43983	1.44220	1.44456	1.44692	1.44927	1.45161	1.45395	1.45629
4.3	1.45861	1.46094	1.46325	1.46557	1.46787	1.47017	1.47247	1.47476	1.47705	1.47933
4.4	1.48160	1.48387	1.48614	1.48840	1.49065	1.49290	1.49515	1.49739	1.49962	1.50185
4.5	1.50408	1.50630	1.50851	1.51072	1.51293	1.51513	1.51732	1.51951	1.52170	1.52388
4.6	1.52606	1.52823	1.53039	1.53256	1.53471	1.53687	1.53901	1.54116	1.54330	1.54543
4.7	1.54756	1.54969	1.55181	1.55392	1.55604	1.55814	1.56025	1.56235	1.56444	1.56653
4.8	1.56861	1.57070	1.57277	1.57485	1.57691	1.57898	1.58104	1.58309	1.58514	1.58719
4.9	1.58923	1.59127	1.59331	1.59534	1.59736	1.59939	1.60141	1.60342	1.60543	1.60744

Table A.5. (Continued)

X	0.00	0.01	0.02	0.03	0.04	0.05	0.06	0.07	0.08	0.09
5.0	1.60944	1.61143	1.61343	1.61542	1.61741	1.61939	1.62137	1.62334	1.62531	1.62728
5.1	1.62924	1.63120	1.63315	1.63511	1.63705	1.63900	1.64094	1.64287	1.64480	1.64673
5.2	1.64866	1.65058	1.65250	1.65441	1.65632	1.65823	1.66013	1.66203	1.66393	1.66582
5.3	1.66771	1.66959	1.67147	1.67335	1.67523	1.67710	1.67896	1.68083	1.68269	1.68454
5.4	1.68640	1.68825	1.69009	1.69194	1.69378	1.69561	1.69745	1.69928	1.70110	1.70293
5.5	1.70475	1.70656	1.70838	1.71019	1.71199	1.71380	1.71560	1.71739	1.71919	1.72098
5.6	1.72277	1.72455	1.72633	1.72811	1.72988	1.73166	1.73342	1.73519	1.73695	1.73871
5.7	1.74047	1.74222	1.74397	1.74572	1.74746	1.74920	1.75094	1.75267	1.75440	1.75613
5.8	1.75786	1.75958	1.76130	1.76302	1.76473	1.76644	1.76815	1.76985	1.77156	1.77326
5.9	1.77495	1.77664	1.77834	1.78002	1.78171	1.78339	1.78507	1.78675	1.78842	1.79009
6.0	1.79176	1.79342	1.79509	1.79675	1.79840	1.80006	1.80171	1.80336	1.80500	1.80665
6.1	1.80829	1.80993	1.81156	1.81319	1.81482	1.81645	1.81808	1.81970	1.82132	1.82293
6.2	1.82455	1.82616	1.82777	1.82938	1.83098	1.83258	1.83418	1.83578	1.83737	1.83896
6.3	1.84055	1.84213	1.84372	1.84530	1.84688	1.84845	1.85003	1.85160	1.85317	1.85473
6.4	1.85630	1.85786	1.85942	1.86097	1.86253	1.86408	1.86563	1.86718	1.86872	1.87026
6.5	1.87180	1.87334	1.87487	1.87641	1.87794	1.87946	1.88099	1.88251	1.88403	1.88555
6.6	1.88707	1.88858	1.89009	1.89160	1.89311	1.89462	1.89612	1.89762	1.89912	1.90061
6.7	1.90211	1.90360	1.90509	1.90657	1.90806	1.90954	1.91102	1.91250	1.91398	1.91545
6.8	1.91692	1.91839	1.91986	1.92132	1.92279	1.92425	1.92571	1.92716	1.92862	1.93007
6.9	1.93152	1.93297	1.93441	1.93586	1.93730	1.93874	1.94018	1.94162	1.94305	1.94448
7.0	1.94591	1.94734	1.94876	1.95019	1.95161	1.95303	1.95444	1.95586	1.95727	1.95868
7.1	1.96009	1.96150	1.96291	1.96431	1.96571	1.96711	1.96851	1.96990	1.97130	1.97269
7.2	1.97408	1.97547	1.97685	1.97824	1.97962	1.98100	1.98238	1.98376	1.98513	1.98650
7.3	1.98787	1.98924	1.99061	1.99197	1.99334	1.99470	1.99606	1.99742	1.99877	2.00013
7.4	2.00148	2.00283	2.00418	2.00553	2.00687	2.00821	2.00955	2.01089	2.01223	2.01357
7.5	2.01490	2.01623	2.01757	2.01889	2.02022	2.02155	2.02287	2.02419	2.02551	2.02683
7.6	2.02815	2.02946	2.03078	2.03209	2.03340	2.03471	2.03601	2.03732	2.03862	2.03992
7.7	2.04122	2.04252	2.04381	2.04511	2.04640	2.04769	2.04898	2.05027	2.05156	2.05284
7.8	2.05412	2.05540	2.05668	2.05796	2.05924	2.06051	2.06179	2.06306	2.06433	2.06560
7.9	2.06686	2.06813	2.06939	2.07065	2.07191	2.07317	2.07443	2.07568	2.07694	2.07819
8.0	2.07944	2.08069	2.08194	2.08318	2.08443	2.08567	2.08691	2.08815	2.08939	2.09063
8.1	2.09186	2.09310	2.09433	2.09556	2.09679	2.09802	2.09924	2.10047	2.10169	2.10291
8.2	2.10413	2.10535	2.10657	2.10779	2.10900	2.11021	2.11142	2.11263	2.11384	2.11505
8.3	2.11625	2.11746	2.11866	2.11986	2.12106	2.12226	2.12346	2.12465	2.12585	2.12704
8.4	2.12823	2.12942	2.13061	2.13180	2.13298	2.13417	2.13535	2.13653	2.13771	2.13889
8.5	2.14007	2.14124	2.14242	2.14359	2.14476	2.14593	2.14710	2.14827	2.14943	2.15060
8.6	2.15176	2.15292	2.15408	2.15524	2.15640	2.15756	2.15871	2.15987	2.16102	2.16217
8.7	2.16332	2.16447	2.16562	2.16677	2.16791	2.16905	2.17020	2.17134	2.17248	2.17361
8.8	2.17475	2.17589	2.17702	2.17815	2.17929	2.18042	2.18155	2.18267	2.18380	2.18493
8.9	2.18605	2.18717	2.18830	2.18942	2.19053	2.19165	2.19277	2.19388	2.19500	2.19611
9.0	2.19722	2.19833	2.19944	2.20055	2.20166	2.20276	2.20387	2.20497	2.20607	2.20717
9.1	2.20827	2.20937	2.21047	2.21157	2.21266	2.21375	2.21485	2.21594	2.21703	2.21812
9.2	2.21920	2.22029	2.22137	2.22246	2.22354	2.22462	2.22570	2.22678	2.22786	2.22894
9.3	2.23001	2.23109	2.23216	2.23323	2.23431	2.23538	2.23644	2.23751	2.23858	2.23965
9.4	2.24071	2.24177	2.24283	2.24390	2.24496	2.24601	2.24707	2.24813	2.24918	2.25024
9.5	2.25129	2.25234	2.25339	2.25444	2.25549	2.25654	2.25759	2.25863	2.25968	2.26072
9.6	2.26176	2.26280	2.26384	2.26488	2.26592	2.26696	2.26799	2.26903	2.27006	2.27109
9.7	2.27213	2.27316	2.27419	2.27521	2.27624	2.27727	2.27829	2.27932	2.28034	2.28136
9.8	2.28238	2.28340	2.28442	2.28544	2.28646	2.28747	2.28849	2.28950	2.29051	2.29152
9.9	2.29253	2.29354	2.29455	2.29556	2.29657	2.29757	2.29858	2.29958	2.30058	2.30158

Table A.6. Percentage points of Bartholomew's test for order when $m = 3$ proportions are compared

C	ALPHA .10	.05	.025	.01	.005
0.0	2.952	4.231	5.537	7.289	8.628
0.1	2.885	4.158	5.459	7.208	8.543
0.2	2.816	4.081	5.378	7.122	8.455
0.3	2.742	4.001	5.292	7.030	8.360
0.4	2.664	3.914	5.200	6.932	8.258
0.5	2.580	3.820	5.098	6.822	8.146
0.6	2.486	3.715	4.985	6.700	8.016
0.7	2.379	3.593	4.852	6.556	7.865
0.8	2.251	3.446	4.689	6.377	7.677
0.9	2.080	3.245	4.465	6.130	7.413
1.0	1.642	2.706	3.841	5.413	6.635

REPRODUCED FROM TABLE A.1 OF BARLOW, R.E., BARTHOLOMEW, D.J., BREMNER, J.M. AND BRUNK, H.D. (1972). ''STATISTICAL INFERENCE UNDER ORDER RESTRICTIONS.'' JOHN WILEY AND SONS, NEW YORK.

Table A.7. Percentage points of Bartholomew's test for order when m = 4 proportions are compared. The table is symmetric in c_1 and c_2

C2	ALPHA	C1 0.0	0.1	0.2	0.3	0.4	0.5	0.6	0.7
	.10	4.010							
	.05	5.435							
0.0	.025	6.861							
	.01	8.746							
	.005	10.171							
	.10	3.952	3.891						
	.05	5.372	5.305						
0.1	.025	6.794	6.724						
	.01	8.676	8.601						
	.005	10.098	10.020						
	.10	3.893	3.827	3.758					
	.05	5.307	5.235	5.160					
0.2	.025	6.725	6.649	6.570					
	.01	8.602	8.522	8.437					
	.005	10.022	9.939	9.851					
	.10	3.831	3.760	3.685	3.606				
	.05	5.239	5.162	5.080	4.993				
0.3	.025	6.653	6.571	6.484	6.391				
	.01	8.525	8.438	8.346	8.246				
	.005	9.942	9.652	9.756	9.653				
	.10	3.765	3.688	3.607	3.519	3.423			
	.05	5.166	5.083	4.994	4.898	4.791			
0.4	.025	6.575	6.486	6.392	6.289	6.174			
	.01	8.442	8.348	8.247	8.137	8.014			
	.005	9.855	9.758	9.653	9.539	9.411			
	.10	3.695	3.610	3.521	3.423	3.313	3.187		
	.05	5.088	4.997	4.898	4.791	4.670	4.528		
0.5	.025	6.491	6.394	6.289	6.173	6.043	5.891		
	.01	8.352	8.246	8.136	8.013	7.873	7.709		
	.005	9.761	9.654	9.537	9.409	9.264	9.092		

Table A.7. (Continued)

C2	ALPHA	C1 0.0	0.1	0.2	0.3	0.4	0.5	0.6	0.7
	.10	3.617	3.523	3.422	3.310	3.183	3.031	2.837	
	.05	5.002	4.900	4.789	4.665	4.524	4.354	4.135	
0.6	.025	6.398	6.289	6.170	6.038	5.886	5.702	5.462	
	.01	8.251	8.135	8.008	7.867	7.703	7.504	7.244	
	.005	9.656	9.535	9.404	9.256	9.085	8.877	8.604	
	.10	3.530	3.422	3.305	3.172	3.017	2.822	2.550	1.987
	.05	4.904	4.787	4.657	4.510	4.337	4.118	3.805	3.137
0.7	.025	6.291	6.166	6.027	5.870	5.682	5.443	5.100	4.346
	.01	8.135	8.002	7.854	7.684	7.482	7.223	6.846	6.000
	.005	9.534	9.395	9.242	9.065	8.853	8.581	8.183	7.279
	.10	3.427	3.296	3.151	2.981	2.770	2.473	1.642	
	.05	4.787	4.644	4.483	4.294	4.056	3.715	2.706	
0.8	.025	6.163	6.011	5.838	5.634	5.375	4.999	3.841	
	.01	7.994	7.832	7.647	7.427	7.146	6.734	5.412	
	.005	9.385	9.217	9.025	8.795	8.500	8.064	6.635	
	.10	3.291	3.110	2.897	2.621	2.166			
	.05	4.631	4.432	4.195	3.883	3.353			
0.9	.025	5.990	5.778	5.523	5.182	4.591			
	.01	7.804	7.577	7.303	6.933	6.277			
	.005	9.183	8.948	8.661	8.273	7.576			
	.10	2.952							
	.05	4.231							
1.0	.025	5.537							
	.01	7.289							
	.005	8.628							

REPRODUCED FROM TABLE A.2 OF BARLOW, R.E., BARTHOLOMEW, D.J., BREMNER, J.M. AND BRUNK, H.D. (1972). ''STATISTICAL INFERENCE UNDER ORDER RESTRICTIONS.'' JOHN WILEY AND SONS, NEW YORK.

Table A.8. Percentage points of Bartholomew's test for order when up to $m = 12$ proportions based on equal sample sizes are compared

M	ALPHA .10	.05	.025	.01	.005
3	2.580	3.820	5.098	6.822	8.146
4	3.187	4.528	5.891	7.709	9.092
5	3.636	5.049	6.471	8.356	9.784
6	3.994	5.460	6.928	8.865	10.327
7	4.289	5.800	7.304	9.284	10.774
8	4.542	6.088	7.624	9.639	11.153
9	4.761	6.339	7.901	9.946	11.480
10	4.956	6.560	8.145	10.216	11.767
11	5.130	6.758	8.363	10.458	12.025
12	5.288	6.937	8.561	10.676	12.257

REPRODUCED FROM TABLE A.3 OF BARLOW, R. E., BARTHOLOMEW, D. J., BREMNER, J. M. AND BRUNK, H. D. (1972). ''STATISTICAL INFERENCE UNDER ORDER RESTRICTIONS.'' JOHN WILEY AND SONS, NEW YORK.

Answers to Numerical Problems

Problem 1.2. We are given the value $P(B) = .001$.

(a) $P(A|B) = .99$ and $P(A|\bar{B}) = .01$. By (1.13), $P_{F+} = .01(1 - .001)/[.01 + .001(.99 - .01)] = .9098$. By (1.14), $P_{F-} = (1 - .99)(.001)/[1 - .01 - .001(.99 - .01)] = .00001$ (i.e., one per 100,000). The false positive rate is too high for most purposes.

(b) Now, $P(A|B) = .98$ and $P(A|\bar{B}) = .0001$. For the new definition of a positive result, $P_{F+} = .0001(1 - .001)/[.0001 + .001(.98 - .0001)] = .0925$ and $P_{F-} = (1 - .98)(.001)/[1 - .0001 - .001(.98 - .0001)] = .00002$ (i.e., two per 100,000). The false positive rate is about one tenth what it was in part *(a)*, and the false negative rate is still very small.

(c) The proportion who are positive on the first test is, by (1.11), $P(A) = .99 \times .001 + .01 \times .999 = .01098$, or 1,098 individuals per 100,000 tested. Therefore, 98,902 per 100,000 will be negative on the first test and will not have to be retested.

Problem 1.3.

(a) (1) 40,000 of the neurotics live alone. (2) 200 of them will be hospitalized. (3) 60,000 of the neurotics live with their families. (4) 360 of them will be hospitalized. (5) $p_1 = 200/(200 + 360) = .357$. (6) $p_1 < P(A|B)$.

(b) (1) 200,000 nonneurotics live alone. (2) 1000 of them will be hospitalized. (3) 800,000 nonneurotics live with their families. (4) 1800 of them will be hospitalized. (5) $p_2 = 1000/(1000 + 1800) = .357$. (6) $p_2 > P(A|\bar{B})$.

(c) p_1 is equal to p_2 even though $P(A|B)$ is much larger than $P(A|\bar{B})$.

Problem 1.4. We are given the values of $n = 100$ and $p = .05$.

(a) By (1.26), $P_L = .01$. By (1.27), $P_U = .15$.

(b) The value of $c_{\alpha/2}\sqrt{pq/n} + 1/(2n)$ is .06, so the lower and upper 99% confidence limits on P using (1.29) are $-.01$ and .11.

(c) The interval in *(b)* is narrower than the interval in *(a)*, but the fact that the lower limit is negative in *(b)* raises doubts about its validity. If the continuity correction in (1.29) were ignored, the limits of the interval would be the same to two decimal places as in *(b)*. The fault does not lie in the continuity correction.

Problem 3.4. We would like to distinguish between $P_1 = .45$ for placebo and $P_2 = .65$ for an active drug with a one-tailed test.

(*a*) For a one-tailed test with a significance level of .01, we use $c_{.01} = 2.326$ in (3.14); for a power of .95, we use $c_{.95} = -1.645$. The value of $\bar{P}$ is .55. Thus $n' = [2.326\sqrt{2 \times .55 \times .45} - (-1.645)\sqrt{.45 \times .55 + .65 \times .35}\,]^2/.2^2 = 191.85$. By (3.15), $n = 201.73$, or 202 patients per treatment.

Alternatively, look in Table A.3 under $P_1 = .45$, $P_2 = .65$, $\alpha = .02$ (recall that we are considering a one-tailed test) and power $= .95$.

(*b*) If $\alpha = .05$ (still for a one-tailed test) and $1 - \beta = .80$, 85 patients per group will be needed.

(*c*) If $n = 52$, the value of $c_{1-\beta}$ from (3.17) is $-.20$. The corresponding power is .58. (Table A.2 gives the value $P = .8415$ as corresponding to $z = .2$. The area under the normal curve below the value $-.20$ is $P/2 = .42$, and the power is $1 - P/2 = .58$.)

Problem 3.5. We wish to distinguish between the probabilities $P_1 = .25$ and $P_2 = .40$.

(*a* and *b*) We apply (3.19) and (3.18) for each value of r, using $c_\alpha = 2.576$ and $c_{1-\beta} = -1.645$, to find the value of $n_1 = m$. The value of n_2 is then rm, so that the total sample size is $n_1 + n_2 = m(r+1)$. The total cost is $10n_1 + 12n_2 = 10m + 12rm = m(10 + 12r)$. The complete table is

Ratio of Sample Sizes (r)	n_1	n_2	Total Sample Size	Total Cost
.5	530	265	795	$8,480
.6	473	284	757	8,138
.7	432	302	734	7,944
.8	401	321	722	7,862
.9	377	339	716	7,838
1	357	357	714	7,854

The total cost is minimized when $r = .9$ (i.e., when $n_1 = 377$ and $n_2 = .9(377) = 339$).

(*c*) If the total cost must be $6,240 and if the investigator decides to employ the value $r = .9$, the value of m is $m = 6240/(10 + 12 \times .9) = 300$. The value of m', from (3.20), is $m' = 300 - (.9 + 1)/(.9 \times .15) = 285.93$. To solve (3.19) for $c_{1-\beta}$, note that, for $r = .9$, $\bar{P} = .32$ and $\bar{P}\bar{Q} = .2176$. Therefore,

$$c_{1-\beta}=\frac{2.576\sqrt{(.9+1)\times .2176}-.15\sqrt{.9\times 285.93}}{\sqrt{.9\times .25\times .75+.40\times .60}}=-1.17.$$

The power of the test is found, by interpolating in Table A.2, to be approximately .88.

Problem 5.4.

(*a*)

	Hospital Diagnosis		
	Schizophrenia	Affective	Total
New York	82	24	106
London	51	67	118
Total	133	91	224

	Project Diagnosis		
	Schizophrenia	Affective	Total
New York	43	53	96
London	33	85	118
Total	76	138	214

	Computer Diagnosis		
	Schizophrenia	Affective	Total
New York	67	27	94
London	56	37	93
Total	123	64	187

(*b*)

Source	Odds Ratio from (5.16)
Hospital	4.49
Project	2.09
Computer	1.64

The odds ratios for the project and computer diagnoses are close, and both differ substantially from the odds ratio for the hospital diagnosis.

(*c*)

Source	o	o'
Hospital	4.49	4.41
Project	2.09	2.08
Computer	1.64	1.63

Problem 5.5. For the values of n_{ij} in Table 5.1 and of N_{ij} in Table 5.5, the value of $\chi^2 = \Sigma\Sigma(n_{ij} - N_{ij})^2/N_{ij}$ is 3.32, which is close to the value 3.25 found in (5.51).

Problem 5.6. For an initial approximation of $\omega_U^{(1)} = 5.37$, the value of X is 527.75 and that of Y is 401.48; the value of N_{11} is therefore 14.45 and that of W is .1993. The value of F, with the continuity correction taken as $+\frac{1}{2}$, is $-.73$, so the iterative process must be undertaken.

The value of T is .9345, that of U is .0049 and that of V, with $+\frac{1}{2}$ replacing $-\frac{1}{2}$, is 1.5428. The second approximation to the upper limit on the odds ratio is therefore, from (5.58),

$$\omega_U^{(2)} = 5.37 - \frac{-.73}{1.5428} = 5.84.$$

For this value of the odds ratio, $N_{11} = 14.87$ and $W = .2016$. The value of F is .01, which is sufficiently close to zero to stop the iterative procedure. The upper 95% confidence limit for the odds ratio underlying the data of Table 5.1 is $\omega_U = 5.84$.

Problem 5.7. The expected cell frequencies are

	B	$\bar{B}$
A	14.87	35.13
$\bar{A}$	10.13	139.87.

For these frequencies, the value of the phi coefficient (and therefore the upper 95% confidence limit on the parameter) is .30. The upper 95% confidence limit on the relative risk is $(14.87/50)/(10.13/150) = 4.40$.

Problem 5.9.

(*a*) For nonwhite live births, the estimated risk of infant mortality attributable to low birthweight is

$$r_A = \frac{.0140 \times .8625 - .1147 \times .0088}{.0228 \times .8713} = .557,$$

which is equal, to two decimal places, to the value for white live births.

(*b*) By (5.79), the standard error of ln $(1 - r_A)$ is

$$\text{s.e.}[\ln(1 - r_A)] = \sqrt{\frac{.1147 + .557(.0140 + .8625)}{37{,}840 \times .0088}} = .043.$$

A 95% confidence interval for the parameter is the interval from .518 to .593. It overlaps well with the interval found in (5.85) for white live births, but is somewhat wider. The number of nonwhite live births is less than the number of white live births, however, so the difference in interval lengths is expected.

Problem 6.3.

(*a*)

	Smokers	Nonsmokers	Total
Cases	26	8	34
Controls	73	141	214
Total	99	149	248

$$\text{Uncorrected } \chi^2 = 21.95$$
$$\varphi = .30$$

(*b*)

	Smokers	Nonsmokers	Total
Cases	163	51	214
Controls	12	22	34
Total	175	73	248

$$\text{Uncorrected } \chi^2 = 23.60$$
$$\varphi = .31$$

The phi coefficients in (*a*) and (*b*) are close.

(*c*)

	Smokers	Nonsmokers	Total
Cases	94	30	124
Controls	42	82	124
Total	136	112	248

$$\text{Uncorrected } \chi^2 = 44.03$$
$$\varphi = .42$$

The phi coefficient in (*c*) differs appreciably from those in (*a*) and (*b*). The percentage difference between the coefficients in (*a*) and (*c*) is $100(.42 - .30)/.30 = 40\%$.

Problem 6.4.

(*a*) The required value of $n_{..}$ is 807.

(*b*) The required value of N_P is 702. The percentage reduction from $n_{..}$ to N_P is $100(807 - 702)/807 = 13\%$.

(*c*) The required value of N_R is 398. The percentage reduction from $n_{..}$ to N_R is $100(807 - 398)/807 = 51\%$. The percentage reduction from N_P to N_R is $100(702 - 398)/702 = 43\%$.

Problem 7.1.

(*a*) The value of the critical ratio (with continuity correction) is 5.54. The difference between improvement rates is significant in the second hospital.

(*b*) The simple difference between the two improvement rates is $d_2 = (.75 - .35) = .40$. Its standard error is $\text{s.e.}(d_2) = .06$. The critical ratio for testing the significance of the difference between $d_1 = .20$ and $d_2 = .40$ is $z = 2.17$. The two simple differences are significantly different at the .05 level.

(*c*) The relative difference between the two improvement rates in the second hospital is $p_{e(2)} = (.75 - .35)/(1 - .35) = .62$. The value of L_2 is $-.97$, and the estimated standard error of L_2 is .19. The critical ratio for comparing L_1 and L_2 is

$$z = \frac{|-.69 - (-.97)|}{\sqrt{.28^2 + .19^2}} = .83,$$

so that the two relative differences do not differ significantly.

Problem 7.2.

(*a*) For the sample of patients given the treatments in the order *AB*, $n = 20$ and $p_1 = 15/20 = .75$.

(*b*) For the sample of patients given the treatments in the order *BA*, $m = 15$ and $p_2 = 5/15 = .33$. (Recall that p_2 is the proportion out of m who had a good response to the treatment given first, i.e., to *B*.)

(*c*) The value of the critical ratio for comparing p_1 and p_2, with the continuity correction, is 2.14. The effects of treatments *A* and *B* are significantly different.

Problem 8.1. For comparing the proportions of patients diagnosed affectively ill, the value of McNemar's statistic is $(|20 - 10| - 1)^2/(20 + 10) = 2.70$. The difference is not statistically significant.

For comparing the proportions diagnosed something other than schizophrenic or affectively ill, the value of McNemar's statistic is $(|15-5|-1)^2/(15+5)=4.05$. Because this fails to exceed 5.99, the difference is not statistically significant.

Problem 8.2.

(*a*) The value of the Stuart-Maxwell chi square statistic is 10.43 with 2 degrees of freedom. The two outcome distributions differ significantly.

(*b*) The value of d_1 is $70-50=+20$, that of d_3 is $10-20=-10$, and that of d_1-d_3 is $+30$. The new treatment seems to be superior to the standard in that it is associated with a greater net improvement. The value of the test statistic in (8.20) is 9.57, so that the new treatment is significantly better than the standard.

Problem 9.3.

(*a*)

Sample	n	Proportion Affective
1	105	.019
2	192	.068
3	145	.166
Overall	442	.088

The value of chi square is 18.18 with 2 degrees of freedom. The proportions diagnosed affective differ significantly at better than the .01 level.

(*b*) $\bar{p}_{1,2}=.051$ and $n_{1,2}=297$. The value of chi square for the difference between $\bar{p}_{1,2}$ and p_3 is 16.06; the difference is significant at better than the .01 level. The value of chi square for the difference between p_1 and p_2 is 2.03, so the difference is not significant.

(*c*) $\bar{\chi}^2=\chi^2=18.18$, and

$$c=\sqrt{\frac{105\times 145}{297\times 337}}=.39.$$

The hypothesized ordering is significant at better than the .01 level.

Problem 9.5.

(*a*) The mean ridit for group *B*, with *A* as the reference, is .963. The probability is .963 that a randomly selected member of group *B* will experience an injury at least as serious as that experienced by a randomly selected member of group *A*.

(*b*) The mean ridit for group *A*, with *B* as the reference, is .037, exactly the complement of the value found in (*a*).

(*c*) The standard error of the mean ridit from (9.40) is equal to .040.

(d) The standard error of the mean ridit from (9.45) is equal to .041, only slightly larger than the value found in (c).

Problem 10.1.

(a) The value of $\chi^2_{2\,\text{vs}\,3}$ is .02, confirmation, if any was needed, of the virtual equality of the odds ratios in studies 2 and 3.

(b) The weighted average of L'_2 and L'_3 is $\bar{L}_{2,3} = 0.862$. The value of $\chi^2_{1\,\text{vs}\,(2,3)}$ is 9.40, indicating that L'_1 is significantly different from $\bar{L}_{2,3}$ at the .01 level (the appropriate critical value for chi square is 9.21).

(c) The sum of the chi square values in (a) and (b) is 9.42, equal except for rounding errors to the value of χ^2_{homog} found in (10.18).

Problem 10.2.

(a) The mean log odds ratio for groups 2 and 3 is $\bar{L}_{2,3} = .862$ [see Problem 10.1 (b)], with an estimated standard error of

$$\text{s.e.}\left(\bar{L}_{2,3}\right) = \frac{1}{\sqrt{w'_2 + w'_3}} = .160.$$

The value of χ^2_{assoc} for these two groups is

$$\chi^2_{\text{assoc}} = \left[\frac{\bar{L}_{2,3}}{\text{s.e.}\left(\bar{L}_{2,3}\right)}\right]^2 = 29.03,$$

so the mean log odds ratio is significantly different from zero.

(b) An approximate 95% confidence interval for the underlying log odds ratio is

$$\bar{L}_{2,3} \pm 1.96\,\text{s.e.}\left(\bar{L}_{2,3}\right),$$

or the interval from .548 to 1.176.

(c) The mean odds ratio is equal to antilog (.862) = 2.37. An approximate 95% confidence interval is the interval from antilog (.548) to antilog (1.176), or, the interval from 1.73 to 3.24.

Problem 11.1.

(a)

Hospital	n	Proportion Catatonic
1	112	.286
2	154	.506
3	31	.419
4	151	.364
5	124	.298
Overall	572	.376

(*b*) The value of chi square for comparing these five proportions is 18.51.

(*c*) The critical value of chi square with 4 degrees of freedom for the .001 significance level is 18.47. The standards for the differential diagnosis of catatonic and paranoid schizophrenia are significantly different ($p <$ .001) across the five hospitals.

Problem 11.2.

(*a*) The value of P_L is .50, but that of p_L is (.75)(.50) + (.05)(.50) = .40.

(*b*) The value of P_B is .40, but that of p_B is (.9)(.40) + (.1)(.60) = .42.

(*c*) The value of $D = P_L - P_B$ is .10, but that of $d = p_L - p_B = -.02$. They are of opposite sign.

(*d*) The value of the odds ratio as a function of P_L and P_B is (.50)(.60)/(.40)(.50) = 1.50, but the value of the odds ratio as a function of p_L and p_B is (.40)(.58)/(.42)(.60) = 0.92. They are on opposite sides of unity.

Problem 12.1.

(*a*) $p_B = 60/200 = .30$, and the value of the odds ratio is (.44)(.70)/(.30)(.56) = 1.83.

(*b*) The value of n_{00}/n_0 is 18/18 = 1, and that of n_{10}/n_1 is 2/32 = .06. The resulting value of P_B is (1)(.30) + (.06)(.70) = .34. The value of the odds ratio for the adjusted rates of smoking is (.51)(.66)/(.34)(.49) = 2.02. The association seems to be stronger than in (*a*).

(*c*) The value of n_{00}/n_0 is 16/18 = .89 and that of n_{10}/n_1 is 7/32 = .22. The resulting value of P_B is (.89)(.30) + (.22)(.70) = .42. The resulting value of the odds ratio is (.51)(.58)/(.42)(.49) = 1.44. The association seems to be weaker than in (*a*).

Problem 13.3.

(*a*)

Study	n	p_e	$\hat{\kappa}$	A	B	C	s.e.($\hat{\kappa}$)
1	20	.59	.39	.0742	.0784	.0009	.21
2	20	.71	.48	.0820	.0393	.0123	.25
3	30	.54	.35	.0714	.1196	.0000	.17

The value of the numerator of (13.21) is $.39/.21^2 + .48/.25^2 + .35/.17^2 =$ 28.63, and the value of the denominator is $1/.21^2 + 1/.25^2 + 1/.17^2 =$ 73.28. The overall value of kappa is 28.63/73.28 = .39.

(*b*) The value of the chi square statistic in (13.22) is $(.39 - .39)^2/.21^2 + (.48 - .39)^2/.25^2 + (.35 - .39)^2/.17^2 = 0.18$ with 2 degrees of freedom. The three estimates of kappa do not differ significantly.

(*c*) An approximate 95% confidence interval for the common value of kappa is $.39 \pm 1.96 \times \sqrt{1/73.28}$, or the interval from .16 to .62. The overall value of kappa is significantly different from zero (the confidence interval does not contain the value 0), but the magnitude of kappa indicates little better than fair chance-corrected agreement (even the upper 95% confidence limit, .62, is low).

Problem 14.1.

(*a*) The rates of abnormal lung functioning in service industries are greater than or equal to those in manufacturing industries for employees aged less than 50 years, but are smaller for employees aged 50 years or over.

(*b*) The only gain would be the simplicity of comparing just two adjusted rates. The major losses would be the failure to describe the crossover phenomenon and the strong dependence of the direction of the difference between the two adjusted rates on the age distribution of the standard.

(*c*)

	Adjusted Rates		
Standard	Manufac.	Service	Difference
1	3.98%	4.40%	Service > Manufac.
2	8.37%	6.92%	Manufac. > Service
3	3.87%	3.84%	Approximately equal

(*d*)

	Adjusted Rates		
Age Interval	Manufac.	Service	Difference
20 – 49	2.58%	3.37%	Service > Manufac.
≧ 50	8.23%	7.14%	Manufac. > Service

Author Index

Numbers in *italics* indicate pages where references appear.

Adelstein, S. J., 8, *18*
Allison, T., 224, *235*
Altham, P. M. E., 56, *80*
Althouser, R. P., 134, *135*
Ament, R. P., 238, *255*
Anderson, R. L., 27, *32,* 194, *200*
Anscombe, F. J., 64, *80,* 107, *109*
Anturane reinfarction trial research group, 100, *109*
Armitage, P., *110,* 145, *158,* 216, 217, *234*
Arnold, J. C., 203, *210*
Arostegui, G. E., 97, *99*
Assakul, K., 194, *199*

Bahn, A. K., 240, *255*
Bakwin, H., 192, *199*
Baldwin, I. T., *159*
Barlow, R. E., 149, *158*
Barnard, G. A., 20, *32*
Barron, B. A., 196, 198, *199,* 204, *210*
Bartholomew, D. J., 147, 149, *158*
Bartko, J. J., 218, 224, *234*
Bartlett, M. S., 67, *80*
Bass, H. E., 201, *210*
Bayrakci, C., *199*
Bennett, B. M., 26, *32,* 114, 133, *135,* 216, *234*
Berger, A., 60, *80,* 132, *135*
Berkson, J., 9, 10, *17,* 91, 92, 93, *98*
Berry, G., 253, *254*
Bertell, R., 94, *99*
Bhapkar, V. P., 120, 132, *135*
Billewicz, W. Z., 133, *135*
Birch, M. W., 175, *186*
Birley, J. L. T., *211*
Bishop, Y. M. M., *viii,* 161, 164, *186*
Blendis, L. M., 216, 217, *234*
Bowman, J. E., 191, *199*
Brauninger, G., 189, *200*
Bremner, J. M., 149, *158*
Breslow, N. E., *110,* 244, 253, *254*
Brewer, G. J., 188, 190, *199*
Bross, I. D. J., 112, *135,* 151, *158, 159,* 177, *186,* 195, *199*
Brown, C., 164, 179, 180, *187*
Brown, G. W., 9, *17*
Brown, S. M., 131, 132, *137*
Brunk, H. D., 149, *158*
Bryson, M. R., 203, *210*
Byar, D. P., 105, *109,* 164, 179, 180, *187*

Canner, P. L., 107, *109*
Carroll, J. B., 60, *80*
Carter, J. L., *199*
Casagrande, J. T., 41, 42, *48*
Castellan, N. J., 140, *158*
Chalmers, T. C., *110*
Chambers, D. S., 96, *99*
Chapman, D. G., 145, 146, *158*
Chase, G. R., 133, *135*
Chassan, J. B., 147, *158*
Checkoway, H., *199*
Chiacchierini, R. P., 203, *210*
Chiang, C. L., 247, *254*

Cicchetti, D. V., 223, 224, *235*
Cobb, S., 201, *211*
Cochran, W. G., 128, 131, 133, 134, *136*, 145, *158*, 174, 177, 178, *186*, 240, *254*
Cochrane, A. L., 5, *17*, 192, *199*
Cohen, G., 189, *200*
Cohen, J., 33, 43, 46, *49*, 208, *211*, 218, 219, 221, 223, 224, *235*, *236*
Colton, T., 107, *109*
Committee for the Assessment of Biometric Aspects of Controlled Trials of Hypoglycemic Agents, 100, *110*
Conover, W. J., 27, *32*
Cook, P., 239, *255*
Cooper, J. E., 78, *80*, 156, *158*, 165, *186*, *211*
Copeland, J. R. M., *80*, *158*, *186*
Copeland, K. T., 194, 195, *199*
Cornfield, J., *xii*, 37, *49*, 52, *55*, 64, 69, *80*, 89, 92, *98*, 107, *109*, 115, *136*, 168, *186*
Cox, D. R., *viii*, *xi*, 67, *80*, 92, *98*, *110*, 123, *136*, 161, *186*
Craddock, J. M., 56, *80*
Cutler, S. J., 52, *55*
Cuzick, J., 228, *235*

Davidow, B., 201, *210*
Davies, L. G., 192, *199*
Day, N. E., 244, 253, *254*
Delmore, T., 9, *18*
DeMets, D. L., *109*
Deming, W. E., 203, 206, *210*
Densen, P. M., 201, *210*
Dern, R. J., 188, 190, *199*
Derryberry, M., 192, *200*
Diamond, E. L., 195, *200*
Dice, L. R., 214, 233, *235*
Discher, D. P., 253, *255*
Dixon, W. J., 56, *80*
Doll, R., 96, *98*, 126, *136*, 239, *255*
Dorn, H. F., 139, *158*
Dunnett, C. W., 103, *110*
Dyke, G. V., 67, *80*

Ebel, R. L., 218, *235*
Eberhardt, K. R., 30, *32*
Ederer, F., 38, *49*
Edwards, A. L., 114, *136*
Edwards, A. W. F., 67, *80*
Edwards, J. H., 64, *80*
Edwards, W., *xii*
Efron, B., 54, *55*, 106, *110*
Eisenstein, R. B., *199*
Ejigou, A., 115, *136*
El-Badry, M. A., 239, *255*
Ellenberg, J. H., *109*
Elveback, L. R., 239, 248, *255*
Endicott, J., 208, 211, 223, *236*
Everitt, B. S., *viii*, 56, 67, *80*, 120, 121, 123, 129, *136*, 161, *186*, 219, 221, 224, *235*

Feigl, P., 46, *49*
Feinberg, H. C., 253, *255*
Feinstein, A. R., 91, 95, *98*
Fienberg, S. E., *viii*, *ix*, 67, *80*, 161, 164, 176, *186*
Fieve, R. R., 189, *200*
Finney, D. J., 26, *32*, 180, *186*
Fisher, B., 206, *210*
Fisher, R. A., 25, *32*, 37, *49*, 50, *55*, 58, *80*
Fleiss, J. L., 8, *18*, 42, 45, *49*, 67, 74, 76, *80*, *81*, 120, 121, 123, 128, 129, 133, *136*, 159, 175, *187*, 189, 192, *200*, 208, *211*, 216, 218, 219, 221, 223, 224, 228, 230, 231, *235*, *236*
Fletcher, C. M., 192, *200*
Fligner, M. A., 30, *32*
Flood, C. R., 56, *80*
Fox, T. F., 100, *110*
Francis, R. E., *xi*
Free, S. M., 104, *110*
Freedman, L. S., 106, *110*
Freireich, E. J., 105, *110*
Friedewald, W. T., *109*
Frieman, J. A., 100, *110*
Frischer, H., *199*

Gabrielsson, A., 132, *137*
Gail, M. H., *109*, 247, 253, *255*
Galen, R. S., 6, 8, *18*
Gambino, S. R., 6, 8, *18*
Ganz, V. H., 106, *210*
Garland, L. H., 192, *199*, *200*
Gart, J. J., 64, 67, 74, *81*, 104, *110*, 165, 166, 169, 175, 185, *186*

Gehan, E. A., 105, *110*
Gent, M., 103, *110*
Gladen, B., 8, *18*
Glynn, M. F., 188, 190, *199*
Gold, R. Z., 132, *135*
Goldberg, I. D., 216, 217, *236*
Goldberg, J. D., 194, 195, 196, *200*
Goodman, L. A., 56, 60, 61, *81*, 175, *186*, 215, 217, *235*
Gordon, R. S., 100, *110*
Graham, P., *211*
Graham, S., 196, *200*
Gray, P. G., 96, *98*
Green, L. M., 8, *18*
Greenberg, B. G., 92, *98*
Greenhouse, S. W., 8, *18*, 27, *32*, 41, *49*, 52, *55*, 107, *109*
Greenland, S., 96, *98*
Grizzle, J. E., 27, *32*, 67, *81*, 105, *110*, 120, *136*
Guilford, J. P., 213, *235*
Gurian, J., 38, *49*
Gurland, B. J., *80*, *158*, *186*, 206, *210*

Haenszel, W., 9, *18*, 95, 96, *99*, 115, 123, 125, *136*, 173, *187*
Haldane, J. B. S., 67, *81*
Halperin, M., 38, *49*, 107, *109*
Hammond, E. C., 95, *99*
Hankey, B. F., 175, *187*
Harman, H. H., 60, *81*
Harper, D., 203, *211*
Hartley, H. O., 26, *32*
Haseman, J. K., 41, 42, *49*
Hemphill, F. M., 238, *255*
Herrera, L., 96, *99*
Hill, A. B., 50, *55*, 96, *98*, 100, *110*, 112, 126, *136*
Hirayama, T., 151, *159*
Hochberg, Y., 203, *211*
Holbrook, R. H., *199*
Holland, P. W., *viii*, 161, 164, *186*
Holland, W. W., 6, *17*
Holley, J. W., 213, *235*
Horwitz, O., 193, *200*
Howard, S. V., *110*
Hsu, P., 26, *32*
Hubert, L. J., 224, *235*

Ibrahim, M. A., 96, *99*, 151, *158*
Ireland, C. T., 120, *136*
Irwin, J. O., 25, *32*, 140, *158*
Isaacs, A. D., *211*

Jackson, G. L., 104, *110*
Jick, H., 96, *99*
Jones, E. W., 201, *210*

Kalton, G., 240, *255*
Kantor, S., 151, *158*
Kastenbaum, M. A., 140, *158*
Keeler, E., 8, *18*
Kendell, R. E., *80*, *158*, *186*
Kernohan, W., *159*
Keyfitz, N., 247, *255*
Keys, A., 194, 196, *200*
Kihlberg, J. K., 194, 196, *200*
Kilpatrick, S. J., 244, *255*
Kimball, A. W., 140, *158*
Kitagawa, E. M., 238, 239, 247, *255*
Klemetti, A., 96, *99*
Knoke, J. D., 140, *158*
Koch, G. G., 120, 126, *136*, 194, *200*, 218, 224, 225, 226, 229, 232, *235*
Koran, L. M., 192, *200*
Kramer, M., 41, *49*
Krippendorff, K., 218, *235*
Kruskal, W. H., 56, 60, 61, *81*, 215, 217, *235*
Ku, H. H., 120, *136*
Kuebler, R. R., *110*
Kullback, S., 120, *136*
Kupper, L. L., 247, *255*

Lachin, J. M., 38, *49*
Lancaster, H. O., 140, *158*
Landis, J. R., 218, 224, 225, 226, 229, 231, 232, *235*
Latscha, R., 26, *32*
Lee, J., *159*
Lehmann, E. L., 85, 88, *99*
Levin, M. L., 76, *81*, 94, 96, *99*, 206, *211*
Lieberman, G. J., 26, *32*
Light, R. J., 225, 233, *235*
Lilienfeld, A. M., *98*, 195, 196, *199*, *200*

Lindman, H., *xii*
Llambes, J. L., 97, *99*
Lord, F. M., 60, *81*
Lysgaard-Hansen, B., 193, *200*

McClave, J. T., 233, *236*
McHugh, R., 115, *136*
McKinlay, S. M., 134, *136,* 175, 176, 177, *186, 187*
MacMahon, B., 83, *99*
McMichael, A. J., *199, 255*
McNeil, B. J., 8, *18*
McNemar, Q., 114, *136*
McPherson, K., *110*
Mainland, D., 9, *18,* 96, *99,* 100, *110*
Mantel, N., 8, 9, *18,* 27, *32,* 95, 96, *99, 110,* 115, 123, 125, 133, *136,* 146, 151, *158,* 164, 169, 173, 174, 175, 179, 180, 181, *187,* 241, 245, 249, 251, 252, *255*
Marks, H. H., 246, *255*
Markush, R. E., 76, *81,* 192, *200*
Massey, F. J., 56, *80*
Mausner, J. S., 240, *255*
Maxwell, A. E., *xi,* 67, *81,* 120, *136,* 213, 218, *235*
Medical Research Council, 208, *211*
Meier, P., 100, 104, *110*
Miettinen, O. S., 87, *99,* 112, 114, 124, 125, 133, 134, *136, 137,* 177, *187,* 238, *255*
Miller, R. G., 121, *137,* 141, *158*
Morrow, R. H., 124, *137*
Moses, L. E., 53, *55*
Most, B. M., *255*
Mosteller, F., 64, *81,* 114, *137*
Mote, V. L., 27, *32,* 194, *200*

Nam, J., 145, 146, *158*
National Center for Health Statistics, 143, *158*
Navarrette, A., 97, *99*
Naylor, A. F., 166, *187*
Nee, J. C. M., 231, *235*
Newell, D. J., 193, *200*
Neyman, J., 8, *18*
Nissen-Meyer, S., 8, *18*
Novick, M. R., 60, *81*
Nunnally, J., 60, *81*

Oakford, R. V., 53, *55*
Odoroff, C. L., 175, *187*
Owen, D. B., 26, *32*

Palmer, C. E., 96, *99*
Pasternack, B. S., 181, *187*
Patterson, H. D., 67, *80*
Pavate, M. V., 27, *32*
Peacock, P. B., 93, *99*
Pearl, R., 8, 9, *18*
Pearson, E. S., 26, 27, *32*
Pearson, K., 22, *32*
Peto, J., *110*
Peto, R., 105, 106, *110*
Pike, M. C., 41, 42, *48, 110,* 124, *137*
Pilliner, A. E. G., 218, *235*
Plackett, R. L., 27, *32*
Pocock, S. J., 54, *55,* 106, *110*
Press, S. J., 203, *211*
Proctor, C. H., 194, *199*
Pugh, T. F., 83, *99*

Radhakrishna, S., 173, *187*
Rand Corporation, 53, *55*
Rao, P. V., 233, *236*
Reinfurt, D. W., 125, *136*
Remington, R. D., 38, *49*
Rimmer, J., 96, *99*
Robbins, H. E., 30, *32,* 107, *110*
Roberts, R. S., 9, *18*
Rogan, W. J., 8, *18*
Rogot, E., 38, *49,* 194, *200,* 216, 217, *236*
Rosenbaum, J., 201, *211*
Roth, H. P., 100, *110*
Rubin, D. B., 134, *135, 137,* 176, 177, *187*
Rubin, T., 201, *211*

Sackett, D. L., 9, *18,* 96, *99*
Sandifer, M. W., 8, *18*
Savage, L. J., *xii*
Saxén, L., 96, *99*
Schaaf, W. E., 192, *200*
Schlesselman, J. J., 96, *99, 109,* 177, *187*
Schneiderman, M. A., 52, *55*
Schork, M. A., 38, *49*

Scott, W. A., 218, *236*
Seeger, P., 132, *137*
Seigel, D. G., 192, *200*
Selvin, S., 153, *159*
Sharpe, L., *80, 158, 186*
Sheehe, P. R., 165, 185, *186*
Sheps, M. C., 36, *49,* 91, 92, 93, *99,* 102, 103, *110*
Shimkin, M. B., *98*
Simon, R. H., 54, *55,* 106, 107, *109, 110, 111*
Simon, R. J., *80, 158, 186*
Smith, H., *110*
Smith, P. G., 41, 42, *48, 110*
Smoking and health, 91, 93, *99*
Smyllie, H. C., 216, 217, *234*
Sneath, P. H. A., 233, *236*
Snell, E. J., 151, *159*
Sokal, R. R., 233, *236*
Somes, G. W., 132, *135*
Spiegelman, M., 246, *255*
Spitzer, R. L., 151, *159,* 192, *200,* 208, *211,* 218, 223, *236*
Spitzer, W. O., 9, *18*
Stark, C. R., 241, 245, 249, 251, 252, *255*
Starmer, C. F., 120, *136*
Stevens, W. L., 69, *81*
Stuart, A., 114, 120, *137*
Symons, M. J., *255*

Tate, M. W., 131, 132, *137*
Taves, D. R., 106, *111*
Taylor, J. W., 94, *99*
Tenenbein, A., 202, 203, *211*
Thomas, D. G., 74, *81*
Tytun, A., 42, 45, *49*

Underwood, R. E., 114, *135*
Ury, H. K., 42, 45, *49,* 125, *137*

Vessey, M. P., 96, *99*

Wackerley, D. D., 233, *236*
Walter, S. D., 44, *49,* 76, *81,* 94, *99,* 103, *111*
Ware, J. H., *109*
Weinstein, M. C., 105, *111*
White, C., 9, *18*
White, S. J., 106, *110*
Williams, G. W., 233, *236*
Wing, J. K., 208, *211*
Winkelstein, W., 151, *158*
Winsor, C. P., 67, *82*
Wood, C. L., 146, *159*
Woolf, B., 67, *82*
Woolsey, T. D., 239, 240, 247, *255*
Worcester, J., 133, *137*
Wynder, E. L., 97, *99,* 151, *159*

Yates, F., 20, 26, *32,* 143, *159,* 178, *187*
Yerushalmy, J., 4, *18,* 96, *99,* 192, *200,* 248, *255*
Youkeles, L. H., 133, *137*
Yule, G. U., 62, *82,* 244, 248, *255*

Zelen, M., 107, 108, *111*
Zubin, J., 19, *32,* 192, *200*
Zweifel, J. R., 64, *81*

Subject Index

Adaptive clinical trial, 54, 107
Adjustment of rates, *see* Standardization of rates
Agreement: chance-expected, 216-217, 219
  indexing magnitude of, 212-216
  see also Kappa; Reliability
Analysis of variance on categorical data, 151, 218, 225-227
Association: average degree of, 162
  bias in, 8-13, 95-97, 194
  confounding factors in, 112, 177
  in cross-sectional sampling, 22-23, 26, 56-58
  definition of, 2
  in different populations, 160
  as function of proportions, 20, 58
  in general contingency table, 56
  generalizing, 13, 160
  homogeneity of, 163-164, 175, 178
  measurement of, *see* Measure of association
  need for control group in estimating, 95
  in prospective study, 83-87
  in retrospective study, 87-90, 92
  in selected samples, 8
  test for significance of, *see* Chi square test; Critical ratio test; Normal curve test
  between two explanatory variables, 241-242, 249
Attributable risk: from cross-sectional study, 76-77
  definition of, 75-76
  from retrospective study, 93-95

Bartholomew's test for order, 147-149
Bayes' theorem, 3, 4-5, 11
Berkson's fallacy, 8-13, 16
Bias: in assignment to groups, 188
  in association, 8-13, 95-97, 194
  in clinical trial, 52-53
  control of, 50, 52-53, 96, 134, 176-178, 205, 207-209
  in estimated difference between rates, 194-196
  in estimated odds ratio, 196-198
  overcompensation for, 50, 193
  in prospective study, 96
  in retrospective study, 95-97
  sources of, 8, 95-96, 134, 191-192, 206, 209
  see also Confounding factors
Biased coin randomization, 54, 106
Binomial distribution, 13
Blind evaluation, 53, 206, 209

Carry-over effect in crossover study, 105
Case-control study, *see* Retrospective study
Case rate, 6-7, 15
Changes in study conditions over time, 53
Chi square statistic: as basis for measure of association, 60
  continuity correction in, 19, 22, 24-26
  equivalence of formulas for, 22, 31, 139, 150
  as function of total sample size, 58
  as measure of association, 58-59
  partitioning of, *see* Partitioning of Chi square

Pearson's, 22
Chi square test: for average degree of association, 163
Cochran's pooled, 174, 179
for comparing independent samples, 139, 147
for comparing independent values of kappa, 222
for comparing two frequency distributions: from independent samples, 150
from matched pairs, 119-123
effects of misclassification errors on, 193-194
for fourfold table, 19, 22, 24
for general contingency table, 56, 139
for homogeneity of association, 163
for hypothesized value of odds ratio, 68, 70
for linearity, 145-146
on logarithm of odds ratio, 165-166
Mantel-Haenszel's, *see* Mantel-Haenszel procedure
for matched pairs, *see* McNemar's test
for matched samples, *see* Cochran's Q test
for method I sampling, 22, 58
for method II sampling, 23-24
for multiple controls per case, 124, 125, 135
one-tailed, 28
for ordered samples, *see* Bartholomew's test for order
power of, 26, 85, 86, 88, 90, 149
significance level for, 19
for slope of line, 124-125, 145-146
with small frequencies, 56
suggested by data, 123, 142, 185
on table of total frequencies, 184-185
theory of, for comparing independent measures, 160-164
as two-tailed test, 27
Clinical trial: adaptive designs for, 54, 107
comparing two independent, 108-109
crossover design for, 104-105, 109
difference between independent proportions in, 24, 101, 108-109
double blind, 206
dropouts in, 38, 105
ethical issues in, 100
as example of method III sampling, 24
informed consent in, 107
layout of data from 100, 104
length of, 105
matching in, 53-54, 104, 112, 116-119, 123, 126, 133
problems in executing, 100, 107-108
randomization in, 50, 52-55, 105-106, 126
relative difference in, 102-103, 108-109, 117, 119
reluctance to participate in, 107-108
sample sizes in, 43, 47, 48, 100
serial entry of patients in, 106
specifying important difference in, 35-36
stratification in, 53-54, 106
Zelen's design for, 107-108
Cochran's method for pooling data, 174, 179
Cochran's Q test, 128-133
Cohort study, *see* Prospective study
Comparative mortality figure, 248
Comparative mortality rate, 248
Comparisons suggested by data, 123, 142, 185
Conditional probability: definition of, 2-3, 11
in evaluating screening test, 4
with two conditions, 10-11
Confidence interval: for attributable risk: from cross-sectional study, 76-77
from retrospective study, 94-95
for difference: between independent proportions, 29-30, 101-102
between proportions with matching, 117, 119
for general measure of association, 164
for kappa, 221, 222, 234
for logarithm of odds ratio, 72, 167
for odds ratio: from cross-sectional study, 71-74, 79
with matching, 116
from prospective study, 86
from retrospective study, 89
for overall log odds ratio, 167
for overall odds ratio, 167-168, 171-173
for phi coefficient, 74-75, 79

for relative difference: between independent proportions, 103
between proportions with matching, 118-119
for relative risk, 74-75, 79
for single proportion, 14-15, 17
theoretical basis of, 14
Confounding factors: controlled by matching, 112, 133-134, 176-177
controlled by stratification, 134, 176-177
example of, 112, 176-177
if left uncontrolled, 177
see also Bias
Continuity correction, 13, 70, 79
critique of, 27
sample sizes with, 42, 45
sample sizes without, 41, 44
Continuous distribution: as approximation to discrete distribution, 26
dichotomizing, 67
underlying ordered outcome variable, 151, 152
Control group: in determining relative difference, 92
for estimating association, 95, 96, 205-207
matched with cases, 112, 113, 123, 176-177
more than one, 96, 126
sources of, 95-96
Controlled trial, *see* Clinical trial
Cornfield-Gart procedure, 168-173, 176
Correlation coefficient, 60
Critical ratio test: for average degree of association, 162
for comparing two clinical trials, 108-109
for comparing two independent proportions, 23, 30-31, 45, 101-102
continuity correction in, 13, 22
for kappa against hypothesized value, 221, 225
for kappa against zero, 219, 224, 228, 232
for method II sampling, 23
for method III sampling, 101-102
in ridit analysis, 154-156
for single proportion, 13
see also Normal curve test
Crossover design, 104-105, 109
Cross-product ratio, *see* Odds ratio
Cross-sectional study, chi square test for, 23
compared to prospective study, 60, 83, 84-85, 88, 98
compared to retrospective study, 60, 88, 89, 90, 98
description of, 20-21, 56
layout of data for, 21
see also Method I sampling
Crude rate, *see* Overall rate

Diagnostic errors, 78, 191-192, 198, 206, 208-209
Diagnostic test, 4, 7-8, 15-16, 126
see also Screening test
Difference between proportions, in clinical trial, 24, 102, 108-109
comparing two independent, 108
confidence interval for: in independent samples, 29-30
in matched pairs, 117-119
corrected for misclassification error, 209-210
effect of misclassification errors on, 195-196, 199
errors in inference about, 33-34
important magnitude of, 34-38
lack of invariance of, 36, 92
layout of data for, 23
in matched pairs, 113-114, 117-118
as measure of association, 90-93
as measure of treatment effect, 102
with multiple controls per case, 124
one-tailed test of, 27-28, 46, 47
sample sizes for detecting, 38-42, 44-46
small in magnitude, 33, 34
standard error: from independent samples, 29, 30, 40, 101-102
from matched pairs, 114, 117-118
with several matched controls per case, 124
test for significance: from independent samples, 23-24, 101-102
with several matched controls per case, 124
two-tailed test of, 27-28
Direct standardization of rates: comparison with indirect standardization, 244, 245

with consistent differences between specific rates, 245
description of, 244-247
with inconsistent differences between specific rates, 247, 253-254
standard population in, 245-247, 254
for two factors, 250
Dropouts: and length of trial, 38, 105
sample sizes to anticipate, 38

Equal sample sizes: deliberate departures from, 44, 48
in design of study, 38
for ordered samples, 149
and sensitivity of statistical comparisons, 52
Equivalent average death rate, 248
Errors of inference, 33-34
see also Power; Significance level; Type I error; Type II error
Errors of misclassification: effects of, 188-192, 193-196
estimation of, 202-203
experimental control of, 205-209
sources of, 188, 193, 208-109
statistical control of, 201-204
ubiquity of, 188, 192
in variables being studied, 196-198, 204-205
Ethical problems: in clinical trial, 100
in eliciting information, 206
in one-tailed versus two-tailed test, 28
Excess risk, 91
see also Relative difference
Expected number: in chi square test, 25
in comparing independent samples, 23, 31
definition of, 1
for hypothesized value of odds ratio, 67-69
minimum for analyzing single proportion, 13
minimum for chi square test, 24
minimum for Mantel-Haenszel test, 175
in testing for independence, 24, 31

Factor analysis, using phi coefficient, 60-61
False positive/negative rate, 5-8, 15
Fisher-Irwin test, 25-26
Follow up study, *see* Prospective study
Fourfold table: in case of independence, 21, 25, 26
combining data from more than one, 160
"exact" analysis of, 24-26
layout of, for matched pairs, 113, 117
for method I sampling, 20-22, 57
for method II sampling, 23
for retrospective study, 88
model example, 19
significance test for, 19-20

General contingency table: association in, 56
chi square test for, 56, 139
identifying sources of significance in, 141
small sample sizes in, 56
Goodman-Kruskal index of agreement, 215

Historical controls, 105-106
Homogeneity of association, need to examine, 164
see also Association, homogeneity of
Hospitalized samples: association estimated from, 8-10
bias in studying, 8, 10, 13
control for bias in, 96
obtaining data on, 205-207
Hypergeometric distribution, 25-26

Identifying sources of significance: for general contingency table, 141
with independent samples, 141-143
with many fourfold tables, 185
with matched pairs, 121-123, 134
with several matched samples, 128-131
Independence: and Bayes' theorem, 11
definition of, 2, 15, 20
expected cell frequencies under, 25, 31
in population, not in selected sample, 8-10
and proportions in fourfold table, 21-22, 27
in selected sample, not in population, 16-17
test for, 23
value of odds ratio under, 63
Independent samples: chi square test for, 141, 149
comparisons among, 140-143
layout of data from, 23, 139

ordered, *see* Ordered samples
Indirect standardization of rates: comparison with direct standardization, 244, 245
deficiencies in, 243-244, 251-252
description of, 240-243
mathematical model underlying, 245
for two factors, 250-253
Informed consent, 107
Interview: control of errors in, 206, 208
errors in, 188, 193, 202, 208-209
measuring error in, *see* Kappa
structure in, 205-206, 208-209
Intraclass correlation coefficient, 218, 225-226

Kappa: as chance-corrected measure of agreement, 217, 219
confidence interval for, 221, 222, 234
for dichotomous variable, 217-218, 225-229
independent estimates of, 222, 234
interpreting magnitude of, 218
as intraclass correlation coefficient, 218, 225-227
as measure of similarity, 232-233
for more than dichotomous variable, 218-222, 229-232
range of variation of, 217, 226
ratings per subject: for constant number of, 229-232
for multiple, 225-232
for varying numbers of, 225-229
standard error, 219, 221, 228, 231
in study of matching, 233
test for significance of, 219, 228-229, 231-232
testing hypothesized value of, 221
in unifying indices of agreement, 217
weighted, 222-225

Layout of data: from clinical trial, 101, 104-105
from independent samples, 139
from matched pairs, 113, 117, 120
from matched samples, 127
from method I sampling, 21, 57
from method II sampling, 23, 84, 88
for more than dichotomous outcome, 150
for multiple controls per case, 125
from prospective study, 84
from quantitatively ordered samples, 143
from retrospective study, 88
Linear trend in proportions: for independent samples, 143-146
for matched samples, 131-132
Logarithm of odds ratio: confidence interval for: from independent studies, 167
from single study, 72
estimation of, 67
from independent studies, 165-168, 175-176
under logistic model, 67
under normal model, 67
standard error of: for independent studies, 165
for single study, 67
for testing hypothesized value of odds ratio, 70
Logistic model: alternative to, 67
applications of, 65, 151, 161
description of, 65-66
odds ratio under, 65-66
logarithm of, 66
Long-term trial: dropouts in, 38, 105
sample sizes for, 38

McNemar's test, 114, 116, 121, 135
as special case of Mantel-Haenszel test, 175, 186
Mantel-Haenszel method: compared to Cochran's method, 174
compared to Cornfield-Gart method, 170
compared to other methods, 175-176
with matched pairs, 115
with multiple controls per case, 123, 125
overall odds ratio estimated by, 169, 173, 186, 189
sample size requirements, 175
for testing significance of association, 171, 174, 186
Mantel-Stark method of standardization, 252-253
Matched pairs: in clinical trial, 53, 104, 112, 117-119, 122-123, 134
confidence interval for: difference between proportions with, 117, 119

odds ratio with, 115-116
relative difference with, 117, 119
difference between proportions with, 114, 117, 118
identifying sources of significance, 121-123, 134
layout of data from, 113, 117, 120
Mantel-Haenszel method, 115
more than dichotomous outcome with, 119-123
with ordered outcome variable, 122-123, 134-135
randomization with, 53
relative difference with, 118, 119
in retrospective study, 113
significance test for, *see* McNemar's test; Stuart-Maxwell test
standard error: of difference between proportions with, 114, 117, 119
of odds ratio with, 115
of relative difference with, 118, 119
Matched samples, 126-133
Matching: advantages of, 133, 134, 177
alternatives to, 134, 176-177
of cases and controls, 112, 113, 123, 126
in clinical trial, 53, 104, 112, 116, 119, 123, 126, 133
for controlling confounding factors, 96, 112, 133-134, 176-177
on date of event, 134
gain in efficiency with, 133
limitations of, 134, 177
with many matched sets, 176
with more than dichotomous outcome, 119-123
with multiple controls per case, 123-126, 233
number of characteristics for, 133, 134
odds ratio with, 115, 125, 126
precision of comparisons with, 112, 133
in prospective study, 112, 116, 126, 133
randomization with, 53, 112, 126
in retrospective study, 112, 113-116, 126, 133
with time limit on study, 133
Measure of agreement, *see* Agreement; Kappa
Measure of association: chi square as, 58-60
conclusions depending on choice of, 93
critical ratio as, 180-182
effects of misclassification errors on, 193-196
effects of unreliability on, 212
function of chi square as, 58-60
general, 161-164
homogeneity of, 162-164
invariance of, 60, 66, 67, 89, 92
significance of association, 58
standards for selection of, 89, 92
and validity of retrospective study, 91, 92
Yule's, 62
see also Difference between proportions; Logarithm of odds ratio; Odds ratio; Phi coefficient; Relative difference
Method I sampling: association in, 20, 21, 57-58
chi square test for, 22, 58
compared to method II sampling, 60, 83, 84-87, 88-90, 98, 138
description of, 20-22
estimation in, 58
layout of data from, 21, 57, 58
misclassification errors in, 188
odds ratio in, 62-63
phi coefficient only valid in, 97
randomization in, 50, 51
sample size for, 33
statistical hypothesis appropriate for, 28
see also Cross-sectional study
Method II sampling: compared to method I sampling, 60, 83, 84-87, 88-90, 98, 138
critical ratio test for, 23-24
description of, 23-24
layout of data from, 23, 84, 88
matching in, 112
misclassification errors in, 188
randomization in, 50, 51
sample sizes for, 33
statistical hypothesis appropriate for, 28
see also Prospective study; Retrospective study
Method III sampling: critical ratio test for, 101-102
description of, 24
matching in, 112

misclassification errors in, 188
randomization in, 50, 52-53
sample sizes for, 33
statistical hypothesis appropriate for, 28
see also Clinical trial
Misclassification errors, *see* Errors of misclassification
More than dichotomous outcome: agreement on, 213, 218-222, 229-232
in case of independent samples, 150-153
in case of matched pairs, 119-123
kappa for, 218-222, 229-232
McNemar's test in case of, 119-123, 134
sample sizes for, 38-39
Mortality figure, comparative, 248
Morality index, 248
Mortality rate, comparative, 248
Multinomial sampling, 56
see also Cross-sectional sampling; Method I sampling
Multiple controls per case, 123-126, 232-233
Multiple significance tests on same data, 121, 142
Multivariate confounder score, 177

Naturalistic sampling, *see* Cross-sectional study; Method I sampling
Normal curve test: one-tailed, 29, 46, 47
power of, 26, 27
sample sizes for, 38-42, 44-46
two-tailed, 26-29
see also Critical ratio
Normal distribution, as alternative to logistic model, 67
Numerical grading applied to ordered categories, 151

Odds: derived from ridit analysis, 153
equality of two, under independence, 77
under logistic model, 66
as measure of risk, 61
Odds ratio: advantages over other measures, 95-96
as approximate relative risk, 37, 60, 78
in case of independence, 63
compared to relative difference, 95-96
comparison of procedures for analyzing, 175-176
from cross-sectional study, 71-75, 79
with matched pairs, 115-116
in prospective study, 84
in retrospective study, 88
confidence interval for: in several fourfold tables, 167-168, 171-173
Cornfield-Gart procedure for, 168-173
corrected for misclassification errors, 209-210
criticisms of, 90-93
derivation of, 61-63
effects of misclassification errors on, 189-192, 195, 196-197
estimate of: with matched pairs, 115
in method I sampling, 63
with multiple controls per case, 125
in prospective study, 84-85
in retrospective study, 88, 97
for small cell frequencies, 64, 166
homogeneity of, 166, 169-170
invariance of, 66, 88, 93
logarithm of, *see* Logarithm of odds ratio
under logistic model, 65
Mantel-Haenszel estimate of, 115, 125, 170, 173, 176
as measure of discrepancy between proportions, 31
precision of estimated, 64, 85-86, 89-90
in sample size determination, 36-37
significance of pooled estimate of, 170-171, 174, 175
standard error of: with matched pairs, 115
in method I sampling, 63
in prospective study, 85
in retrospective study, 90
testing hypothesized value of, 67-71
One-tailed test: description of, 27-29
ethical problem in, 29
power of, 28
sample sizes for, 46, 48
significance level for, 28
Ordered outcome variable: derived from underlying continuum, 151, 152
with independent samples, 150-156
logistic model for, 151
with matched pairs, 122-123
Ordered samples: of equal size, 150
qualitatively ordered, 147-149

quantitatively ordered, 131-132, 143-146
Overall proportion of agreement, 213-214, 219
Overall rate: in case of independence, 2
difference between two, 237-238
limited value of, 2
ratio of two, 237
as weighted average of specific rates, 5, 11-12

Partitioning of chi square: for homogeneity of association, 185
for independent samples, 140-143, 145-146
for matched samples, 129-131
for several fourfold tables, 161-164
suggested by data, 141, 185
Phi coefficient: comparing two, 60
confidence interval for, 74-75, 79
as correlation coefficient, 60
deficiencies in, 60, 88-89
definition of, 59
in factor analysis, 60
lack of invariance of, 88, 97
in prospective study, 86
in retrospective study, 89, 97
validity only in method I sampling, 97
Power: of Bartholomew's test, 147, 149
under budgetary restrictions, 34
of chi square test for linearity, 145
with continuity correction, 26-27
of critical ratio test, 30-31
definition of, 34
with equal sample sizes, 42, 138
of McNemar's test, 114
with matching, 133
with multiple controls per case, 126
of one-tailed test, 27-29
for predetermined sample sizes, 42
of prospective study, 84, 85-86, 87, 90
of retrospective study, 87, 90
in sample size determination, 43-44
selection of value of, 43
with stratification, 177
with unequal sample sizes, 45-46
see also Type II error
Precision: with equal sample sizes, 138
of estimated odds ration, 86, 89-90, 176
as function of standard error, 161
with matching, 112, 133
with multiple controls per case, 126
sample size necessary for specified, 98
of specific rate, 239, 250-251
with stratification, 177
Precoded responses, 208-209
Predictive value, 8
Prevalence rate, *see* Case rate
Proportional mortality rate, 248
Proportion of specific agreement, 215
Prospective study: association in, 83, 84-85
compared to cross-sectional study, 60, 83, 84-85, 88, 98
compared to retrospective study, 88, 90, 95-97, 98
description of, 23
designed to replicate retrospective study, 37-38
invalidity of phi coefficient in, 84
layout of data from, 84
matching in, 112, 116, 126, 133
method II sampling underlying, 83
odds ratio in, 85-86
power of, 84, 85-86, 88, 89-90
precision of, 85, 86, 89-90
sources of bias in, 96

Qualitatively ordered outcome variable, 122-123, 150-156
Qualitatively ordered samples, 147-149
Quantitatively ordered samples, 132, 143-146

Randomization: in adaptive clinical trial, 54
alternatives to, 105-106, 107
in assignment to treatments, 50, 52-53
criticisms of, 105
in crossover study, 104
imbalance with, 54
within intervals of time, 52
for matched pairs, 53

for matched samples, 112, 126
necessity of, 50, 105
in selection of sample, 50, 51, 57
with stratification, 53-54
Random sample: simple, 50
systematic, 57
Rank order analysis, 153-154
Relative difference: in clinical trial: with independent samples, 102-103, 108-109
with matched pairs, 118, 119
compared to odds ratio, 92-93
comparing two independent, 108-109
confidence interval for: with independent samples, 103
with matched pairs, 117-119
as modified simple difference, 92
in sample size determination, 35-36
standard error of: with independent samples, 103
with matched pairs, 117
Relative mortality index, 248-249
Relative risk: confidence interval for, 74-75, 79
as component of attributable risk, 76
definition of, 64
see also Odds ratio
Reliability: and accuracy, 212
and biased estimation, 201, 202
and cost, 201
of diagnosis, 191-192, 208-209
of ratings, 218
of recollections, 95
of subjective grading, 149
see also Agreement; Kappa
Replication of previous research, sample sizes needed in, 36
Retrospective study: compared to cross-sectional study, 60, 88, 89, 90, 98
compared to prospective study, 88, 90, 95-97, 98
confidence interval for odds ratio in, 90
confounding factors in, 112
criticisms of, 92-93
description of, 23, 87
important difference between proportions in, 36
invalidity of phi coefficient in, 88, 97
layout of data from, 88
matching in, 112, 113-116, 126, 133, 176-177
method II sampling in, 83
odds ratio in, 89, 97
power of, 88-90
precision of, 90
sources of error in, 95, 96
stratification in, 176-177
superiority over other study designs, 89-90, 98
validity of, 92-93, 98
Ridit analysis, 151-156
Rogot-Goldberg index of agreement, 216

Sample size: for clinical trial, 35, 43, 48
for comparing several treatments, 38
for cross-sectional study, 31
for detecting specified difference, 38-42, 44-46
for differential costs, 48
with drop-outs, 38
effect on statistical significance, 58
equal in two groups, 38-41
large, 26, 29, 33, 34, 47
for long-term trial, 38
for more than dichotomous outcome, 38
for one-tailed test, 46, 47-48
prespecified, 34, 42
for replicating previous research, 36-38, 44
required for analyzing single proportion, 13
required for chi square test, 25
required for Mantel-Haenszel procedure, 175
small, 25, 101
for specified precision, 98
unequal in two groups, 44-46, 48
without continuity correction, 38-41, 45
Screening test: error rates of, 4-8
predictive value of, 8
repetition of, 8, 15-16
sensitivity of, 4, 6, 8
specificity of, 4, 6, 8

Selected samples, association in, 8-13
Sensitivity, 4, 6, 8, 194, 195
Serial entry of patients over time, 52, 106
Significance level: for comparisons suggested by data, 121, 123, 142-143, 185
  in controlling Type I error, 33
  for one-tailed test, 28, 46, 47-48
  selecting a value of, 43
  and value of power, 43
  see also Type I error
Single proportion: confidence interval for, 13-15, 17
  standard error of, 13
  test of hypothesis for, 13, 14
Slope of straight line: with independent samples, 143-146, 156
  with matched samples, 131-132
Small frequencies: chi square test with, 56
  Cochran's Q test with, 133
  estimated odds ratio with, 63, 166
Source of sample, bias in, 8-9, 95-97
Specificity, 4, 6, 8, 194, 195
Specific rate: in case of independence, 2
  as component of overall rate, 4-5, 11
  consistencies in, 245
  definition of, 1
  inconsistencies in, 247, 253-254
  need for comparing, 239
  precision of, 239, 250-251
  ratio of, 64, 248
  unavailability of, 238, 251-252
Standard error: of attributable risk: from cross-sectional study, 77
    from retrospective study, 94
  of cell frequency in general fourfold table, 69, 168
  for comparing two differences between proportions, 108-109
  for comparing two relative differences, 108-109
  of difference: between mean ridits, 155
    between proportions from independent samples, 23, 29, 30, 39-40, 108
    between proportions from matched, pairs, 114, 117
    between proportions with matched controls per case, 124
  of kappa, 219, 221, 228, 231
  of logarithm of odds ratio, 67, 70, 165
  of mean log odds ratio, 165
  of mean measure of association, 163
  of mean ridit, 154, 155, 157
  of odds ratio: in fourfold table, 63, 64, 85, 90
    with matched pairs, 115
    with multiple controls per case, 125, 126
  and precision, 161
  of proportion corrected for misclassification error, 203
  of relative difference: from independent samples, 103
    with matched pairs, 117
  of single proportion, 13
  of standardized difference, 179
  of standardized rate, 247
Standardization of rates: criticisms of, 239
  direct method, *see* Direct standardization of rates
  indirect method, *see* Indirect standardization of rates
  miscellaneous methods, 247-249
  reasons for, 239-240, 246
  for two factors, 249-253
Standardized difference, 178-180
Standardized mortality ratio, 247
Standardized rate: magnitude of, 246
  standard error of, 247
Standard population: in direct standardization, 244-245
  effect on standardized rate, 245-247, 254
  in indirect standardization, 240
  in ridit analysis, 152-154, 155, 156, 157
Stratification: as alternative to matching, 160, 176, 177
  in clinical trial, 106
  imbalance with, 54, 106
  randomization with, 53-54
  into small number of strata, 175
Stuart-Maxwell test, 119-123
Subject serves as own control: for comparing experimental conditions, 126

in two-period crossover study, 104
Summation observed *vs.* summation
expected procedure, 182-184
Summation of chi procedure, 180-182

Tabulation of data, *see* Layout of data
Tenenbein's double sampling scheme,
202-204
Time limit on study, and matching, 133
Total rate, *see* Overall rate
Two-tailed test, 27-29
Type I error, 33, 39, 43
see also Significance level
Type II error, 33-34, 43
see also Power

Unequal sample sizes, 44-46, 47

Weighted average: in Bartholomew's
test, 147-149
overall rate as, 5, 11
standard error of, 163
standard errors used in, 161, 163
weights in, 5, 185
Weighted kappa, 222-225

Yates correction, *see* Continuity
correction

Zelen's design for a clinical trial,
107-108

Applied Probability and Statistics (Continued)

FLEISS • Statistical Methods for Rates and Proportions, *Second Edition*
GALAMBOS • The Asymptotic Theory of Extreme Order Statistics
GIBBONS, OLKIN, and SOBEL • Selecting and Ordering Populations: A New Statistical Methodology
GNANADESIKAN • Methods for Statistical Data Analysis of Multivariate Observations
GOLDBERGER • Econometric Theory
GOLDSTEIN and DILLON • Discrete Discriminant Analysis
GROSS and CLARK • Survival Distributions: Reliability Applications in the Biomedical Sciences
GROSS and HARRIS • Fundamentals of Queueing Theory
GUPTA and PANCHAPAKESAN • Multiple Decision Procedures: Theory and Methodology of Selecting and Ranking Populations
GUTTMAN, WILKS, and HUNTER • Introductory Engineering Statistics, *Second Edition*
HAHN and SHAPIRO • Statistical Models in Engineering
HALD • Statistical Tables and Formulas
HALD • Statistical Theory with Engineering Applications
HARTIGAN • Clustering Algorithms
HILDEBRAND, LAING, and ROSENTHAL • Prediction Analysis of Cross Classifications
HOEL • Elementary Statistics, *Fourth Edition*
HOLLANDER and WOLFE • Nonparametric Statistical Methods
JAGERS • Branching Processes with Biological Applications
JESSEN • Statistical Survey Techniques
JOHNSON and KOTZ • Distributions in Statistics
Discrete Distributions
Continuous Univariate Distributions—1
Continuous Univariate Distributions—2
Continuous Multivariate Distributions
JOHNSON and KOTZ • Urn Models and Their Application: An Approach to Modern Discrete Probability Theory
JOHNSON and LEONE • Statistics and Experimental Design in Engineering and the Physical Sciences, Volumes I and II, *Second Edition*
JUDGE, GRIFFITHS, HILL and LEE • The Theory and Practice of Econometrics
KALBFLEISCH and PRENTICE • The Statistical Analysis of Failure Time Data
KEENEY and RAIFFA • Decisions with Multiple Objectives
LANCASTER • An Introduction to Medical Statistics
LEAMER • Specification Searches: Ad Hoc Inference with Nonexperimental Data
McNEIL • Interactive Data Analysis
MANN, SCHAFER and SINGPURWALLA • Methods for Statistical Analysis of Reliability and Life Data
MEYER • Data Analysis for Scientists and Engineers
MILLER, EFRON, BROWN, and MOSES • Biostatistics Casebook
OTNES and ENOCHSON • Applied Time Series Analysis: Volume I, Basic Techniques
OTNES and ENOCHSON • Digital Time Series Analysis
POLLOCK • The Algebra of Econometrics
PRENTER • Splines and Variational Methods
RAO and MITRA • Generalized Inverse of Matrices and Its Applications
RIPLEY • Spatial Statistics
SCHUSS • Theory and Applications of Stochastic Differential Equations
SEAL • Survival Probabilities: The Goal of Risk Theory
SEARLE • Linear Models
SPRINGER • The Algebra of Random Variables
UPTON • The Analysis of Cross-Tabulated Data
WEISBERG • Applied Linear Regression
WHITTLE • Optimization Under Constraints